MORGAN THE TRAVEL

For Scoob & Liliana

Warmest wishes

Ronald D Mayes.

MORGAN
THE
TRAVEL

From the Topic of Cancer
to the Tropic of Capricorn

Ronald D Morgan

Matador
9 Priory Business Park,
Wistow Road, Kibworth Beauchamp,
Leicestershire. LE8 0RX
Tel: 0116 279 2299
Email: books@troubador.co.uk
Web: www.troubador.co.uk/matador
Twitter: @matadorbooks

ISBN 978 1 8004 6442 1

British Library Cataloguing in Publication Data.
A catalogue record for this book is available from the British Library.

Printed and bound by CPI Group (UK) Ltd, Croydon, CR0 4YY
Typeset in 12pt Adobe Garamond Pro by Troubador Publishing Ltd, Leicester, UK

Matador is an imprint of Troubador Publishing Ltd

*For my family and all our wonderful friends
who have filled our lives
with love, fun and laughter.*

Never swim against the tide as the expansive blank canvas of the newly washed 'fresh start' beach promises that discovery and opportunity await as the sands of time fall through our fingers.

Morgan the Travel

The Traveller's Poem

As we dream of far-off lands
Of oceans lapping shore
Of cultures strange passing us by
Inquisitive for more
We pack our bags
We wave goodbye
Passport and plans we grip
We start our wondrous journey
By plane and train and ship
We know not what awaits us there
But the world we must explore
Like a drug it drives us on
To travel more and more
And when we return with travellers' tales
To prompt laughter and enthrall
Of mishaps that occurred whilst there
And sights like China's wall
We will count ourselves so humbled
As our photographs will show
Of the wondrous world that we live in
And of the chance we had to go!

Contents

CHAPTER ONE

———

Good health, the greatest asset you will ever possess

March 2015. I felt like a peach being hit by a sledgehammer. I was sitting with my wife in a small, impersonal, sparsely furnished clinic office in my local hospital. A worn wooden desk bereft of any adornment and an empty office chair that had seen better days sat before us. We were waiting to hear the results from a recent colonoscopy, an examination that I regularly had every two years and had done so for two decades, but we knew the results of this one was not favourable. I'd been told I had what appeared to be a tumour in my sigmoid colon.

We'd been waiting in no man's land, trying to be upbeat as we had not had official confirmation that I had cancer, not until we heard the biopsy or CT scan results. On this rollercoaster I had even imagined being put in the mortuary but not being dead.

Now we were back at the hospital to find out just what the prognosis was. After what seemed an eternity the colorectal consultant at Royal Shrewsbury Hospital appeared from an ante room, muttered a greeting as he shuffled a few papers and did not beat around the bush. He confirmed I had cancer. He pulled no punches: there was nothing they could do to cure my disease, which was quite advanced. The consultant told me he was recommending palliative care only and suggested I go home and get my affairs in order. As

1

I stared past the soothsayer of doom through the window to the courtyard beyond, my mind went blank, I was numb!

We were told that not only was the tumour cancerous but the cancer had metastasised to my liver and unfortunately the tumours in the liver were grouped around my vena cava and other important blood vessels, which made it impossible to operate. We were stunned. I heard my wife Dianne saying, 'No, he's too young, too fit, he runs most days, there must be something you can do.' On Dianne's insistence the consultant finally agreed to refer me to the liver specialists in Birmingham, but 'we do not think it will change anything,' he warned.

This fifteen-minute consultation was followed by a colorectal nurse taking us into a small room and unceremoniously and with even less compassion than the consultant giving us a Macmillan compendium of information in a brown paper bag. Our world was in tatters and we were absolutely adrift in an ocean of doubt, fear and isolation.

When you are told you have cancer and you may not win the fight it brings a certain amount of perspective! Material things are never more immaterial at this time, whimsical fancies are replaced with simplicity, like just wanting to listen to the dawn chorus or the sounds of the waves lapping on the beach one more time. Your senses seem keener to smells like freshly cut grass and sounds seem more prominent. You actually take time to taste your food and eat more slowly, almost as if time is in slow motion. You cherish every second.

You lie awake in the morning thinking about the people in your life and the loved ones you will leave behind, you think about your funeral and what hymns or songs or other laments may be apt, such as 'When I'm dead and gone' by McGuinness Flint, which sprang to mind. But, seriously, there are so many songs that mean things in my life and remind me of people, classics like 'Danny Boy' for my father and Alison Krauss singing 'When you say nothing at all', 'White Sandy Beach of Hawaii', which all mean a lot to Di and I and, of course, 'Somewhere over the rainbow' by IZ Kamikawiwo'ole. As I write this I start to fill up because I do not want to go! It's not yourself you feel sorry for it's the thought of leaving your family and friends.

Why me? I was fit. Three years before I trekked to the top of Mount Kilimanjaro, the highest free-standing mountain in the world, a daunting but wonderful experience. Last year I was in Tibet at Everest base camp and Rongbuk monastery in the footsteps of George Mallory and Sandy Irvine. I have spent a good deal of my life pushing the parameters and stretching the limits of endurance, but now I had a new challenge.

When I was first told it was like walking into a pitch-black room devoid of light. I could see nothing. Then, as I stood still and waited, my eyes adjusted and soon I could see a way forward. We gradually started to regroup and say to ourselves, 'This cannot be the end, what can we do? Who can we ask?'

This is where technology came in handy. We started to research to try and understand what was being said to us and establish what our options were. Indeed, did we have any or were we just in denial? We researched liver consultants at Queen Elizabeth Hospital Birmingham and made an appointment to see a colorectal consultant Robert Sutcliffe privately. One Friday evening shortly afterwards we sat in the reception waiting room of this old building at the Priory Hospital in Edgbaston, full of hope but naturally fearing the worse. We extolled the virtues of the splendid machine dispensing free hot chocolate as we tried to distract ourselves from the fear and appreciate the small things in life.

We were shown upstairs into a small room and very quickly invited into a consultancy room, where a surprisingly young and engaging consultant immediately put us at our ease. 'I have reviewed the scans and confirm the original diagnosis,' he began. 'However, if chemotherapy can shrink back the tumours from just one of the main blood vessels in the liver, I will operate.' It was the lifeline we had been looking for.

We told friends and family. My youngest son decided to fly back from Australia to be with me for whatever time was needed. We headed to a weekend with friends in the Peak District in a more positive mood and we were able to put our problems to one side for a day or two before heading home. We set about engaging with the oncologists at Shrewsbury Hospital, which, despite chasing for appointments, proved so difficult to achieve.

Our recent optimism was turning into pessimism as we became demoralised by NHS waiting times. Finally, fifty-three days after the tumour was first located and after persistent badgering and pressure from us, we had an appointment and we saw the oncologist a week later.

We started the gruelling chemotherapy at the end of May and five-and-a-half cycles of chemo drugs oxilaplatin and 5fu later, lo and behold, a scan showed that the main tumour had shrunk by two-thirds and Mr Sutcliffe confirmed he would operate.

Monday morning, 5 October 2015 came. After an anxious night alone in a side room on Ward 28 at the Queen Elizabeth Hospital Birmingham, I was full of trepidation. I had barely slept, scared that I wouldn't survive this large open-liver operation. However, the terrific staff soon put me at ease. They were brilliant as they wheeled me down to theatre. As they prepped me, I could see the surgeons and the team through the small windows in the doors to the theatre. I felt a small cool trickle into my vein, then nothing.

Six days of being uncomfortable and sore later, I was discharged and left hospital, very weak, with just 25 per cent of my liver remaining, but the operation had been successful. I was told the liver would regenerate back to about a third its original size and hopefully stay cancer free. It was a long and uncomfortable two-hour journey home in our car. Dianne drove as carefully as she could, but I felt every small bump in the road and every gear change and braking action.

Life was now about recuperation, but as relieved as I was to be home, when I tried to climb our stairs' I found my feet and lower legs and hands were all pins and needles. I could barely feel them. It transpired I had developed a peripheral neuropathy.

Within a few days I was back in hospital with pains in my shoulders and chest. I could barely breathe. It turned out I had a 'collection', a residue of blood and other fluids by my liver that needed draining. I had to wait in Shrewsbury Hospital for a week as they could not do the procedure and I needed a transfer back to Birmingham. I could not go home or I would be out of the system and have to start again at A & E.

Finally, I was transferred by ambulance. I had not realised how uncomfortable these vehicles are. I figured they would have cushioned, state-of-the-art suspension, but it was like being in a Russian tank or like riding a bicycle down a cobbled street. But I was sharing the ambulance with a lovely young lady who was awaiting a liver transplant, which put my fears and pain into perspective.

I waited four days in Birmingham for the procedure, which ended up being one of the most painful I have encountered. There was no general anaesthetic. A huge needle was put through my ribs to the location of the collection tracked using ultrasound. Then belatedly, as I was in such pain, a barely adequate dose of Novacaine was used as a local anaesthetic. The initial entry into my body cavity left me with my toes curling and a nurse holding my hand, then a drain was inserted and left in. The pain was still so bad in my chest that every time I tried to inhale, I almost passed out with the pain. I was in so much discomfort they put out a call for the clinician who had done the procedure. I was still on the table he had performed the incision on. He checked me out and said, 'Yes! It will be sore but get on with it,' dismissing my condition as if I was complaining over nothing.

I was bounced back to the ward, lying on a trolley, feeling every bump and jolt. Up in a lift and along corridors, I could barely speak to tell the ward staff of the pain I was in. Fortunately, my wife arrived and within fifteen minutes I had two types of pain relief and could then begin to cope with the excruciating soreness. After a few days the drain could finally be removed and I was allowed home.

Whilst this was just something else debilitating to be added to my daily burden, I concentrated on trying to recover from the liver operation. I knew the mother tumour was still in my colon and this had to be removed for me to stand a chance.

On Christmas Eve I went to see my colorectal surgeon in Shrewsbury to discuss the plan for my care going forward. I had expected a resection, just a small part out of my large bowel, as per my last conversation with him. We were dumbfounded to be told that, following a meeting of the surgeons and others in the Multi-Disciplinary Team, it had been decided to remove the whole of my large bowel and create an ileostomy, which is basically my small bowel shoved through my abdomen and an external bag added to cope with the body waste. I cannot tell you how shocked I was. This was another setback for me in my already weak state, and I reeled from the emotionally charged distress.

It put a huge pall of angst over the festive period. I just could not think and Dianne, who was trying to comfort and care for me, was beside herself with worry. 'Please help me, God! Are you there?' came the inner prayer for salvation.

Christmas came and went. My eldest son came over from Australia for the second time in six months and soon the New Year arrived and a late January date for the bowel operation with it.

In the meantime, I went privately to see a top neurologist in Birmingham about my peripheral neuropathy, but he was not very helpful, just confirming what we already knew. Then he damaged my Achilles by hitting it too hard checking reflexes. He confirmed it was the platinum-based chemotherapy drug 'oxilaplatin' that had caused the neuropathy and that there was unlikely to be an improvement. Oh! he also threw in the little tidbit, 'it may get worse.' I still felt it was the tight DVT stockings they gave us in hospital, where my toes would get stuck and go numb and blue through the constricted hole in the stocking toe, but no one would listen. I felt I was deteriorating mentally and physically.

I continued to try and eat well and exercise by walking each day and resting as appropriate. The time came for my huge open-bowel operation, but as bad luck had it I had contracted a cold and they would not operate. All the waiting creates havoc with your mind and well-being as the stress is unbearable. Surgery was delayed until 2 February. I was still not feeling sharp, but the operation took place.

It had been hard trying to get used to the idea of the ileostomy both from a physical and emotional point of view. How would it affect the way I would live, play sport or do everyday things that once were simple but now I would need to plan more to not embarrass myself or those around me? However, the overriding and sole driver was that this might eradicate the cancer. That was the goal!

We had enjoyed a few days in Borth-y-Gest on the Welsh coast, trying to relax before this surgery, but nothing could have prepared me for the ensuing trauma. Post-surgery I felt darn awful for a week. I just wanted to be beamed up. Incredibly the human body and mind is nothing but resilient. The ensuing weeks involved an 'ileos', which meant the bowel was not working, having a gastric tube forced up my nose and into my stomach, making me vomit litres of foul swamp-like green bile. Together with the post-operative pain, it really tested your endurance and willpower.

A huge piece of apparatus like a medical plumber's plunger was shoved into my bowel as I lay awake on my hospital bed, its task to remove any kinks in the bowel. Luckily, it worked, and I overcame this bowel inertia and it suddenly exploded into life. Finally, I could eat and get stronger, the hiccoughing and nausea ceased, the tubes came out and I had turned a corner, one I never believed in those dark early days would ever be possible.

I began to walk, dealing with the indignities of carrying huge bags of my own excretion around with me. Occasionally the bag split, and I was covered in it. All through this, the staff were brilliant and good-humoured cleaning me up, reassuring me, showing me how to care for my stoma. I cannot speak highly enough of them. As I started to improve it was good to share in the camaraderie of the ward too and I met some very pleasant folk indeed.

Although I was terribly sleep deprived. I finally left hospital on 16 February 2016, two weeks after the operation, very weak, but alive! It was so good to be in my own bed. I gradually regained my strength and desire for life. After a week I walked the short distance to the end of the drive, then the second week to the local shop, the third week I did this twice a day, all the time dealing with the pain of the surgery and especially the derriere, which had been sewn up and which was very painful to sit on. Such was the price of life!

A good mate brought me a reclining chair, so I could shift my weight to cope with the discomfort. By week five I was much stronger, I was eating well and adjusting to everything. I was now excellent at sleeping on my back even though my back became sore too.

More appointments came and went and a date for more chemo loomed: 23 March 2016, for another six cycles. The operation had rid me of the huge colon mother tumour and eighteen lymph glands, six of which had shown cancerous activity, so the chemo was a must. Chemotherapy is brutal, making you feel so sick and fatigued but when it stops you can feel better quite quickly. Initially you go from drinking red wine from the nectar of the gods to chemotherapy the nectar of the devil, which ironically can bring redemption and longevity.

Surgery is very tough, but mentally you know it has improved your chances of life, so you live with it. People talk about battling cancer and others more positively talk about living with cancer. I prefer to think I am 'dancing with cancer', covering every part of the dance floor I can, and listening to the music of life and responding. My senses have been heightened to the lyrics and the melodies of every day and I intend to keep on dancing the slow numbers and when I can, the twists and rock and roll tunes, some Charleston and Bollywood until I can dance no more, then I will be ready for home, knowing I have enjoyed the dance of life!

Cancer is indiscriminate, it affects the wealthy and the poor, the academic and the blue-collar worker, the grafter and the shirker, all religions, all colours of skin. My mantra was 'resilience and brilliance', be as tough and bright as you can and surround yourself with comedy and wit, self-deprecation, music and most of all love!

I was born on 24 June under the sign of Cancer. Perhaps I was always going to be a 'cancer dancer'! It is easy to see why people feel a hopelessness at first, a resignation that this is their fate and lose the will to live, but in fairness most people when the shock has subsided look at how they can overcome this dreadful diagnosis. It does not take away your fears. The fear of dying is not actually the biggest concern, it is more the fear of leaving the ones you love, the fear you will leave something unfinished or not achieve what you had planned to do. You can think you are constantly living on death row, but aren't we all?

You are immediately appreciative of everything and everyone around you, of what you have and especially aware of that precious commodity 'time' that drains with every second. There are silver linings: hope and faith flourish. Your sensibilities are heightened, the goodness of friends and family comes to the fore, but you see beyond your self-absorption how your predicament also affects them, probably more than you. You appreciate a celebration and experience joy of what you have and see more clearly your blessings. Cancer is just a wake-up call that endows us with the clarity to see, and appreciate what we have done and what we have, and what we can do and who we can be. We are lucky to have had the dance!

Here is an entry from my diary as an addendum to the above:

7 June 2016. I may have to have a seventh cycle of chemo but also been back to Birmingham to be advised by a haemotologist that I have a protein in my blood called MGUS that needs monitoring, as it can lead to myeloma leukaemia through the bone marrow. Happy days. I feel wretched, weak, nauseous and lacking any desire to do anything but I try to exercise, walk, that is, and interact, but it's not much of a life at present.

Boy, I was frightened at times, full of trepidation, but I was also uplifted by positivity, faith and hope. It was so different to the year before!

I recovered well and even got back to a little running going on one of the alternative runs with the Shropshire Shufflers, which essentially is like cross country. I love them, only four miles or so, mind. One weekend I went down to support Dianne and Kerry at a local race, the Shrewsbury 10K, from the Quarry Park in Shrewsbury and started to run with the back markers to encourage them but found myself doing the whole distance. We all finished together tired but fulfilled.

Then another scan showed that I had two more small tumours on the liver, so I went to see a specialist and they said they could deal with them by ablation, meaning they blast the tumours by entering my abdomen laparoscopically and thus I would recover more quickly. So, in for a penny in for a pound. On 28 September 2016 the procedure was done and after an overnight stay I went home and recovered very quickly, as promised.

Of course, you are always only as good as your next scan so on 4 November I had a CT scan and kept my fingers crossed. two weeks later I went to Birmingham to see the consultant radiotherapist Mr Roberts, who said they had reviewed the scan and deemed the operation a success, and my main consultant had added that there was no evidence of residual disease, meaning for the time being they thought I was cancer free. Just like when I was diagnosed, I was not quite sure what I was hearing. Was I really cancer free? Finally, it sank in that I had a reprieve and for the first Christmas in two years I could celebrate with my family and friends, who all rejoiced with me. Sure, it's only until the next scan but life is life!

In 2017 my next scan showed more tumours on my liver and I had these ablated at the end of March 2017. This led to some urinary problems but I managed to overcome them. In the summer and Autumn following Dianne and I managed to holiday all over the UK, then ventured overseas to Spain, Tenerife and Cyprus, returning to as normal a life as possible. I was blessed with three consecutive clear scans. As of January 2018, the myeloma was in check so on to the next scan at the end of April 2018. All good there, too, so I kept dancing! Thereafter I was on six-month scans and check-ups.

Cancer heightened the senses and the importance of enjoying every day. We vowed to pursue a time of exploration of this beautiful planet while we could. Cancer was just another fork in the road on my safari through life, another peak to be conquered, another adventure.

CANCER DANCER RAP

I'm a cancer dancer
I'm on a mission
to cure the body
with good nutrition
to spread the word
on this expedition
Cancer free is the ambition
So, eat and drink right
detox and relax
use those essential oils
you know what's best
you have to de-stress
to take away life's toils

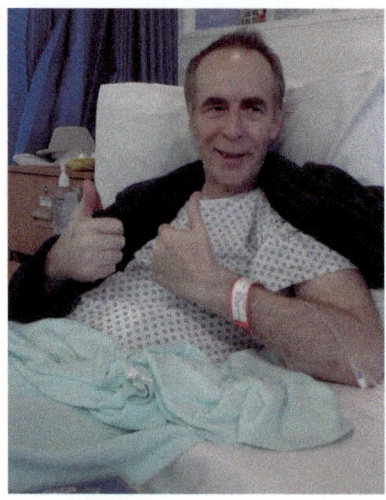

So change your lifestyle
put your body in remission
be positive and brave
change your mindset and position
With your loved ones around you
add in hope and faith
you can overcome the spectre
of cancer's wraith
Though the medical fraternity
may feel its sedition
through determined research
find a sympathetic physician
Whatever you do hold onto life
each day is a joy
forget all the strife
silver linings will prove an enhancer
as you survive in good shape
as a true cancer dancer
So, cancer TB4
take a hike
there's nothing in my body
that you will like

CHAPTER TWO

———

Formative years

I was always a worry for my dear old folks. Having fractured my skull twice by the time I was eleven it was always going to be a challenging and tough life. The mould was set! But hey, it's not all catastrophes … read on!

Wow, what a fantastic fifty years plus I have had travelling this wonderful world. How could I have foreseen as a seventeen-year-old just leaving school where life's rollercoaster would take me?

I was born on Thursday 24 June 1954 at 16 Corndon Road, Sundorne in Shrewsbury, in the front bedroom of my maternal grandparents' house. Little did I know the adventure I was embarking on. Rationing was just finishing following the Second World War and money was tight.

In the words of the old English nursery rhyme 'Thursdays child has far to go'. I guess I have put my own interpretation on that!

I recall being asked as a child what I wanted to be when I grew up and replying I wanted to be a missionary like David Livingstone, no doubt influenced by the books I had browsed through from my grandparents' bookshelf, my grandfather's church ministry in the community and the kindness of my dear Nan Morgan, who was always helping someone 'worse off' as she would put it!

My childhood was full of love, family and music. Isobel, my eldest sister who is almost nine years older than me, proved to be a brilliant and very talented singer. She sang with a well-known folk group called the Spinners

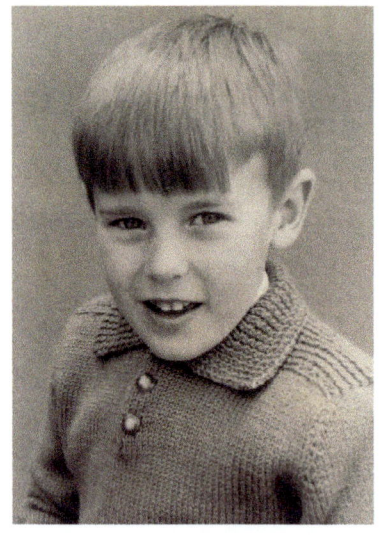

and a number of well-known artists like Peggy Seeger, Ewan McColl and others. Dad would occasionally get on the piano at home and could hold a tune and my brother Chris was also a good singer. My sister Sue and Mum were able to carry off a melody or two, but singing was not really my forte. I loved music, though, and of course joined in. Mum taught us the songs from both World Wars, especially the Vera Lynn tunes like 'We'll meet again' and old marching songs like 'Pack up your troubles in your old kit bag.' I was in Meole Brace Church Choir with my brother Chris for a number of years.

We loved going to see live music and through the years have been fortunate to see many top stars in concert, even meeting one or two of them. A fond memory was meeting Julie Felix in deepest Shropshire at a village hall in the little village of Clee St Margaret. It was so wonderful to listen to all her hits and say hello in such intimate surroundings rather than a big concert hall. Needless to say she sang 'If I could' (El Condor Pasa).

In our early twenties my pals and I used to go to the Drum and Monkey pub at Bromlow in Shropshire, which was often be frequented by well-known musicians like Ronnie Lane and Eric Clapton to name but two, who might end up jamming into the small hours.

We met the lovely Diana Jones at Shrewsbury Folk Festival, a fabulous annual event which attracts great folk and indeed a lot of Americana and country and bluegrass singers. We listened to her concert and as it again is so easy to chat to performers we approached and had a lovely chat with the very easy-going Diana and told her how we came to listen to her after hearing her restorative track 'Better times will come'.

Over the years we have been to many concerts, amongst them Alison Krauss, whose track 'When you say nothing at all' Dianne and I played at our wedding in New Zealand. Bruce Springsteen, Tina Turner, The Who, Bob Dylan, Van Morrison, David Bowie, Argent, Skippinish and so on, we've seen them all, and sometimes got the T-shirt!

We travelled down to Empire Pool Wembley in early May 1976 to see

David Bowie on his 'Isolar' tour supporting his 'Station to Station' album. I wasn't a big Bowie fan but what a brilliant concert. He was a fantastic artist and the consummate showman!

We loved the pop festival at Knebworth in 1976, the hottest summer on record at the time, set in a natural bowl of countryside perfect for acoustics. The Rolling Stones were the headliners. It felt like a medieval gathering as we all congregated. Many had flags on long poles to signify where they were so friends could find them. The performing stage was in the shape of a long tongue which curled into the crowd, like a fashion model's runway. There were lots of buskers, circus acts and clowns around the grounds. The Stones came on late and played a very long set, so much so the hours of darkness arrived and the lighting of the band's performance looked amazing as the event became a sound and light show.

We witnessed all sorts of irreverent behaviour at the festival, including lots of drugs of course. A black guy made love to a woman in the back of Land Rover, both buck naked and in full view of everyone. Topless women were abundant and nudity was just commonplace. A guy jumped on stage between acts totally naked and shall we say enjoying himself in front of the

The Morgan Children: Ron, Chris, Susie and Isobel

whole crowd before he was removed sharpish. Other bands included 10 CC, Todd Rundgren and Lynyrd Skynyrd. Some of the latter were killed in an air crash shortly afterwards and the band was destroyed.

As much as I love music, participation for me was more tuned to sport. Football in-particular was more my bag, I always remember my sister Isobel had a boyfriend called Sid Evans who used to bring me football programmes. I liked him very much, of course.

Over my childhood years, school holidays were mostly spent with Nan Morgan in Ironbridge then Madeley, especially if Dad was working away on a building contract.

Awareness of the larger world was not as full-on as it is today, with no social media or 24-hour rolling news, but I can remember that in the early sixties I didn't think I was going to get much past eight years of age when the Cuban missile crisis brought the world to the brink of disaster. One evening watching television at our family home in Mary Webb Road on the Meole Estate in Shrewsbury, the news programme broadcast doom and gloom, the end was nigh. I remember Dad walking to the living room window, surveying his beautiful and cherished garden planted with roses and other English flowers. As he talked to Mum, they were pondering, indeed fearing, the outbreak of a nuclear war and the destruction of the earth. Back then in October 1962 I had no idea where Cuba was, nor what a nuclear war entailed, but I knew it wasn't good!

Mum and Dad were very aware of international news, thanks to the BBC. I recall a piece of homework where I had to write a poem. I was about nine and my Mum flagged up as a possible theme the march on Washington in Summer 1963 when people went to listen to Martin Luther King. It had made an obvious impression on Mum, and I have since appreciated the momentous time in history this was as the seventeen-minute-long speech became known as the 'I have a dream speech', delivered on the steps of the Lincoln Memorial. The 250,000 marchers who had descended on Washington thronged the whole area all the way up to the memorial and beyond, calling for racial equality and an end to discrimination. The march was also for jobs and freedom and American civil rights, all reflected in my poem, helped by Mum. When many years later I saw the Lincoln memorial for myself, I had shivers down my spine, never thinking I would ever be here. I stood where there is an inscription on the spot commemorating this historic speech. I also went to the Lorraine Motel in Memphis where Martin Luther King was murdered a few years later, shot by James Earl Ray. It's depressing how even in 2020, fifty-seven years later, racism and discrimination still abounds, especially in the USA.

The furthest we travelled when we were kids was Wales. We had some wonderful childhood holidays in Tenby, Saundersfoot and Newgale on the Pembrokeshire coast. For the most part we travelled down by train from Shrewsbury, changing at Whitland. I loved the whole adventure, the smell of the steam trains, the corridors and carriage compartments with their overhead laced luggage racks, the pictures of seaside destinations on the wall and the sound of the engine hissing and spluttering, then the noisy whistle.

We loved to go to Monkstone Bay, between Saundersfoot and Tenby, where we would stay in a caravan on land belonging to Trevayne Farm. There was no bathroom in the caravan, just a little stove and sleeping quarters which were seats during the day. I fondly recall the smell of the Calor gas and the little gas lintels lighting the caravan at night. We used to kick a football or play tick in the huge field, or if it was raining stay in playing cards, I Spy or a board game.

The footpath down from the caravans was exciting, steep, winding and rocky. It was exhilarating to look down over the beach to the waves beyond. We spent many happy hours playing in the rock pools with small fishnets atop pea sticks trying to catch crabs in our tin buckets, then running into the waves with Dad holding our hands.

Mum used to hate the toilet block on the caravan site, which was part of the old farm outbuildings and draughty and cobwebbed. I always remember Mum wiping the toilet seat and then laying toilet paper on the seat before she would sit, whilst debating the attendance of spiders and the possibility of mice and worse … God forbid rats!

I recall staying in a caravan at Newgale during the summer of 1966. Dad couldn't get time off, but Mum took us kids on her own. We had lovely weather, bright sunny days and a wide sweep of beach to play on, but we used to get spooked at nights when the wind used to blow things around the outside of the caravan, especially at that time as Harry Roberts was being sought for the murder of three policemen in West London. He was on the run and sleeping rough, so this heightened our imagination, fearing he was outside.

The owners of the caravan had tied a knot in the gas pipe to restrict the flow. Those were the days! We used to sit in this small sandpit, which became our pretend fox hole, and use pieces of wood as ack-ack guns. Every time a plane took off from R.A.F. Brawdy we had to dive for the sandpit and take aim, pretending it was an attacking bomber. You can tell it wasn't that long since the Second World War. It was ingrained in our thinking, probably from things we heard from our parents and other folk.

In 1966 the World Cup came to the UK. What a wonderful time for a twelve-year-old boy! I was full of it, collecting all the posters and magazine articles and filling out the chart as the games were played. Even the World Cup, the Jules Rimet trophy as it was known, was stolen, then found by a guy out walking a dog called Pickles. You couldn't make it up. My mum along with Sid Evans took me to Villa Park and I watched West Germany despatch Switzerland 2–1 with a fabulous goal from their winger Lothar Emmerich. I had been picked up from school and still had my blazer and cap on, the latter of which I lost on the Holte End. What a brilliant day out, though. All those fantastic teams of countries from around the globe from Europe to Brazil and Argentina. Of course, watching England in the final at Wembley, albeit on television, was the culmination of a fascinating tournament with games like the tough one versus Argentina when their Captain Antonio Rattin was sent off and we won 1–0. Also games like South Korea versus Portugal where South Korea took a 3–0 lead and lost 5–3 and of course England beating Portugal in the semi-final to stop Eusebio and his team getting to the final 2–1

What an occasion the Cup Final was! I have the football programme for this famous occasion somewhere in the attic. But to lose a goal just before full time and go to extra time was agony to watch. Then came the infamous 'is-it-a-goal-or-not' Hurst shot, which hit the underside of the bar and bounced down on or behind the line in extra time. The game finished off with my favourite commentator Kenneth Wolstenholme's immortal words 'people are on the pitch. They think it's all over …' and then, after a long ball from Bobby Moore to Geoff Hurst who crashes the ball into the roof of the West Germany net and the closing statement from Wolstenholme 'it is now' as England won 4–2!

I enjoyed my time at the Wakeman School, which we all called the 'Tech' after its previous name 'Shrewsbury Technical College 'but was disappointed in some ways that I passed my eleven-plus exam as a lot of my mates went to Meole school, a secondary modern school back then, and I wanted to be with them. My folks were proud and I had a few friends like John Nicholls who went to the Wakeman, but there was only a handful of us off the estate.

I was not the best of scholars. I didn't really apply myself. I was all sport, football and cricket, and in the last year or two, girls became more of a distraction. My favourite subjects were geography and history. Mr Drury was my geography teacher. He had lost one of his arms after D-Day when he was injured at Caen. He, like my history teacher Miss Picken, inspired and engaged me. Miss Picken used to use a book called *1066 And All That*, which made history far more interesting. I recall spilling a bottle of ink all over my exercise book and having to copy up notes for a week to catch up, but she was very understanding. There were chances at school to go overseas and on trips nearer home like Arthog in Wales, but as a family we could never afford it. I guess I missed out on that experience and the opportunity of bonding with other pupils and teachers.

I did okay at cricket, but football was another matter. I had passion and desire but I guess I was just not good enough. In fairness we had one of the best school football teams ever at that time, which even went to the English Schools finals and did very well. My only appearance for the first Xl was at Oswestry Boys school in my final year courtesy of the main squad of players being away at the English Schools Tournament. John Hawksworth (Nodge) captained what was essentially the second XI masquerading as the first XI. Our games teacher John Evitts said 'Right chaps, tomorrow I want you to arrive with your kit washed and ironed, your boots polished and when you enter that arena to play, put your shoulders back and be committed.' Very sound advice, if he hadn't added, 'It will take them a good fifteen minutes to work out they're much better than you.'

Looking back, he was spot on. We looked the business and for the first fifteen minutes we surprised them and even went 1–0 up thanks to our captain Nodge. Alas, the occasion must have got to us, as we eventually ended up losing 8–1. Still, it was a great experience and at least I could say I played first X1 football and cricket for the Wakeman!

Whilst I was not a great eleven-a-side footballer, I did enjoy a lot of success in the 1970s in a five-a-side team in tournaments in Shrewsbury, usually up at Sundorne Hall representing Anchor Youth Club, which provided long-sleeved white shirts with a blue and amber round neck. There were about eight of us in the squad, which varied slightly over the three or four years I was involved: Peter 'Mags' Maguire, Graham 'Wobble' Croft, Roger Lee, John Brake, Keith 'Doog' Ivison and myself with a few younger ones, like Roger and Graham's brothers Kev Lee and Dave Croft joining the side. We won the league and cup competitions quite a lot over the years and if we didn't win, we were often runners-up. It certainly kept us fit!

There were the usual school altercations and shenanigans, fights over nothing with classmates, just for posturing most of the time, all handbags-at-dawn-no-knife-fights in those days, as I recall. But I made some good friends at the Wakeman school. We used to kick a tennis ball around on the car park area by the Gay Meadow (the quaintly named old Shrewsbury Town ground next to the school) and watch the Shrewsbury players practice on the cinder pitch at the back of the ground. Witnessing fast trains pass by on the railway line above us was all part of schooldays. In fact, the deputy head's son and I were avid trainspotters for a while, even going to some relatives of his in Darlington for a weekend so we could visit the train sheds. We of course saw the football players in our break time at school and collected autographs. The team had added interest for me as my cousin Trevor Boucher was playing for Shrewsbury at that time. Because Tottenham Hotspur, or Spurs, were winning everything in the early sixties and I used to watch them on the television in cup finals I supported them as my 'big' team.

Miss Leahy, my English teacher, was excellent. She shook her head at some of my efforts, but nurtured and improved me. Mr Withers, English Literature, used to love putting on a gramophone record for us all to listen to and discuss. Dylan Thomas's *Under Milkwood* was his favourite.

Basil Stott was another of my favourite tutors, our Mathematics teacher. I particularly remember him coming into class after first break one morning in

First Year at Wakeman Ron is top left!

October 1966 with a very solemn, almost stern demeanour. He then proceeded to tell us about the Aberfan disaster. We did not do much maths that day and the whole episode shocked the nation and stays imprinted on our minds even today. I have since been to the graves at Aberfan, a very poignant pilgrimage.

Alan Jones, John Evitts and Peter Roebuck were my games masters. Roebuck was a tough guy, who used to say after PE that we had five minutes to dress or he would lock us in the changing room, which he did more than once. Then we had to explain why we missed the next class and often had detention for it, so a double whammy.

We used to take the school bus to Monkmoor, where we played football and did cross country, or to London Road, where we played cricket, hockey and enjoyed athletics. There was a seat two-thirds of the way down the bus that was on its own facing into the aisle. If you sat there, the masters sitting in a double seat sideways on would slap your bare legs when you were not ready for it. We all tried to avoid that seat, seen as sport by the masters and feared as the torture seat by the pupils.

Who could forget Jenny Gleave and Miss Handley, or the art teacher Miss Burton? They were all attractive women and a number of stories come to mind about them. I remember Miss Gleave preparing us for a sort of music-and-movement mime performance during PE lessons. We actually staged a mimed football match with no sound or ball, ready for a parent's evening. It was like football in slow motion but very effective. Some days we had to do old-time or country dancing and learned dances like the 'Gay Gordon'. Suffice to say not my cup of tea, I'm afraid!

John Waddington-Feather was an English Literature teacher I only had for a short time but he made a positive impression on me. He encouraged all of us to read and developed my interest in English literature and reading especially.

When I was about fifteen years old I went with the family (except Dad) to Northern Ireland, taking the Belfast Steamship Company ferry MV Ulster Queen from Liverpool to Belfast. We stayed at a place called Holywood, pronounced 'Hollywood', with my sister Isobel's in-laws Paddy and Jean. Isobel had married an Irish lad called Christopher Randolph Stone, who had joined the marines and was serving all over the world and was usually based at Arbroath or Lympstone in Devon when back in the UK. We had a great time going over on the big ferry and going fishing in Bangor Bay and walking on the mountains of Mourne. On the fishing trip we rowed out into the bay and Paddy, my eldest sister's father-in-law, put feathers on the line with lots of hooks but it kept breaking because the line was rotten. Eventually we caught one mackerel, only to see as we rowed back in a host of anglers catching loads of fish just sitting on the end of the pier. We didn't even get to eat our sole mackerel as Paddy's dog seized it and ran off into the garden with it!

On the way back to Holywood from our fishing we were diverted from our planned route at one of the many British Army checkpoints, as a bomb had exploded in a pub. Such was the active sectarian violence back then. It was such a shame as everyone we met was so kind and friendly it was hard to believe this aggression and hate existed.

I was full of trepidation waiting for my GCE 'O' level results while in Northern Ireland, but thought I was safe on holiday away from the knowledge of the outcome. That was until my Dad called and told Mum, who was not very impressed. I had only achieved two 'O' level passes and some okay GCSE grades. Nevertheless, I was allowed into the sixth form to do retakes of 'O' levels and start some 'A' levels. This meant wearing a black blazer and essentially you had a bit of cred as an upper school prefect. In reality, I retook subjects and studied with the year below and loved it. I met a chap who was to become lifelong friend, David Turner, who had come from the Belvedere school.

In my spare time I used to read the *Victor* comic and especially 'Tough of the Track' Alf Tupper, who always seemed to have a bag of chips before breaking some athletics track record. I loved getting the *Soccer Star* each week too. Saturday nights in Shrewsbury meant the technology of the day brought us the *Sporting Pink* newspaper with all the up-to-date sports results of the day! *Star Soccer* was the ITV programme for footy highlights, but *Match of the Day* on the BBC was streets ahead. I enjoyed Subbuteo, which was a football game played on a large baize green cloth by flicking model footballers on a wobbly base at a ball. Others preferred to build things using Meccano and Lego which were popular at the time.

When I went to Shrewsbury Town matches I used to wear the blue and white, later blue and amber, 'bar scarf', but later decided not to wear one because it identified you at away grounds and made you a target for the hooligans of the seventies. Umbro soccer kits were popular if you could afford one. I wanted an England one but my mum bought me one with the badges of the four home nations on, which, to my shame, instead of being thankful, I was disappointed with.

Rosettes and bobble hats were big in the sixties, rosettes more so for cup days. I used to take a large home-made wooden football rattle with me to matches, which I still have to this day. It made a real racket and this was whizzed around whenever the Town scored, had a near miss or did something exciting. When my dear old mum first took me to matches, I recall the smell of tobacco from the briar pipes lit up by the men as they waited for the match to start. Half time signalled another opportunity to get the Ogden's flake out. Many blokes wore a jacket and tie and flat caps but more casual wear was about to take over.

There were lots of choices for the fashion conscious, but I was not really one of them. Sure, I had a Parka coat and also some Levi jeans, which I shrunk to size by sitting in the bath. Ox-blood or black Dr Martens (DMs) were all the rage at that time, but my folks would not let me have any, so I went for brogues with segs in the soles instead. You could hear me coming a mile away. As well as Wranglers, Dockers were in vogue, not forgetting Ben Sherman's shirts. Bleached jeans vied with French-flared trousers. I went for the latter in a velvet loon look as I was a bit of a hippy for a while. In due course after my dalliance with the long-haired rebel look, fur trimmed hooded Parkas gave way to donkey jackets, then Crombies.

As for snacks, Wagon Wheels and Jammie Dodgers were my favourite tuck-shop foods. I am sure the Wagon Wheels were bigger in those days. I liked my third of a pint of milk when it was free at school, although in the winter when the teacher put the milk on the radiator to thaw, it curdled and tasted awful!

Lyons Maid ice lollies called Goal and Rocket. Those who could afford them had a Mivvi with an ice-cream centre. Locally in Shrewsbury we loved Sidolis Bros, that familiar cream- and maroon-coloured ice-cream van that brought us expensive but delicious creamy ice cream. Vanilla was my favourite flavour.

Lucky Bags were a treat as were sticks of licorice, although packs of football cards with a thin sheet of bubble gum inside were brilliant, and addictive as you tried to collect the set, taking swaps to school to achieve this. For a short while American civil war cards rivalled the football cards and I learned a lot about history from these.

Smith's crisps were my choice over 'Golden Wonder' if only for the novelty of finding the blue bag of salt to unwrap, pour over the crisps and shake the packet! We didn't tend to have any food at the football, perhaps just a Kit Kat or a mug of Bovril in the early days, unlike some grounds that had a hot dog or burger stall. We saved our money for the journey home and some pie and chips at Kerr's fish shop by Belle Vue Bridge.

I collected loads of football programmes, mostly from Shrewsbury but there used to be a packet of assorted match programmes you could buy at the local shop and I did that for a while, swapping some with friends at school as kids do.

Televised sports programmes on a Saturday were *Grandstand* on BBC 1 with David Coleman or *World of Sport* on ITV with Des Lynham. Most Saturdays I used to watch *Football Focus* and catch the bus to Sundorne, the area I was born, to visit my Auntie Doris and Uncle Harry on Allerton Road. They had a dog called Dash, a mongrel and so loving, probably because I took him Goody Boy chocolate drops for dogs! I used to do some chores when I was there, perhaps helping with the mangle to get the clothes drier. I would do some gardening, then have some lunch. Auntie Doris always bought me *Commando* comics or JT Edson cowboy books.

I was a sort of surrogate son to Auntie Doris and Uncle Harry as they lost their son Barry when he was just sixteen. Life had been tough before that as my uncle was a prisoner of war in the Second World War and had fought in Iraq and then been taken prisoner at Mersa Matruh in North Africa by the Italians. He was moved to Italy and then, as the allies invaded Italy, he was moved to a German POW camp before being liberated. He had a mass of blond hair but came home completely bald following harsh treatment and incarceration conditions. He regularly had to cope with recurring bouts of malaria. He was a gentle giant, said little and carried the unseen scars of war. I watched him die of cancer on the settee at his home in Sundorne, this once powerfully built man reduced to skin and bone.

We would often watch rugby league with Eddie Waring the famous commentator with his sayings, 'eeeh up and at em' and 'he's a big lad'. Auntie Doris loved wrestling. At about 4pm we would sit down and watch the stars of the day like Mick McManus, who everybody hated, and Jackie Pallo and Giant Haystacks. I liked the tag wrestling where it was two against two but only one at a time in the ring and you had a get to the rope to tag your partner to swap over. The culmination of the Saturday afternoon sport was the exciting teleprinter on the BBC showing the final football scores as they came in. It was a slow teaser as the names of the teams were printed first, then the score. You started to check your pools coupon., Littlewoods, Vernons or Zetters, for the chance of becoming a millionaire and changing your life!

Approaching evening just before I went home, Uncle Harry would have a shave and put on a Marty Robbins album as a prelude to them going down

11th birthday party

Meole Brace Choir 1967

the Heathgates pub for a night out. I would catch the bus home, always with my books and five shillings pocket money. Happy days!

When we were growing up my mum ran a Freemans catalogue, from which she earned commission and concessions. Catalogues were ideal for people to have goods up front, usually clothes, and pay so much a week when they couldn't afford to buy something from the high street.

Mobil, Esso and other fuel companies always had advertising campaigns and some giveaways to encourage you to buy petrol from them. One I remember best was' 'Put a tiger in your tank' and another was 'Keep going well, keep going Shell, you can be sure of Shell, Shell, Shell'. And Esso's was 'Esso sign means happy motoring, happy motoring, happy motoring, the Esso sign means happy motoring'.

Television advertisements had a big influence on what products you bought and for that reason Dad disliked ITV with all its commercial breaks. As there was only two channels in the early days, we always watched the BBC when he was around because it had no advertisements. When Dad went out we switched to ITV, where I remember lots of jingles like: 'Murray Mints, Murray Mints, too good to hurry mints' and 'Don't forget the fruit gums mum', 'Snap, crackle and pop Rice Krispies', 'go to work on an egg', 'Don't be vague, ask for Haig' and 'A Double Diamond works wonders, works wonders, works wonders, a Double Diamond works wonders so drink one today'. I think it was Double Diamond who coined the phrase 'I'm only here for the beer, it's Double Diamond'.

A beer that didn't take off in Britain but did okay in the States was COLT45. It's the jingle that sticks in my mind sung to a marching rhythm. 'Any Tom, Jack or Walt who likes the taste of malt will like the taste of a Colt 45.' Then there was this jingle from Davenports:

Beer at home means Davenports!
That's the beer!
Lots of cheer!
The finest malt with hops and yeast,
Turns a snack into a feast.
Straight from breweries to your home,
Why collect?
We'll deliver!
Soon you'll know why folks all say:
'Beer at home means DAVENPORTS'!

CHAPTER THREE

Time for a real job!

In 1971 the band T-Rex was big in the pop world and Rod Stewart was singing Maggie May. Hot pants were all the rage for young ladies, Arsenal beat Liverpool 2–1 in the FA Cup final and a certain George Osborne was born. Decimilisation confused a few for a while when it came into being on the 15 February. By July that year when I started work in the travel industry, we were pretty au fait with it. Most people paid with cash and cheque back then, as the credit card revolution had not yet got going.

One day after school in early July 1971 I went to the Youth Employment Exchange in Shrewsbury and asked about a job, my folks and teachers oblivious of this course of action. I was shown to a small, uninspiring, sparsely furnished room. It had an old wooden desk which was topped with a large telephone, covered with many buttons to connect to different extensions within the building. I sat on an uncomfortable, worn leather chair. I was joined by a bespectacled lady, my job advisor, who came in armed with a new-fangled Rolodex filing system and a box of file cards. No computers in those days!

She said in a somewhat irritated manner, 'I believe you are looking for a job. What do you have in mind?'

'I have no idea, I just want a job,' I replied. With a dismissive sniff and a roll of the eyes intimating I was wasting her time and she had better things to do, she asked me about my qualifications. She was not impressed with my two 'O' levels in History and Geography and the grade two CSEs in Maths and English. However, I optimistically added that I hoped to have more passes from the examinations I had just taken.

Looking over her glasses, perched precariously on the end of her long roman nose, she viewed one card after another as she spun the Rolodex filing system. Firstly, I was made aware of a job vacancy in a local Barclay's Bank branch but then she added I needed a minimum of four 'O' level passes, maybe five, and the same for the Abbey National Building Society. Then strangely she pulled some cards out and looked at me and put them back without telling me what was on the them.

Finally, she extracted a card from what appeared to be the reserve box (the last-chance saloon) of wonderful job opportunities. Tantalisingly, she held the card in her hand, perusing it, as if deciding whether to share the information with me.

'Well, there is a Junior Clerk required at Luxitours' and before I could answer, she continued with the caution '… but the work is poorly paid.'

Just as she was about to return the card to the box of cards I intervened: 'Could you tell me more about that job, please?'

She couldn't tell me a lot more from the details on the card, but with a sigh she asked, 'Would you like to have an interview?' She appeared to be getting bored of the whole scenario now.

I replied excitedly, 'Yes, please' and an appointment was duly arranged. That same afternoon I walked the short distance to the High Street and had an interview with Mr Brian Bass, the Managing Director at Luxitours. I was of course still in my school uniform albeit with my sixth-form black blazer showing my upper school status.

I recall the interview as a short, straight-to-the-point affair. It was my first ever interview, so I had no clue what to expect or if I had messed up or impressed with this tall imposing man who was thinning on top but had a distinct air or authority. He gave the impression he would take no messing and I was not wrong! The interview took place in the far from salubrious surroundings of the stockroom, which was off a narrow shut over the road from the travel office at 44 High Street. It all felt rather surreal. The stockroom was piled high with brochures and had a table resplendent with what would become my nemesis, a printing machine with drum for creating a mass mail shot.

I went home not at all optimistic and full of trepidation. Later that afternoon our newly installed home telephone rang in the hallway. Before I could reach it, my mother had answered. 'Ron, you have a call from a Mr Bass.' Her bemused voice echoed up the stairs to my bedroom, where I had been playing some music on my record player.

To my pleasant surprise, the job was mine and I was to start my travel career on the following Wednesday, 14 July 1971. All I had to do now was explain to my parents that I was leaving school and not going on to do A levels.

Initially my decision to look for a job rather than continue to University was not too well received by my parents, until I enlarged on my ambition to earn some money and that I had a great opportunity, explaining that the phone call from Mr Bass was news that I had a full-time job.

Mum did not know quite what to say to me. She was clearly disappointed I was not going to get my 'A' levels and that I was foregoing the chance of a university education. However, time soon showed the decision to be the right one. I soon found out I had only achieved one extra 'O' level pass from my examination retakes! I now had amassed the grand total of three 'O' levels and a smattering of CSE grade two passes. So, travel it was!

Luxitours was a small group of travel agencies, what would later be referred to as a 'miniple', which revolved around a Blackpool head office, probably because two of the directors, Mr and Mrs Sleigh, came from there. There were also offices in Colwyn Bay, Wellington, Aberystwyth and Llandudno Junction. Luxitours also operated the Severn Travel Bureau in Butcher Row, which I recall was run by Beryl Pugh, who I worked with for a few years. In later years Beryl had an elderly sick mother and she was often late arriving for work because of that. It wasn't until later in life I appreciated her sacrifice when I was faced with similar family responsibilities. Roy Gandy had started as junior clerk the week before me, so we were both novices together.

With money from my first pay packet, I booked a driving lesson with Keith Challinor. A driving lesson cost £1.75, a fair chunk out of a take-home pay of £6.53. I paid my mum £3 a week housekeeping on payday, which was Thursday, and was borrowing it back by Friday … didn't we all? Beer was about 16p a pint.

When I started at Luxitours it was a male-dominated industry, with just two lady travel clerks and a couple of female typists. It was a manual recording and filing system so everything was written long hand. I was taught my trade for about 18 months before I was allowed to deal with the public and actually sell. During my training, I would assist all the other clerks as they progressed a booking and learned my craft from their experience.

As junior clerks we were essentially town runners and general errand and small task chaps, going to Burt Dann's to get the tuck-shop goodies and Beddard's for pork pies – but only when the boss Brian Bass wasn't around. One morning on one of our errands, Roy and I found Mr Bass passed out by his motor scooter in Peacock's passage, where he always parked. I believe he'd had a heart attack. Fortunately he survived and taught me my trade brilliantly along with his professional staff like Peter Duckers and Fred Wilson.

As mentioned earlier, there was a stockroom where we kept our travel brochures across the High Street from the travel agency. The bane of my life resided here, the aforementioned duplicator/printer. A skin would be typed out by the typist with a proforma letter and this was attached by punctured margins to the drum and then, taking care to make sure all the air bubbles were out of the contact between skin and drum by affixing the skin tightly to the drum, we then tentatively spun the drum with the attached handle. As the Luxitours headed paper whizzed through, we had to check the quality. If it was not right, the skin would come off and we would try again. The amount of clothes, especially white shirts, ruined by ink on them was horrendous for a young guy, who had little money and attended work for the first six months in his school sixth-form blazer until my mum bought me a smart light brown jacket from Halons, but even this was ink-stained after a while. A pity, as that doubled as my going-out jacket for dancing at Tiffany's.

I remember a large graphic poster in a frame of a new cruise ship being put in above our accounts section so that could be seen over the counter. Brian Bass was very proud of this picture and to this day I do not know who drew the shark fins circling the ship, although Mike Dodd the accountant and John Startin-Field were in the frame, the former being the most likely!

We sold a lot of journeys by ship, especially cruises and had wooden pigeon holes to hold all the brochures of the shipping lines. I had never heard of the likes of the Washington Line, but there were more widely known marine companies like Union Castle, Sitmar and Chandris. Union Castle had boats like the 'Reina Del Mar' but I particularly remember the 'Windsor Castle' with line sailings to Cape Town, which regular clients used each winter to travel south to the better weather.

Thomas Cook Continental Rail was a big seller too. We had to learn the timetable almost by heart, despite having the paperback book with a distinctive orange cover to use. Clarksons, Horizon and Horizon Midlands, Arrowsmith , Thomsons, Lyons, Global, Cosmos and Hickie Borman were just some of the well-known travel companies.

After eighteen months training I was allowed to serve a client. By then I knew off by heart many of the schedules, especially the Thomas Cook Continental Rail timetable. With rail services via Paris, it was popular as a major gateway to Europe.

The most popular destination at the time was Spain, especially the Costa del Sol, Costa Brava and Costa Blanca. Benidorm, Lloret, Calella and Torremolinos were all prominent resorts, not forgetting the Balearic Islands, Majorca with the commercial resorts of Palma Nova and El Arenal and later Ibiza and Menorca.

Back in those days in travel agencies, the clerks stood at counters to serve people. There were no real comforts for the clients such as sitting at a desk to make their travel plans. We issued rail tickets for the UK and continent, and of course air tickets, which we hand wrote, being very careful not to make a mistake, as it was like spoiling a cheque and had to be accounted for. Sometimes we issued 'MCOs' which were miscellaneous charges orders, used for ancillary services such as excess baggage on aircraft.

Jet aircraft had started to change people's habits as you could get to places more quickly. Jets chartered by tour companies brought down the cost of air travel. Some airlines, though, had a poor reputation. Dan Air was often ironically referred to as 'Dan Dare' after the *Eagle* comic book character. It was alleged some of their safety procedures were not always the best.

But these cheaper air costs, combined with deals with overseas hotels at advantageous rates by buying rooms in bulk over a season and some airport transfers to get clients between the airport and hotel meant the mass market package holiday was born!

Cruise lines were not really that big a market at this time. P&O operated educational cruises for schoolchildren on ships like the SS Uganda and did introduce a new flagship called the *Canberra*, which in time became a favourite with clients. The *Canberra*, alongside the *Queen Elizabeth II*, were later used as hospital ships in the Falklands conflict in 1982. CTC was a Russian cruise line which was cheap but whose idea of entertainment was initially showing you a film of the Moscow May-Day parade, or so the story goes. Cruise staff multi-tasked as stewards, entertainers and maintenance staff, although to be fair some cruise members still do that today.

As far as revenue was concerned the rapid increase in package holidays was predominantly during the summer, meaning May to October in the sixties and early seventies. Winter sun package holidays took a while to take off, as a number of socio-economic factors needed to take hold, not least of which were longer paid holidays from work and the wider use of credit cards. Winter ski holidays, of course, developed first but winter sun holidays were a different ball game, but gradually thanks to industry-leading companies like Thomsons, Intasun and Clarksons they gained traction.

By the late seventies and early eighties I had moved on to other companies, where we sold a lot of mini-cruises from Harwich and Newcastle to Esbjerg with DFDS, but in reality, cruising in this case meant we used a smarter ferry than the channel ferries and there was good musical entertainment and a smorgasbord offering in the restaurants. These three- and four-day cruises went down well, especially the longer one, which involved a night's stay in port and a visit to the twelfth-century town of Ribe in Denmark, where we offered a short tour and afternoon tea.

As time went on destination choice grew with the addition of Tunisia, Romania and Bulgaria, for example. The latter two were exceptionally cheap but quite austere and accommodation was typical of the Eastern bloc, bland but functional.

British Airways was yet to arrive. It was still British Overseas Air Corporation known to all as BOAC for the long-haul flights and British European Airways known as BEA for short haul. Clarksons was the flag bearer for the cheaper charter flight package holiday companies, whereas Thomsons was a market leader in good quality and companies like Horizon tended to be a little more upmarket yet competitive.

How times have changed. I still remember those days when a blackberry was a fruit you picked to put in a pie and an I Pad sounded like something you covered your eyes with to enhance sleep. Email and computers were still to come, everything was manual filing and a lot of long-hand writing. Anything to be typed was dictated to our typists.

One day I was in the office and two police officers came in. As I went to serve them at the counter I wondered why we always feel so guilty in front of a policeman. One of the officers said to me and another lad behind me, 'Right you two, at the police station this morning at 11am.' We were speechless. He continued, 'Identity parade.'

'Okay, I'll have to check with my boss,' I replied, finding my voice.

'Eleven o'clock on the dot. Your boss will understand,' he said with a smirk as he walked out. It seemed we had no choice but to be there. So we told our boss and, though annoyed, he confirmed we had to go. We reported to the old Victorian police station on Swan Hill. We were put in a line-up of about fifteen blokes, then a lad came in and tapped a few of us on the shoulder. I thought, blimey, that's it, I'm guilty of something. But we were then escorted away by a police officer to the front desk, asked to sign a piece of paper and given five shillings for our trouble.

Well, I was relieved and made up all in one go. Mind you, I said to my mate Clive who was also released, don't say anything to the boss as he'll want the money for the time we've been missing! As it happened on the walk back to the travel agency, I found a wristwatch so in my lunch hour I took that back to the police station. They recorded the information and put it in lost property, saying that if it wasn't claimed after a certain time it was mine. It was never claimed but I didn't go back for it. Somehow it didn't feel right.

CHAPTER FOUR

———

My first overseas holiday

I planned my very first overseas holiday with Lyons holidays, chosen from a bright red brochure. We managed to get a good discount based on the commission my company would have received and we paid about £50 for a two-week half-board holiday, put in perspective about six weeks wages. We, John Brake, Dave McCormick, Neil Williams and I booked to go to the Hotel Alejandria, Playa De Palma Majorca, a cheap and cheerful two-star hotel, in July 1972.

We had agreed to travel to Luton airport in Neil's Wolseley Hornet, which was similar to the famous mini in size. No surprise, then, that the lads were aghast when they came to pick me up and I had the family-size, expanding suitcase packed to bursting with every bit of clothing I owned. This included a large, sleeveless winter fur and leather jacket which had just arrived from the Freeman's catalogue and a heavy white wool trouser suit with huge penny round collar and flared bottoms, both of them far too warm for a Majorcan summer where temperatures would reach well over 30°C on a daily basis, but I thought I was the bee's knees. Of course, it just proved I had no fashion sense or even common sense.

Just as we were ready to leave, having secured my suitcase on the top of the car with rope, a young lady arrived in a taxi. We didn't see many taxis in our street and I was amazed to see the young girl place her suitcase outside our gate, look at an envelope, then at our house. She collected her luggage and proceeded to lift the latch on the cottage-style wooden gate made by my Dad, hopefully not noticing the central strip of wood was broken at the top after a drunken neighbour had fallen over the gate a few days earlier after a heavy night on the town. She proceeded, case in hand, up our path.

Puzzled, I followed her and soon realised it was my German pen friend Margarethe. I was shocked but not half as much as poor old Mum. Pen friends were quite common in those days, we actually wrote letters to each other. The trouble was it took a long time for letters to reach one another by 'snail mail'.

Apparently, Margarethe had written to me to say she was coming to stay for a week, and I must have written back and said casually that would be great. No dates were agreed and she just appeared. I had to explain to her I was off on a two-week holiday in the sun. Always one to grasp the positive, I pointed out to Mum my room was free for Margarethe to use. I had not long moved into this room, as the pecking order changed when my older sister left home and I now had a single room instead of sharing with my younger sister and brother, so Margarethe was in luck!

Margarethe was, shall we say, rather demanding and needed looking after and taking places, which Mum duly did, but it was difficult for my family as we didn't have a car and nobody could drive. Also, Margarethe had one heck of an appetite being a farmer's daughter. My sister told me at Sunday teatime my mother put a bowl of tomatoes on the table for all to share but Margarethe just picked up the bowl and ate the lot!

Well, not to worry, the lads and I were off on our holidays. What could I do, bar getting on with things, stopping en-route to Luton at a pub, as you did in those days, and then flying out to the sunshine?

Checking in and getting on the aircraft I was aware of this chap in the queue who was in a heavy woollen sweater, thick trousers and walking boots. He sat across the aisle from us and we engaged him in conversation. He told us that he had somehow boarded the flight to Majorca instead of his planned destination in Austria. We laughed at his predicament, but he did okay, as the tour company, admitting it was their fault, gave him £15 on arrival to buy some shorts, flip flops, T-shirts and toiletries. In reality his Austria flight was probably overbooked and they redirected him.

On arriving at Palma airport we were duly transferred to our hotel. We were informed our rooms were not ready even though it was the evening.

We were shown through to the restaurant and given a meal and copious amounts of beer and sparkling wine, which they described as champagne. Before long we, along with a group of Irish and German lads, were told the hotel was overbooked.

By now an alcohol-fuelled bonhomie had developed and we banded together in protest and negotiated free drinks at evening meals as part of the deal for the rest of our stay plus they promised us an upgrade to a three-star hotel from the two star we had originally booked.

So, a little worse for wear, we transferred, feeling smug with the negotiation outcome especially as we were upgraded to a brand new hotel, the Bahia de Palma, in the touristy party resort of Arenal. On arrival we ascended the marble steps and entered the attractive and rather grand portico into the marble-floored reception, before realising the down-side of the deal was that they had not finished building our new abode.

Lift shafts were empty chasms with no doors, just a red and white warning strip of polythene was all that stopped anyone falling down the shaft. Gradually over the two weeks these building projects were completed and we had a great time. These were the days when hotels were being built so quickly that many brochures just had artist's impressions, so it was no

Four long haired guys having dinner in a restaurant

surprise when people arrived only to find many hotels were not complete!

We had a blast, the first of many holidays to come. We explored the island, travelling up to sedate Formentor with its cocooned, famous hotel and tree-fringed beach. It was located in one of the most beautiful areas of the island, facing the idyllic Formentor Beach, the perfect place for a few days of relaxation and a place of retreat for novelists, poets and musicians in search of inspiration, a fabulous hotel which has inspired the artistic for many years. We visited Manacor for its pearl factory and of course the caves of Drach at Porto Cristo. I seem to recall stopping at a Bodega for some wine tasting on the way home, but I don't recall much about that day, if you know what I mean.

During our stay we spent a great evening going over to Barbarella's night club by boat from Arenal on a so-called 'Champagne Cruise'. We drank and drank and despite the crew rocking the boat on purpose and many being ill, we walked down the narrow gangplank, all still drinking from a bottle (think sparkling cheap plonk), which was duly extracted from our grasp as we hit land. We had a great time but were so wasted in the club, which was massive and billed as the biggest subterranean disco in Europe. It was filled with tourists and sailors from the American sixth fleet. Gary Puckett and the Union Gap were the live band playing tunes such as their famous hit song 'Young girl'. We eventually got a taxi home all bleary-eyed for breakfast about 6am having missed our coach transfer.

We challenged the waiters to a game of football one day and agreed to get an 11-a-side team ready for the weekend. We and four Irish guys and a few others turned up only to find out that the waiters actually played for the local football team Arenal FC. To say they were pretty useful was an understatement, not to mention the fact that we were playing in 35°C temperatures, to which they were far better acclimatised and we lost 5–3, a lesson learned! We could only cope with about twenty minutes at a time and had roll-on roll-off subs.

It was so hot we spent most days at the beach or around the pool and of course partying at night. One thing that sticks in my mind is seeing the packed beach the first day we were there. What I thought was a very old woman, brown as a berry, her face very wrinkled like the corrugated terrain on a relief map, turned out to only be in her late fifties. She was so wizened from the constant exposure to the sun, it made me make sure I had smothered my skin with plenty of the sun lotion of the day, Ambre Solaire, or the Spanish equivalent, which was cheaper, and some of the sights like this lady put me off prunes for life!

I'm not sure how we arrived back in one piece from this holiday. I watched one of the lads come down the staircase towards the restaurant one day and go towards what he thought was an open patio door. Unfortunately it was not open and he bounced back into Reception and ended up laid out on the marble floor, slightly dazed. Sunlight plays some strange tricks. Being a travel virgin myself, I did all the wrong things. I drank too much cheap booze and got sunburned badly. But what a hoot!

CHAPTER FIVE

———

Learning my trade

In 1973 Luxitours were taken over by Hill Samuel Hunting Lambert, a shipping company. We moved to the top of Mardol in Shrewsbury, to a huge three-storey building. We felt the bee's knees in such a prestigious office, but I myself was starting to get a little ambitious, thinking bigger and wanting a change.

During this time, I went on an educational, my first familiarisation trip ever, to the nearby Channel Island of Jersey. Modern Line were the company who invited us, and it was clear I was offered the 'fam' trip because none of the senior staff at Luxitours wanted to go, preferring a more exotic destination. I was happy to have the short break. Our party of travel agents stayed at the Mayfair Hotel in St Helier and the company showed us around the attractions and various beauty spots and beaches that Jersey had to offer.

The whole idea of fam trips is to influence you in your choice of tour company or airline that you might recommend to your clients and also to educate you in what destinations had to offer from sightseeing to the quality and location of accommodation. Tourist boards, airlines, tour companies and hotels and other travel affiliates all contributed to make the cost of such an educational experience low for the travel agent and sometimes it was even free.

We were taken to the German underground hospital, a fascinating, cold and frightening subterranean environment which was built by slave labour, many of them Russian prisoners of war. We thought the wide sweeping sands of St Brelades bay were beautiful, with a lovely little chapel and churchyard nestled at one end.

We headed out one night during our stay for some evening entertainment. The headliner act at a local club was the Bay City Rollers, a popular group of the time, though not really my bag, and we stayed after their gig and had a drink with them. They were typical of the touring musician stereotype, flirtatious with the ladies and a little arrogant and full of themselves. I was a little surprised how prickly they became when I wound them up about copying the tartan scarf image from Rod Stewart, who at the time was my favourite pop artist.

In April 1974 I left Luxitours for another company, benefiting from a promotion. It was a pity, really, I loved my time there and learned a lot from everyone and especially Pete Duckers, Fred Wilson and Brian Bass. Fred was a typical Londoner, from Acton or Southall, I think, with lots of lines if you asked him things. If I said, 'You look busy, Fred, but can I ask you something?' he would say, 'I am busy, the ship is sinking, the mast is below the waterline' and so on, but he would coolly advise and answer my question.

Not long after we were taken over by Hunting Lambert, they introduced luncheon vouchers, about 75p a day, I seem to remember. We used to go and get a sandwich or pie from Wimbush's with them, but then one lunchtime I went with Pete and an airline rep to the Plough pub in the Square. Pete would go most days but for me I didn't like to drink during the day, especially as I had to face the public. Then Pete shared with me that he had made an arrangement with the landlord where he would take luncheon vouchers for alcohol. Well, that made it very tempting but I managed to resist. Pete was a clever chap. He worked in business house which dealt with local companies and the travelling they did for servicing and gaining contracts. Peter was very worldly. He knew so much off by heart, but that was the way in those days. Without computers, knowledge was king. He would test me on the three-letter airline codes for airports and wind me up with codes like FUK and FUJ for Fukuoka and Fukue just to hear me say them.

My fellow junior Roy didn't escape the odd wind-up, either. When we first started, he was sent down the road to the ironmongers to get a packet of 'skyhooks' to be greeted by a bewildered assistant who smiled and said, 'I think you've been had!'

Pete made up his own meanings for the airlines. Lufthansa was 'let us fondle the hostess and not say anything' (I'm trying to be polite there!), SAS was 'sex and satisfaction' or 'sex after service', BOAC – 'better on a camel', TWA – 'tiddlywinks airlines' or 'try walking across the Caribbean', LIAT – 'late if at all', Polish airline LOT – 'loads of trouble', SABENA – 'such a bad experience, never again', DELTA – 'doesn't ever leave the airport' and BA – 'bloody awful'.

Pete was such a character. He would reel you in with jokes when you thought he was being serious, for example 'I was on a TWA flight last year and the hostess said "Would you like some TWA coffee?" and I said, "No, but I would like some of the airlines' tea"' Think about it (he would make the delivery more obvious).

On leaving Luxitours I joined Gold Case Travel as an assistant manager. I loved it there. I essentially ran the agency as Eileen the manageress often went AWOL to horse-racing events with her bookmaker boyfriend Tony.

Eileen would often leave a message on my desk saying, 'Gone to York races' or similar and I might not see her again for a few days. She would come in late in the day with Tony on her return, often with some interesting characters and ask me how business had been, then disappear down the Three Fishes in Fish Street, where I would occasionally join them for an après-work beverage or two.

I remember them coming to the office early one evening as I was closing up. They were accompanied by the Australian golfer Bob Shearer, who had just won the Piccadilly Medal at Coventry Golf Club and it might have been Sam Torrance with him. I want to say Greg Norman but my memory is dulling and I cannot be sure! Eileen was a great lady, quite stern towards her staff at times, in the old-fashioned autocratic style, but she was always good to me.

One of the staff I worked with there, served a client one day who came in to make a travel enquiry. Our staff member took on board his request and started looking through the train timetable for the rail schedule to Aberdovey, presenting the client with some suggested times. The client smiled, slightly perplexed, looked her square in the eye and said, 'I didn't realise you could go by train to Abu Dhabi.' A little tip here: make sure you know what the client needs and before you start researching always go through the enquiry with your client and establish all the facts and requirements!

Whilst at Gold Case Travel I was invited to an educational a little further afield than my first one, this time to the burgeoning Mediterranean

Gold Case Travel

party resort of Lloret de Mar, situated on the north-east coast of mainland Spain. You must remember that just like my first visit to Spain in 1971, Spain was still run by the dictator Franco, something that was lost on me at the time.

We had a great time, with other travel agents and Thomson tour operator staff like John Gullidge. I made some very good friends, we checked out hotels and attractions in Calella, Santa Susana and Pineda, which boasted the well-known Taurus Park Hotel. We basically partied whilst staying at the lovely Hotel Monte Cristo. At every hotel we went to inspect we were offered a drink and we saw a lot of hotels over five days, our judgement and recollection of rooms and hotel facilities waning by the time the sun went down each day. No doubt we were young but not quite as professional as we might be expected to be in later years.

During one of the small hotel visits in Santa Susana, four of us got into a lift which was very bijou. It jerked alarmingly into operation, reached the first floor, then, instead of the old door opening, there was a creaking and grinding cacophony, before the lift cage stuttered and plunged back down to the bottom floor and the lights went out.. It was probably only about a four-metre drop and fortunately none of us was physically injured. We were just shocked and traumatised. We had to wait for folk to come and get the doors open as we had ended up slightly below the exit zone for the reception and it was pitch black.

I cite that as one of the reasons why on the last night we had a blow-out celebration and about fifteen of us ended up in somebody's room, where we carried on drinking. When morning came I woke up in the bath and someone else was sleeping on the bathroom floor. As I stepped over them, I viewed the carnage in the room, about three people fully clothed on each bed and people asleep on the floor, in a chair and some on the balcony. It was as if someone had thrown a grenade into the room. I tiptoed out and returned to my room, showered and went to breakfast. Fam trips became a lot more professional and objective in the years ahead but there was still the odd party or two.

That summer I holidayed with a good friend David, heading to Germany and France. We assessed our linguistic capabilities and agreed that whilst in France David would speak a little French and I would take care of the German later in the trip. We had a great time in spite of a number of interesting experiences, shall we say.

As usual we were on a tight budget, so, on arrival in Calais we wanted somewhere to camp. With our backpacks on we walked out of the port and David started asking passers-by, 'Avez vous un camping place?' Most people ignored us, but eventually someone said there wasn't really a public campsite they knew of. So we slept the night in the sand dunes before catching a train to Lille.

We then decided we would hitch-hike as our destination was Northern Germany, close to Bremen. This was the farmhouse home of my pen friend Margarethe Erdmann (remember her?). Now rest assured I had told her to expect us!

We stood on the side of a very busy motorway for over two hours without luck and eventually gave up with our thumbs suffering from repetitive strain. We decided to walk back to Lille railway station, even though it meant using some of our limited funds to get a train. We figured we could sleep overnight on a train and get nearer Germany.

As we were heading from the motorway to the railway station, we passed down a rather narrow path and we realised we were being followed by a group of local youths. As these guys gained on us, we decided to hotfoot it and, sure enough, they gave chase. With backpacks on and the adrenaline pumping we tried to stay ahead of them, but they caught up with us just as we entered the square in front of the railway station. Fortunately for us there were some lads from Yorkshire drinking in a bar who saw the little fracas and came over to assist, and the local youths ran off.

Phew … saved by our compatriots, we had a drink with them, then went to catch a train. By now it was evening and we could only get as far as Brussels before changing and heading towards Bremen on a morning train.

On the train we sat opposite a smartly dressed businessman who smiled at us and we returned the gesture. Eventually I suggested to David to engage him in conversation to see if he could advise us on where to stay in Brussels for the night, a cheap hostel perhaps.

However, David was a little tongue-tied as he galvanised himself before letting fly with his opening gambit which was 'il pleut' and pointing to the train window. The gentleman smiled and nodded agreement. Ice broken, David waited a few more minutes then tried the same friendly observation and the guy started laughing. He spoke a little English and when we got to Brussels, he showed us where we could store our backpacks in lockers. He then kindly took us to a local cafe for a snack, which he paid for.

After about an hour he invited us to his place, a club in the suburbs. We hesitantly agreed. David was a six-foot-tall lad and we decided we were both fit and strong enough to cope with any aggravation, so we took the risk.

We took a tram and after just a few stops alighted. After a short walk of no more than a few blocks we arrived at a café, which it turned out was a strip club on some nights and a bar on others, but on this night, it was closed. Our host made us coffee and a plate of bread and cheese. He then proceeded to show us how he could fix the pinball machines to stop people winning, all purely by flicking a switch behind the bar. Seeing we were tired he showed us the bathroom where we freshened up. He then led us up a wrought-iron spiral staircase which led directly into an attic bedroom which contained a double bed. He said, 'You can sleep.' We thanked him and he disappeared.

We were very tired so stripped down to our underwear and jumped in, which felt strange for two blokes sharing a double bed even though we were good friends, but a few minutes after lights out, just as I was dozing off, David shook me and said 'He's got into bed,' which now meant David was in the middle. I simply said in my drowsy state, in a spirit of seventies liberalism, 'He's probably just being kind and sharing his bed. He'll be all right.' A few minutes later, just as I was dozing off again, David shot up in the air, standing in the middle of the bed saying, 'Let's go,' and that was that. Fifteen minutes later he called us a taxi and we headed back to Brussels Midi station.

It was probably all very innocent but for two 'wet behind the ears boys' from Shropshire, we didn't take the chance and after all, it was easy for me to be relaxed, I wasn't the one in the middle! Needless to say, the cafe owner was bemused. By now it was well after midnight, but he even paid the taxi fare to the rail station, fair play to him!

Fortunately, we only had a few hours to wait for the early morning train to Bremen so we could visit Margarethe's family in Schwarme near Bremen. We enjoyed our visit to Bremen and the Erdmann family immensely, they made us very welcome from the word go when we arrived late at their home. We slept in sleeping bags on the floor in one of the downstairs rooms and awoke the mid-morning the next day to be treated to a wonderful smorgasbord of farm fare.

However, no sooner had we finished than the daughters laid the table for lunch and the farmer Herr Erdmann and a farm worker came in to eat having been working since early morning. It was a short while before we realised we were now expected to eat lunch straight away with them. Well, instead of insisting we were too full, we thought it polite to try and eat some more. We soon became somewhat bilious and incapable of eating anything else. It was very difficult to try and do the fine luncheon spread any justice and the soup was the straw that broke the camel's back. It seemed to be a cold, thin, wine-infused dish with what could only be described as polystyrene floating on the top. I'm sure this local speciality was more nutritious than that, but that's how it appeared.

Herr Erdmann was a quiet, unassuming, yet very interesting guy. He had served in the Wehrmacht and survived the Eastern Front in Russia during the Second World War. In the evenings, after dinner, there was no television or heading for a computer as many may do today. We all went into a music room, and in turn the four children sang or played piano or some other instrument. David and I feared we would be asked to contribute, but fortunately the Erdmann family were spared that ordeal!

Bremen was a very interesting city. Its architecture had been restored to its former glory since the war and to all intents and purpose the Rathaus and the Platz looked original. We were fortunate to watch an International Moto Cross grass track event in Schwarme with just one British motorcyclist in

Don Godden and he was brilliant, winning his events. It was also great to go to the local community pool which was open air, quite a novelty to us.

On our return home we stopped off in Paris to explore. This was another opportunity for David to show off and practice his French. One of his less glorious moments was wandering along the boulevards asking passers-by for directions to the iconic Eiffel Tower. His memorable question was 'Qu'est-ce que c'est le Tour Eiffel?' Strangely, many moved on without answering, looking bemused themselves and others just laughed. We eventually realised why when it dawned on us that he was asking *what* the Eiffel Tower was, not where!

Every November as a prelude to the following season's holiday selling, travel agents were invited to promotions by travel companies to highlight their products and thank us for our support. We used to get some great bands playing at these promotions, such as Martha Reeves and the Vandellas at a Cosmos evening. In 1974 Cosmos had again arranged a party in central Birmingham, but the very day before we were due to go, the Birmingham pub bombings by the IRA took place. We were lucky not to have been in that vicinity, as our venue was close to where the explosions took place. Cosmos reacted quickly and our promotional party was transferred to Chateau Impney at Droitwich. Even there, the evening was subject to a bomb scare and an evacuation of the premises had to take place, such were the times we lived in.

It was a tough year economically speaking, who can forget? This was the era of the 'Three Day Week'. To reduce electricity consumption and conserve coal stocks, Prime Minister Edward Heath announced a number of measures on 13 December 1973, including the Three-Day Work Order, which came into force at midnight on 31 December. Because of fuel shortages commercial consumption of electricity would be limited to three consecutive days each week, television finished at 10:30 each night, and many pubs were closed. This went on into February 1974 and we got used to candles for lighting. It was quite exciting in some ways, but the feelings against the miners and other strikers became quite bitter at times instead of them gaining support.

In 1975 a group of us lads decided to go and see Scotland so we set off in our old Oxford Cambridge, the team car we chipped in and bought together. It was an upgrade on the previous team vehicle, a Bedford van with a mattress in the back, but we will not dwell on that, other than to say the van was useful to take a lot of people home from the pub. We stopped off for a drink in Carlisle on the way up and as it was getting late, we pulled over in Gretna Green in the car park of the Old Smithy and we all slept in the car, not the most romantic thing that has ever happened at that location, I'm sure.

We then proceeded north. Our plan was to go up the west coast and down the east coast and this we did. What beautiful scenery! Fabulous shimmering lochs and stupendous mountains and vistas. One place, Gairloch, stood out for us apart from the obvious like Ben Nevis, which we didn't have time to climb. Shame, but it was just a whistle-stop one-week tour.

We sped up the west coast into Wester Ross and stayed on a campsite in the seaside village of Gairloch, a spectacular location on the loch. That evening we popped down to the large modern Gairloch Hotel, but it didn't have much atmosphere. We were recommended to check out an attractive little pub called the Badachro Inn. We drove along a rather tortuous, winding narrow road around the bay. Set on a lovely little cove the low-ceilinged inn with lots of local fisherman and crofters in was full of character.

Did we have a good time? Well, let's just say they had a top shelf of scotch bottles, so many different types it would have been rude not to sample them all. How we returned to the campsite without causing injury to ourselves or others is a miracle and an indictment of the stupidity many people indulged in, not just youths, showed on those pre-drink-driving-law days

Next morning, I awoke in our six-man tent with just one of the guys. To say it took a while to focus and then try to think where we were, is a bit of an understatement. I crawled to the mouth of the tent to see where the others were. I saw another of our party asleep under the car and that left two unaccounted for. I roused my sleeping companions and we searched all over, but no one was found in the field or toilet block. Where could they be?

Eventually we considered going to the police, but before that we climbed to the top of a large sand dune to get a better 360-degree view. As we reached the summit we gazed down the other side and there they were, lying side by side face down, arms out, almost as if they were holding hands. We hurriedly reached them and fortunately they stirred from their drunken coma. Wow, what a lesson in what not to do when imbibing the local tipple!

We lit the primus stove and made a full bacon, egg and sausage breakfast to try and get the systems going. Immediately we had an entourage of seagulls wheeling above and then having the bravado to dive and try and get the bread and even something from the frying pan. Graham solved the problem by coating a ball of bread in cooking fat and sending it skywards. With great dexterity the gulls caught the treat and they left us alone.

We headed further north to Cape Wrath and we could see why it got its name. There was an inn cum hotel not far from here, in the middle of nowhere. It appeared closed until we stumbled across a door at the rear of the property. With a slight push this opened and we literally fell in down some steps. We had stumbled into a snug with just two crofters in it. These local inhabitants just looked up, then returned their attention to the half pints and wee drams in front of them. There didn't seem to be anyone serving at the bar so we called out, 'Hello! Anyone home?' No response, then we saw a bell on the bar and rang it.

After a short while, a small rotund lady in her fifties came through the doorway behind the bar, drying her hands on her apron. She stepped up onto a box or ledge so she could see over the bar (I kid you not) and we think she asked us what we wanted, her brogue being almost intelligible. We all asked for a half pint of beer (we couldn't face whisky again for a while). Our barmaid nodded and then promptly disappeared. Five minutes later she was back with a jug of ale, which she poured and handed a glass to each of us, took the money and left again, pretty much without another word.

We had a lovely drive through the wild narrow road going over to John O'Groats, then down the east coast back through towns and cities, but it's the west coast that left the lasting impression on us.

Back in Gold Case travel, I recall a junior member of staff asking me where Helsinki was and, let's face it, we all have to learn and it was interesting for them to find out it was in Finland and indeed the capital of the country. Nowadays any request is put into Google or some other search engine and personal knowledge is not as great or some may say not as needed as we have become computer reliant.

Being a young guy, one of the joys of working in a travel agency I found was being able to meet lots of people and share in the excitement booking a holiday brings. Of course, there was also a fair share of young ladies booking holidays and one day a couple of very attractive girls came in, one blonde, one brunette. We got chatting and seemed to hit it off, a reservation was made for them and, satisfied, they headed for the office door. In those days at Gold Case Travel we had an old-fashioned cash register and as I was putting away their deposit I was simultaneously watching them leave. Both waved goodbye as I rang up the payment and the till drawer opened to the familiar sound of the bell. As they closed the door the big smile on my face turned to a grimace. I had caught my thumb in the till drawer. It hurt so much I couldn't breathe at first. As I extracted my thumb, blood rushed to the site of the injury causing a huge blood blister under the nail. It kept throbbing and was so unbearable I asked the lass I was working with to mind the shop while I went round the corner to the Royal Salop Infirmary to see if they could alleviate the pain.

This hospital was, shall we say, rather old-fashioned having been built in 1830 and the minor accident and emergency department was very small. After what seemed an age I was invited through behind a curtain and told to sit at a table. Shortly I was joined by a very pleasant Asian doctor, who asked me how the injury had been caused and so on. As he did this he opened up a paper clip he was fiddling with and started heating it on a Bunsen burner. I can remember thinking that he wasn't paying much attention, then all of a sudden he grabbed my hand and forced the heated end of the paper clip into the thumb nail, which immediately released the pressure and relieved the pain.

It may not have been very orthodox but this low-tech solution to my predicament worked wonders! I was more careful with the till after that episode.

For Christmas 1975 David, whose father worked on the railways, obtained a rail pass to travel in Europe. I managed to procure a return ticket to the south of Spain from our rail agents within the company I worked for. I had a stinking cold and was so happy to be heading for the sun. We were both wearing heavy greatcoats, my RAF overcoat courtesy of my Christmas bonus and a sale of army surplus at the Army and Navy store. Our hair was long as was the fashion and looking back we looked a proper rough pair of undesirables.

After a long journey on the train through France our trip took us to the Holiday Inn between Malaga and Fuengirola for two nights. We chose this for no other reason than I spotted a darn good hotel deal without paying any thought to the likely weather we might encounter. We soon learned the temperatures were going to be low and the coast rather breezy, and there was nothing of allure like entertainment n offer. No wonder the price was so good. I was a very young, inexperienced traveller in those days.

Unsurprisingly, there was no one in the hotel except us. It was a ghostly place with a windswept beach leading to a turbulent sea. I had just £9 spending money for the whole trip. It was lucky we had the free rail tickets. We were therefore careful with what money we had. We were only buying bread, cheese and cheap wine, but we soon learned that 20-peseta wine was much better than 10-peseta wine in the supermarkets.

Towards the end of our vacation, with a little money left in the kitty, we had a splurge, catching the train into Malaga. We waited for a Chinese restaurant to open at about 5pm, then ordered a slap-up Chinese meal. David and I were so hungry we gulped down the first meal and ordered it again. Restaurant staff were perplexed and thought we were complaining about the first meal as if we were not going to pay for it. Before we knew it, the chef came out to remonstrate with us. A game of charades ensued as we kept rubbing our stomachs and saying we were hungry, finally it clicked and they all started laughing and we joined in. The joys of trying to make ourselves understood with no language skills!

One of the memorable bits of our holiday was a day trip to Torremolinos, which was a very popular summer destination. We were walking nonchalantly down a very quiet pedestrianised thoroughfare in the tourist area of the town. All of a sudden, we encountered a large group of local youths, who insisted on us having a shoeshine. We protested saying we were wearing trainers and didn't need it. David's casual shoes were the main target and they tried to grab him and polish his shoes. He complied as I walked on a little. Then I heard a commotion, so went back. These chancers were trying to charge him a fortune for the shoeshine, which he didn't want in the first place. Soon they were getting violent, essentially my friend was being mugged. I said, 'Right, David, I am going to throw a load of coins in the air. When I do, run.' The ploy worked, as the youths scrambled for the coins before following our flight. We outran them and nervously laughed our socks off at our escape as we adjusted to the adrenaline rush and narrow escape! This whole episode was witnessed by a couple of police officers in fancy helmets who ignored the whole event.

On the way home we were warm and cosy, if a little stifled in the smoke-filled carriages of the third-class carriage to the last Spanish station before the French border. This is where David jumped off the train as we pulled in and headed to the cantina to buy some refreshments with our last pesetas. The service was slow and the train started to pull out and he had to run alongside and jump on as we were moving, having bought nothing.

So, then we travelled back up through France to Paris, with little food or money between us. We were very cold in Paris, so we just rode around on the underground for a few hours to keep warm until our connection to Dunkirk was due. My ticket had expired, but I managed to get on to the ferry anyway, as they didn't even notice the ticket was out of date.

On the ferry, we met a French girl whose name was Dominique and an American girl called Wendy. We befriended them, chatting as we were listening to the jukebox, which only seemed to have one record in English that I remember, which was the 'You are my first my last, my everything' by Barry White. The girls bought us a beer and when we arrived in the UK Dominique went on her way.

Wendy, however, had time before her train left for Devon, where she was staying with her grandmother in Tavistock. We implied to Wendy we knew London and subsequently showed Wendy the sights. We went to Harrods, where I recall Wendy cleverly operating a marionette puppet, and may even have bought it. We then went to Soho and headed past the Eros Statue in Piccadilly up a side street by the Regents Palace Hotel, to a little Chinese restaurant, where we were amazed how the Chinese scooped the rice from the bowls with their chopsticks.

After that we took Wendy to the railway station. I found a cigarette packet which had been discarded on the pavement and wrote Wendy's address on it, so I could keep in touch. Wend had been studying psychology in Nancy having had to learn French at the same time. We kept in touch and wrote to one another. She was impressed that I lived on an estate, figuring my family must own the estate, without realising it was a council estate ... oh, well! Her illusions were eventually shattered.

Whilst in Nancy learning French in order to study psychology Wend was walked home by a young man one night, she told me, and he asked her if she would like to go on to a bar. She wanted to say she was too tired and needed to sleep but unfortunately managed to say in her pidgin French that she had no underwear. She was strangely puzzled by his immediate excitement at her refusal to go to the bar.

As young lads we had some great opportunities for holidays. One I remember was having a great time in Rimini with a bunch of friends. We arranged an impromptu European games championship against the hotel staff, which included a swimming gala and other sports, which we managed to win. We duly celebrated by drinking the hotel dry. 'Doggo' used to challenge everyone to an arm wrestle and beat all comers, earning a two-litre stein of beer each time he won a contest, which probably led to his eventual come-uppance when a huge chef came out of the kitchen to take his crown!

We met a model called Beverley from Bethnal Green on that holiday. She was lovely and her family were too. We went to see them in Bethnal Green a year later. We all went to a fabulous evening out in the hills behind Rimini, with a promise of copious drink and wonderful entertainment. There was also a 'Miss Rimini' competition, which we thought Beverley would win hands down, but she didn't want to take part.

So, after quite a few beers and other tipples. I was persuaded to dress up as a woman. The things you do for a laugh, eh! It was great fun, and before long another guy was encouraged to do the same from what seemed to be an Italian group of tourists. We had a ball, plenty of laughs and I am pleased to say I came third, beating the other guy but losing to two rather attractive belles.

I worked hard to achieve and progress in the travel business. I was loving the responsibility, and was shocked and devastated, as we all were in the shop, when my manageress was killed in a car crash. Before I could recover from that sucker punch, another one came in the form of a new manager from the Potteries. No consideration had been made for me being made manager. I was astonished, after all I had been running the place. I was told company policy was that you had to be twenty-one years of age to manage an agency. I was just a few months off that birthday.

I soon aired my views on this to my Area Manager at Gold Case Travel, a lovely guy called Harry Evans, who was very old-school. Harry contacted me and said, 'We can offer you a Manager's job, come with me tomorrow.'

Next day, he picked me up from the Shrewsbury office in his rather nice saloon car and we headed to Liverpool. It was clear he was trying to appease me. However, he had miscalculated somewhat. Firstly, we had to go to a coach yard which Gold Case/Salopia Coaches had just bought. As I recall the name was 'Home James'. I'm not sure who did the due diligence on Home James, but looking over the inventory of coaches, it was clear many were not fit for purpose and as we opened one door a bird flew out and we discovered why: there was a nest on the luggage rack with eggs in it.

After this amusing incident which defied common sense, Harry took me on to my proposed management office. To say it was like a war zone was an understatement. The office was in a row of shops and apparently had a flat above to benefit a new manager. The fact the office was shuttered day and night said it all. It was in Halewood, a run-down area in Liverpool at the time. I didn't even go in and have a look. I was quite straight with Harry and said thanks, but no thanks. I decided to move on to pastures new.

My next spell of travelling was after finishing work at Gold Case Travel and taking a gap year to study on one of the first Travel and Tourism college courses in the UK. That summer, accompanied by a few friends, I went Interrailing for a month to Europe and had a great time exploring. We visited some fantastic places, including Rapperswil in Switzerland, where unfortunately there was a proliferation of midges around the lake. One of our party convinced a new friend travelling with us that the midges were a local delicacy and were bred there, and furthermore the Rapperswil breeders pulled off the wings and dipped them in batter!

This same new friend was taken in by the lads once more as we travelled through the snow-capped peaks of the surrounding Alpine mountains. Our train windows revealed a vista of famous peaks including the Eiger and the Matterhorn. All of a sudden one of the chaps went, 'Wow, there's a yodeller.' Like a kangaroo on heat, up jumped our pal, his expensive camera in hand, trying to get to zoom in on this non-existent character. He was so disappointed he had missed him as the train sped by.

In Menton in the South of France we were sleeping on the beach, as were a number of other nationalities, when this one chap woke us up trying to be a bit too over friendly. I guess I responded a little too aggressively given my personal space was intruded upon, and before long a stand-off became a little heated and the local police turned up and tried to arrest us all for public vagrancy. Fortunately, a group of the beach-sleeping community reacted aggressively towards the police and we managed to sidle off to avoid being arrested while the police were dealing with these other guys!

Munich was interesting, set amidst the Bavarian Alps. We had a look around the Olympic Village as the games were taking place that year. We were starting to run out of money and very hungry. We were so desperate we even waited until people left their tables at a park restaurant before we ate their leftovers

Italy was generally a funny old place. We went to Rome, Milan and Pisa, taking in all the usual sights. We stopped off in Rimini for a short while too, where an abiding recollection was going to the toilets mob handed as two gay guys were trying to proposition everyone who entered, so we felt we needed protection. It was a bit of culture shock. Venice was so brief I only came out onto the station's steps, saw the canals and got back on a train. Don't ask!

In Pompeii we camped near Herculaneum but didn't have a groundsheet for the tent. It was so hot we slept on our sleeping bags under our flimsy canvas home, only to wake up the next day covered in a moving black mass. We must have pitched our canvas above an ants' nest. Suddenly, Gra just stood up, taking the tent with him and we rushed to the showers to swill off the offending insects.

In Milan we arrived late, so spent the night sleeping on the floor and benches of the large waiting room in the grand ornate railway station. We had been warned about thieves so an American girl who was with us and who had a really expensive camera wrapped the camera strap around her wrist and tucked the camera into her midriff to go to sleep. In the morning, sure enough when she stirred, the camera was gone, but she still had the cut strap around her wrist.

We went to the police desk at the station to report it to them. They spoke little English but understood enough to give us a form to complete, as if they were used to this sort of complaint We duly filed the report.

We were not very worldly and as the girl filled in the form to report her stolen camera, Graham and I went to the bathroom to use the facilities and freshen up. Graham went to the toilet stall and shouted for me to check it out. I went to the door and we had our first experience of the European hole-in-the-ground toilet, but Graham's first words were 'Blimey, Morg, they've even stolen the toilet...' You just had to laugh!

CHAPTER SIX

———

College gap year and the holiday that changed my life!

Most people take a gap year from studies nowadays, but I took a gap year from work to improve my education! I reverted to study mode for my one-year crammer course at St Austell College in Cornwall. I was taking a travel and tourism qualification which was brand new in the UK. You could go to Bournemouth or St Austell. I chose the latter and also added to my studies GCE 'O' level Accounts and Law alongside an Institute of Linguists course in German.

My good mates took me down to the south west with all my belongings, records, record player, clothes and so on. We stopped off for a night to party with friends at Kingskerwell, as they were working somewhere near Torquay. Following that brief period of frivolity my pals dropped me off at a bed and breakfast, which was a temporary arrangement for a week staying with a Mrs Cock (we can't choose our birth name but nowadays the married name is open to arrangement).

During this 'freshers' week, as new students we signed on to all our courses and familiarised ourselves with our college and local environment. My room was pleasant enough. but the downside was it was Ramadan and there were a lot of Iranian students staying whose sleep pattern changed due to the nocturnal demands of the relief from fasting and their need for sustenance. Therefore sleep for me became a rare commodity.

In that first week I took the opportunity to find some new digs. I found new accommodation a little way out of St Austell by the English China Clay quarries at a place called Carclaze. Therefore I had to catch a bus into college. It was a house with an annexe of about five self-contained rooms on the side of the main house. It was run by a Brummie couple and we boarded on a bed and breakfast basis, but for those who stayed at the weekends we had additional meals. It was only later that it became clear to me the couple's daughter, who worked in the college and recommended this property, obviously had ulterior motives.

I hazily recollect our landlords' name was Cooper. Mrs Cooper was a friendly enough diminutive and pleasantly plump landlady reminiscent of a Dickens character who bustled around, but her husband was a strange fish. Mr Cooper reminded me of someone from the pre-war era who had not adapted to the winds of change the more liberal sixties and seventies had swept in for many but obviously not Mr C, who frowned on any indiscipline or casual approach to life.

The lodgers consisted of myself, a young lad called Darren from north-west Cornwall and a few ladies, all students. Jayne, a blonde attractive lass from Market Drayton had the room next to mine. She was a mature student like me. I was twenty-two, but she was late twenties, perhaps even thirty.

It was the most intense period of study imaginable, but I wanted to make up for the poor showing at grammar school and achieve academically to make my parents proud and gain some self-respect.

Every other weekend I would try and hitch-hike home to see my beloved football team play at home and of course see family and friends. Jayne, my fellow lodger and student who was also on the same course as myself often accompanied me, so that she could see her folks in Market Drayton.

One such weekend in the winter I was hitch-hiking on my own. I got as far as the Gordano services near Bristol on the M5. It was a rash decision to even set out from St Austell as the weather was awful and it had taken me a long time to get this far. It was the early hours of the morning and I was only halfway home.

I called my parents, only to hear that the game against Bristol Rovers (ironic as I was calling from Bristol) had been postponed. I decided it was best to cut my losses and try and hitch back to St Austell as the chances of getting lifts was reduced now given the late hour.

I met this young fellow who was also hitching. We managed to get a lift together as far as Exeter, which was where he lived. He said he shared a house with some other folk and I could stay the night, so I thought fair play how fortunate, that's a bit of luck, that's what I will do.

He was a pleasant enough chap and we chatted as we walked a fair way after being dropped off on the outskirts of Exeter. Finally, we arrived in front of an imposing terrace of dilapidated Georgian buildings. My fellow hitch-hiker tried to enter the door but it seemed stiff. He knocked and we heard some bolts opening and a rather dishevelled guy greeted us with caution. We followed him up the bare wooden stairs with broken stair rods where carpet had once cushioned the sound of footsteps, treading over debris and the occasional missing step, and soon entered a room with about a dozen people sleeping on the bare boarded floor. It was winter and cold and a small fire had been lit in a fire grate. I was in a squat!

That night I could not sleep for fear of what might happen. I had been met with not a little suspicion and not too much bonhomie, my new friend trying to put me at ease by saying, 'They're all alright when you get to know them.' Well, I must have dozed off, to be awoken by a guy who was ranting and raving, obviously not just drunk or full of drugs but positively unhinged. He kept shouting he was going to kill us all! Fortunately, they calmed him down, but the commotion had unnerved me even more.

I watched as a couple of the sleepers stoked the fire and tore laths of wood out of the Georgian property's damaged wall in order to fuel the fire. I waited for about ten minutes, gathered my belongings and stealthily went down the stairs, unbolted the door and, relieved, I escaped into the cold night air and hotfooted it down the road, not knowing where I was going.

Fortunately, I was not far from the railway station and managed to get my head down in the waiting room and catch the first train in the morning to St Austell, reflecting on my interesting expedition!

Another time Jane and myself were hitching back to college. We always used to hold up cards showing a destination. We had worked out that a midway point is always a good tactic rather than putting St Austell or Shrewsbury, as many drivers wouldn't be sure where those towns were.

Jayne was a real character and used to make me laugh. She was also attractive, voluptuous and blonde. I would sit on our bags while she hitched holding the sign for Bristol under her ample bosoms. With its obvious connotations, she never failed to get us a lift!

On one occasion we hadn't made it all the way back from home to St

Austell and were in deepest Somerset somewhere, so decided about midnight to walk from the motorway slip road where we had been dropped off to the nearest settlement to see if we could find a B& B for the night. Amazingly in a small village called Sampford Peverell one such property opened its doors and took pity on us, offering us a single room and a double room. We said we would share a double room as we wanted to keep it as cheap as possible.

There was a slight resistance by the owner to this decision based on moral grounds. We explained we were boyfriend and girlfriend and going to get engaged, so she relented. We laughed about it as Jane had the bed and I slept on a chaise longue and then the floor, which turned out to be more comfortable.

We always managed to get lifts but sometimes there were complications. One time we had a lift in a truck and the driver took us as far as Frankley services near Birmingham. Again, the strategy of Jayne being the carrot for these chaps to pick us up worked, but it almost backfired.

When we pulled into Frankley services I descended from the truck's cab first, and as I did so Jayne handed me her bag. The truck-driver grabbed his chance and tried to drive off with Jayne still in the cab. I leapt onto the footplate, the door still swinging open and a tirade of expletives and displeasure descended upon us as I managed to extract Jayne from the cab and he drove off mad as hell. It's a good job it wasn't a car or he would have accelerated far too fast for me to do anything about it. It was a lesson to us to be more cautious.

It was not until the Spring term of 1977 after exams that I had my first foray into the wild nightlife of St Austell. By then Jayne was going out with an Iranian student called Hossein, tall and handsome but very controlling.

We had a few drinks celebrating one of our group's birthday and all returned to Pitt Cottage, our Carclaze accommodation, slightly enebriated and very happy. We were a bit rowdy, I suspect, but all good clean fun.

Mr Cooper appeared and told us to keep quiet, so we all said good-night and went to our rooms. Shortly afterwards I heard a commotion next door in Jayne's room. Mr Cooper had apparently knocked on her door and barged in to remonstrate with Jayne about something she had said to him a few minutes before but Jayne was in a state of undress.

This wasn't a surprise to me as he used to walk in on the girls in the morning with wake-up calls. So, I entered the room and became assertive with him just pushing him vigorously to leave the room and pushing him up against the hall wall. He retreated to the cottage and called the police. The police did attend but we persuaded them it was all a misunderstanding, endorsed by the Coopers, who I am sure were a little concerned regarding my allegation of improper behaviour by Mr Cooper. The police, a little disgruntled by the callout, warned us to behave, then left.

Awkwardly, the next day the Coopers were driving to Birmingham and gave Jayne and myself a lift which we had been arranged with them earlier in the week. I apologised to the Coopers to try and smooth things out and improve the atmosphere but it was a long, quiet journey to Birmingham.

The very next week I moved out into some new accommodation, staying with a lovely couple Barb and Richard Coombs, who had two pretty little daughters and were a great friendly and fun family. I had a single room in their bungalow, which was more centrally situated in St Austell, and I loved it.

Our end-of-term bash at the Carlyon Bay hotel, examinations over, was a fun night, but my lecturer Frank Squibb who lived in Falmouth had far too much drink to drive home so I smuggled him into my room and he slept on the floor.

Next morning, I got up to tell Barb and Richard. 'I hope you don't mind but I brought my tutor home last night.' Now, as I had pulled their legs a bit on April Fool's Day, they didn't believe me until this six-foot-plus, bespectacled and hungover tutor emerged. Again, they were brilliant, offering him breakfast.

I played cricket for Bugle in the Cornish league while I was in Cornwall and really enjoyed my fellow team players and the experience it afforded me as it meant I went to lots of different venues to play. One evening I was invited to play against a Poldark XI in a charity match made up of some of the cast and crew of the television series and had the great pleasure of playing against Robyn Ellis and also Angharad Rees who joined in the fun. Later in life at a funeral of a distant cousin Gareth Watkins, I went back to the house of Vaughan Rees and learned I was distantly related to this fine actress Angharad Rees who played Demelza in the original Poldark series. Vaughan was her brother. Angharad died in 2012.

That college year flew by, as it was intense and I was snowed under with studying and revising for examinations. As part of the course, we had to create a travel project. For my project I chose to design an itinerary for the United States aimed at people my age trying to keep their costs as low

G.N.A.E. Project folder

as possible. It was fully costed with mileage, accommodation and transport, something that is so easy to do today with the aid of computers and search engines like Google. But this was all manually researched, typed and enhanced with pasted pictures cut out of brochures and magazines.

The upshot was that it was successful and helped me pass the course. Indeed, it was thought good enough by my tutors to be put forward for national scrutiny and eventually judged to be the third-best effort in the country. Now the reason I tell you this is that before starting back to work or indeed looking for a job, the project had ignited a desire to visit North America.

So, myself and three friends planned our vacation based on the content of my project, aptly named the 'Great North American Expedition'.

In July 1977 we took advantage of Freddie Laker's new service, the 'Skytrain' ABC charter service to New York. As a game plan we used my college project itinerary as a template, then set about finding any relations, friends or random people we might call on should we need assistance, sustenance or lodging. For starters, I had an American pen friend who had a brother living in New York. I had been in touch, and they said we could stay with them for a few days on arrival, so the die was set! America, here we come!

But first there was a farewell party at the Golden Goblet function room in Tiffany's, Shrewsbury. We had a great send-off, complete with punk rock dancing, which was starting to show its anarchic face on the music scene.

We set off for the land of the free, home of the brave. It was a very long flight compared to Europe, but we were so excited it passed rapidly. At New York JFK airport we were some of the first passengers off the plane

Going Away party invitation

with our backpacks complete with Union Jacks sewed on them. We all sported long hair, and were immediately targeted by customs, who took our backpacks apart, even unscrewing the tubing, looking for drugs, I guess.

After a thorough search, we were allowed in and exited customs, by now last off our plane. We were met by my pen friend Lydia, her pal Nancy, Larry, Lydia's brother, and Becky, his girlfriend. It was great to see them, as well as a relief. Our eyes were agog as we chattered excitedly whilst gazing out at the skyscrapers and cityscape.

We arrived at the small apartment belonging to Larry and Becky just off 56th Street in Manhattan, which was rather small and easily congested by the sudden influx of weary Brits. Larry and Becky also had a friend staying with them. So it made it impractical for us to stay that night. Their friend was a musician doing a concert at Carnegie Hall.

So, with no spare room available, we were shown to the YMCA around the corner. This was one heck of an experience, as there were crowds of folk of every ethnicity, including an intimidating group of black guys at the entrance. We checked into two rooms and then went for a shower, where a huge black guy was showering. We exchanged a few words of general chat and pleasantry trying to understand his accent and probably he ours, then we headed to our rooms.

A few minutes later, we heard a noisy commotion in the hallway with lots of shouting and screaming. We all came out into the corridor to witness the black guy we had spoken to earlier prostrate on the ground having been stabbed. We rushed back to the room and contacted reception. We heard the ambulance staff and the police in the corridor but kept the door shut. A little while later we peered into the hall we saw the body had gone, the only evidence he had been there was the residue of his blood. No one interviewed us or asked any questions and that night all four of us slept in just one of the bijou rooms, for safety in numbers. Welcome to New York!

Next evening, we all went out to Chinatown with Larry and his entourage of friends, probably about ten of us, but the underground (subway) in those days was not very safe, so we felt better travelling in a large group. We had a good meal and a laugh as we sat around a large circular table with a 'lazy Susan' full of Chinese food in the middle. Our sarcastic and self-deprecating sense of humour, combined with accent differences made for plenty of mirth and puzzled looks.

Later we walked around iconic Time Square, which was certainly an eye-opener. This is an iconic intersection in the centre of Manhattan. Located at the junction between Broadway and 7th Avenue, it is surrounded by well-known buildings displaying bright neon signs at all hours of the day and night. Wide-eyed we proceeded to explore the sidewalks. To our astonishment we were openly offered just about every drug going and as we approached the famous illuminated apex iron-shaped building complete with bright neon lights and changing messages familiar on the New Year's Eve broadcasts and travel programmes, we were just in awe that we were actually there.

We walked around the block coming down the other side as we headed back to the apartment and were then propositioned by a number of prostitutes, before some small boys at the end of this line offered us spliffs. What a first night in New York.

It was all rounded off by observing two police officers at a busy intersection commandeering a cab and saying, 'Follow that car!' as they struggled to jump in, one of them slamming the door on his long 'billy' club and the door bouncing open again with the officer hanging out as the taxi cab careered away. Absolutely priceless!

New York was a lot more dangerous and lawless in 1977 than it is today. The *New York Times* wrote a couple of months before we arrived.

In Central Park, the once-green lawn of the Sheep Meadow is wearing away, gradually becoming a dust bowl with overuse.

At the Bethesda Fountain, drugs are sold routinely, and the Duck Pond at night becomes a receptacle for beer and soda cans. Crime, graffiti, and decay are the buzzwords of 1970s New York City. And just because Central Park was the city's jewel didn't mean park structures and landscapes were immune.

Before heading out of the city, we made sure we saw all the main sights. We were overawed witnessing Central Park, the Empire State building and all these familiar landmarks from films we had watched. We took in Wall Street, the financial centre of the USA before travelling down to the harbour and taking a ferry out to Liberty Island and seeing Ellis Island, the arrival point for many immigrants including some of my relatives. We climbed the narrow staircase inside the Statue of Liberty, to look out from the crown, a real highlight for us! The statue is made of copper and was a gift from the people of France to the people of the United States. Designed by French sculptor Frédéric Auguste Bartholdi, its metal framework was built by Gustave Eiffel of Eiffel Tower fame. The statue was dedicated on 28 October 1886 and certainly would have been an impressive sight for the new immigrants arriving in New York. To many it was the symbol of hope and freedom.

Taking in Greenwich village with its colourful image and clearly high percentage of gay residents and indeed eccentric characters, not least the guy dressed as the sugar plum fairy, was a real eye-opener to us guys from sleepy Shropshire.

We needed to start following my master plan from college, the GNAE, the Great North American Experience, so we headed out of town and went south first, towards Virginia. Nancy had kindly agreed that we could use her car, which she and Lydia had driven down from Indiana. It was cosy on the bench seats in the car as there were six of us, Lydia, Nancy, Graham, Dave Mc, Purky and me, although luckily this was a larger vehicle than we were used to in the UK, as were the roads.

We consulted the list of the few contacts we had collected between us who we might be able to stay with or visit whilst exploring the USA and Canada. Of course, Lydia, with ancestry in the Crow Indian tribe, and Nancy, both being Americans also had some friends they wanted to visit. I recall heading down to Philadelphia and Washington first.

One place we stopped on the way down south was Gettysburg, which had a profound effect on us, hearing the story of the battle of Gettysburg during the American Civil War and the loss of life and injuries incurred. It had been a particularly bloody battle between the armies of the North and South on 1-3 July 1863. In just three days there were a total of over 50,000 casualties from the warring armies, the bloodiest battle of the war and this was also where, on 19 November, President Lincoln used the dedication ceremony for the Gettysburg National Cemetery to honour the fallen Union soldiers and redefine the purpose of the war in his historic 'Gettysburg Address'.

We stayed with a friend of the girls called Cindy Jacobs in Washington. I remember getting a parking ticket outside her apartment, but we were advised that once we were out of state it didn't need paying unless we returned. We were amazed at how spacious, clean and efficient the underground was compared to New York or indeed London and it felt safer too.

We visited all the monuments we came across, and in Washington marvelled at the Smithsonian Museum, which had a biplane suspended from the ceiling as you entered. We ascended the George Washington monument (on the inside, of course), which gave us a wonderful view looking down onto the pools and the Lincoln Memorial.

I have a lasting memory of the Vietnam and Korea war memorials especially the Vietnam one, where you walked down an incline to walls of names set below ground level. We witnessed people holding their fingers on the name of a loved one and old soldiers crying as they witnessed the name of a fellow veteran and no doubt recalled a savage and brutal time in the history of warfare.

As we hitch-hiked later in the trip, we were regularly offered a spliff by the driver, whether it was a city gent, a well-dressed woman or whoever, it was a bit like offering you a beer or a sweet, and was generally accepted and a way of life. It struck us that the influence on society of the returning Vietnam vets and their constant need for 'weed' to help them with PTSD and get through life also crept into society and this escalated to chemical drugs in the following years. Movies like *Coming Home* and *Born on the Fourth of July* with Tom Cruise highlight the problems these guys had adjusting to the experiences they were subjected to in Vietnam, then returning home and how they were not treated as heroes but more like villains.

Going into the House of Representatives was a great experience, as was being photographed by the Lincoln Memorial. We did a tour of the White House but just standing by the surrounding fence looking at the gardens of red and white tulips and lush green lawns was a marvellous experience.

It brought back childhood memories of my mum helping me to write that poem about the equal rights march on Washington in the sixties.

It was pretty warm being July, lots of folks were milling around and lying on the park's grassy areas enjoying the sunshine. Looking towards the Jefferson Memorial we saw so many planes rising into the sky from Washington National Airport, the skies seemed to be constantly full of planes and vapour trails. It was somewhat surreal for a boy from the blue hills of Shropshire.

In those days we were allowed to tour the armed forces headquarters, the Pentagon, which was an amazing experience too. We were met by an assigned guide of military bearing and shown around. Arlington Cemetery had a big effect on us too. It's a huge area and the row after row of white tombstones is some sight. How amazing to see the tomb of the unknown warrior and the grave plinth of President John F Kennedy, which has an eternal flame to mark it, especially as we were all of the era to remember the Cuban missile crisis and JFK's assassination in Dallas.

We then headed north to Philadelphia, home of the 'Liberty Bell'. The bell is the one thing that really stuck in my mind alongside the story of Benjamin Franklin, one of the founding fathers of the United States, a smart, multitalented fellow, as this was the birthplace of freedom in America. We visited Independence Hall, where the Founding Fathers signed the Declaration of Independence and is the building where both the United States Declaration of Independence and the United States Constitution were debated and adopted. The bell is now the centrepiece of the Independence National Historical Park in Philadelphia.

As we travelled along our way, we seemed to home in on the various fast-food outlets at the comfort stops, in those days a novelty to us and certainly not ubiquitous worldwide. Their names, McDonalds, KFC, Dunkin Donuts and so on were amazingly unfamiliar back then. It amused us that we would often order and be asked to repeat it, purely because they loved to hear our accents ... 'Gee I just love your accent.' To which we would respond, 'We love yours too.' I guess time has changed all that. McDonalds always had a running total of the number of burgers they had served throughout the country, flashed up on electronic billboards outside many of their outlets. It was in the millions even back then!

In my GNAE college project I had listed all the things we needed to see and do, wherever we visited and explored, so we could get the best out of everywhere we went. As we were trying to cover a lot of places and distance in just six weeks and we were driving most of it thanks to our friends and their car, we needed to be organised.

On to Boston. We loved this attractive city, there appeared to be lots of parks and commons. I remember walking Paul Revere's Trail and also going onto the tea cutter, where we ceremonially threw over a bale of tea to commemorate the Boston Tea Party. We visited Faneuil Hall and wandered around Quincy market, full of food stalls and visited the preserved Beacon Hill with its gas lanterns illuminating the cobbled streets and the attractive brick period townhouses.

We cruised into Toronto, Canada and made a beeline for the CN Tower down on the waterfront. We took the great glass elevators up the highest freestanding structure in the world at 1,800 feet/550 metres. This fantastic structure had only opened the year before and looked down upon the harbour front and the baseball stadium below. Views from the Observation Deck were astounding. We trod warily over the transparent glass floor decking, looking at the people who appeared like ants far below our feet … magical!

After the tower, we visited Yonge Street, billed at the time as the longest in the world, until the 1999 Guinness Book of Records put paid to this claim. We started to get the gist that there was always a claim to the largest, widest structure somewhere in North America as we travelled around as the marketeers tried to out-do their tourism competitors.

We then drove onto Montreal in Quebec. It was about the time of the rise of a guy called René Lévesque with his 'Quebecois' party, who wanted Quebec to be independent from Canada, so a lot of Quebec independence pride was being displayed.

In fairness all I can recall about Montreal on this trip was the view of the cityscape from Mount Royal, the Olympic Park that had been used for the 1976 Olympic Games here, and also the rather impressive Notre-Dame Basilica in the historic old quarter of Montreal. We stayed with some folk the girls knew and they had a fabulous house with a swimming pool, which we truly used to the max. What a treat!

One contact we had in North America was a lady I met at college in St Austell, a mature student called Sandra Jukes. Her mother lived in St Catherines, Ontario, and she very kindly gave us her address. We had driven through the night and perhaps inappropriately, we knocked at this poor lady's door at just after 8am one morning.

For a short while we heard nothing and we questioned why we were doing this so early, then before we could walk off, she came to the door and the mosquito netted door creaked open. She looked rather perplexed, as she wrapped her housecoat over her nightclothes and was quite astonished to see five young, bedraggled individuals.

We explained to her we knew Sandy and to her credit she didn't baulk at all, and invited us in. This wonderful lady gave us breakfast and some information and ideas for places to visit. Amazing hospitality!

St Catherines was quite a quaint little place but we soon headed for the iconic Niagara Falls. As we drove along the road towards the falls, we parked up then walked. We could firstly hear this natural phenomenon, then feel the spray, before we actually saw them. Absolutely awesome.

We ventured under the falls wearing blue 'pacamac'-type capes. We all got togged up in a changing room area, then walked through tunnels peering through the occasional roughly hewn window in the rock at the back of the falls. It almost took our breath away as the mass of H2O cascaded over the Horseshoe Falls, so named because of their shape.

We thought we could not leave without attempting to see the falls on the American side, up close and personal. We boarded the *Maid of the Mist* tourist boat and floated downstream past the rather benign-looking falls on the American side with their crumbling rocks and less dramatic water plunge. It was a great experience. From the Canadian side we ploughed on towards the vicious eddies and whirlpools of the resultant Niagara River after its plunge from the iconic precipice above. We were drenched in spray as the powerful engines of the boat fought to stop us being dragged under the huge wall of water with its resultant vortexes caused by the volume of water and height of the drop.

This does beg the question how on earth did a chap from Dawley in Shropshire called Captain Matthew Webb attempt to go over Niagara in a barrel, and what possessed him to do so? Sure, he was a strong swimmer and had swum across the English Channel, but as he found out too late, this was a different barrel of fish, so to speak, in which he ultimately lost his life.

We journeyed on, heading west through some pretty grim industrial areas like Gary Indiana, until we reached Elkhart and then Bristol, Indiana. This was Lydia's home town area and very rural. We were going to stay at her father's smallholding on one of the rural routes.

The stay proved to be very educational. Lydia's father, dressed in baseball cap and dirty dungarees over a well-worn check shirt with fraying collar, took us on a tour of his fields, which seemed to contain a certain plant with a very distinctively shaped leaf. In this era the leaf adorned tee shirts of the youth of the day. Sure enough, when we went into the barn, he had all this marijuana drying on racks and hanging from the rafters. I am sure he was readying the crop for medicinal purposes.

The small farmhouse itself was not very well maintained. Its clapboard exterior had the paint peeling off and the interior was untidy and shabby, with well-worn furniture.

Lydia spotted some ground beef (minced beef to us) in the fridge and said she'd make some hamburgers, but when she broke open the meat, maggots wriggled out. Needless to say, we passed on that. We went down to the local bar with Lydia's dad instead, it was clear he was well known and a real country boy and character. In fairness to Mr Abbott, apart from his social relaxants, he was very generous. He gave us an old estate wagon which we insured and got the sticker (road tax) for and we drove that for a long while, with a few repairs en-route until it eventually gave up the ghost in Alabama.

We continued our expedition west to Chicago which was a neat place and known as the 'windy city'. We could see why, with its location on Lake Michigan. In the winter it was bitter, apparently, as the icy winds drove in off the lake and whistled around the skyscrapers. People would scurry to reach shelter from the chilling onslaught, but this thankfully was summer.

Chicago means 'wild garlic' according to the old Indian name. Prior to arriving we had this vision of the city in the 1920s ruled by Al Capone and the scene of the famous St Valentine's Day Massacre. But the modern skyscraper-filled city amazed us. We loved all the cycling and running tracks along the lakeside. We revelled in summiting Sears Tower, now called Willis Tower. I believe it was the highest in the world at 110 storeys and 1,454 feet high when we were there in 1977 and it is still the highest in the USA today. I have a vivid memory of having a photograph taken at the top of this tower, which was an optical illusion and looked like we were walking outside onto a steel beam perched at a great height with nothing but balance to keep us from falling!

We loved the way the river ran through Chicago and all the iconic buildings, like the John Hancock Tower and the Wrigley Building, owned by the producers of Wrigley's chewing gum, a name familiar to us from childhood. We visited Wrigley Field, home of the Chicago Cubs baseball team. We drove to the north of the city and hit the blues bars, like Kingston Mines around the Lincoln Park area. Not that we met Muddy Waters and Howling Wolf, but we could not but fail to hear their music and get down to the rhythm.

Following our list of contacts, we deviated south to Davenport, Iowa, which was the home town of some relatives of Dave McCormick and another address we had collected to visit. Yet when we called them up on arrival in Iowa, they were a little overwhelmed and initially reticent if not suspicious, especially as there were still five of us in our party.

In fairness to them when they regrouped and acknowledged Dave Mac as a distant family member, they took us out for a meal and allowed us to sleep on their floor. Although not the most friendly visit, they were kind to do that! Davenport is nothing to write home about, with some manufacturing companies like Deere Tractors. It was clearly once an agricultural town and back in the day flourished on the Mississippi River.

Go west, young man, as the saying goes and so, we headed out west. We progressed westwards towards Badlands National Park. On the way we had the odd flat tyre, which took some changing as the wheels were a lot bigger on those American vehicles than our small saloon cars back in the UK. After a long journey we finally arrived at Mount Rushmore.

Mount Rushmore National Memorial was popular in the seventies, visited by nearly three million people each year, we were told, as they came to marvel at the majestic beauty of the Black Hills of South Dakota and especially to seek out the huge carved faces of famous presidents Washington, Jefferson, Lincoln and Roosevelt, created by sculptor Gutzon Borglum initially, but continued after his death by his son Lincoln. In all, the work took fourteen years.

What we didn't realise until we arrived was that nearby another massive mountain sculpture of Crazy Horse was being created, all carved from granite. This famous Indian chief has been worked on by sculptor Korczak Ziolkowski since 1948 and it is still not finished, with a target size of 641 feet (195 m) wide and 563 feet (172 m) high. The head of Crazy Horse will be 87 feet (27 m) high; by comparison, the heads of the four US Presidents at Mount Rushmore are each 60 feet (18 m) high.

I grew up loving westerns and of course Custer's Last Stand was legendary and at Little Bighorn in Montana we were able to roam the valley sides to the top of the bluffs above and wander around the Little Bighorn River. Apparently, the Indians called it the 'Battle of the Greasy Grass'. Great Indian chiefs fought here, notably Sitting Bull and Crazy Horse. This area memorialises the US Army's 7th Cavalry and the Lakota. Northern Cheyenne, and Arapaho in one of the Indians' last armed efforts to preserve their way of life. Here on 25 and 26 June 1876, 263 soldiers, including Lt. Col. George A. Custer and attached personnel of the US Army, died fighting several thousand Lakota and Cheyenne warriors.

I still have a facsimile copy of the newspaper reports of the massacre from the time, which I bought in the tourist information centre there. There are gruesome details of how the soldiers' bodies were mutilated after the battle and what amazed me were the crosses placed on the battlefield where the soldiers fell. It seemed such a tranquil site nowadays as grass-covered bluffs led down to the lazy trickle of the Little Bighorn river.

Our journey onwards took us to Cody, Wyoming, named after Buffalo Bill Cody, another western hero of mine. There is a historical centre dedicated to him here. It talked about Buffalo Bill's Wild West Show, which apparently toured the British Isles too and featured Annie Oakley.

We managed to see a rodeo here! It turned out it was a couple of weeks too late for the Cody Stampede Rodeo festival but they have Rodeo every night, which gave us a real sense of cowboy country. Sure, a lot of amateurs or those learning the trade performed but it was very entertaining. Even children, dressed like mini cowpokes, were involved, roping calves. What amazing feats these riders and rodeo stars pull off, such as bull riding, bareback bronco riding, calf roping, barrel racing. The music and pizzazz and sense of occasion and excitement is great fun!

We now headed west to Yellowstone National Park, the first American national park founded in 1872. A vast wilderness, it was full of wild animals. Sulphur springs, giving off that distinct smell of rotten eggs, and terrific geysers certainly test the senses. No geyser was more impressive than 'Old Faithful', which they reckon reaches 135 feet high when it erupts. It was certainly high and went with a dramatic whoosh. It seemed to go off about every hour but folk have told me it's nearer to every ninety minutes now so I guess the pressure is weakening.

There were more than 3,500 square miles to explore, full of bison and elk, moose, fox, wolves, antelope and yes, bears!!! The sight of a white fox loping along was terrific. It was rather humbling seeing such beautiful creatures in their natural habitat protected by man. Cascading waterfalls, such as the 94-metre-high Yellowstone Lower Falls, were so dramatic and beautiful.

A change of direction took us north back into Montana and across the border into Canada and its wide-open prairies. Calgary, whilst not being anything to write home about, is a leaping-off point for the Canadian Rockies, which you see rising from the plains in the distance towards the west.

Of course, there was yet another tower to climb in Calgary before heading for 'them thar hills'. We had missed the extravaganza that is the 'Calgary Stampede' by just two weeks, which was a shame. However, we did our own whirlwind city tour, including the Heritage Park which was a sort of living museum with lots of old frontier buildings and folk in period costume, and Fort Calgary, site of the first Mounties' fort.

Our real excitement, though, was focused on the majestic peaks of the Rockies. We headed out along the strip of bitumen highway shimmering in the sunshine, full of great anticipation as the road led like a black version of the yellow brick road towards the distant mountains!

The first place we stopped was the small town of Canmore, typical of what we imagined a frontier town would be like. With lots of low buildings and local stores Canmore is just outside the national park limits. Not yet overrun with souvenir shops and excessive tourism, it has a wonderfully attractive and testing golf course nowadays. It has some good restaurants and plenty of bed and breakfasts.

Road cruising on to Banff, we reached the more commercial destination for most tourists, especially tour groups. You could see this was a tourist honeypot. We drove on through the main street and out of town looking for a campsite in the forests that thrived on the surrounding mountainsides. We reached a rather remote but designated area to park and camp. Rather strangely, we found we found no other campers in this clearing.

We erected our tents surrounded by the coniferous trees towering above the pine-needle and twig-laden forest floor. We were soon visited by a park ranger, who advised us to string up any food we had in a tree and not to keep any in the tents. He pointed out a garbage bin which was bear proof, where we could discard any leftover food or other materials. This certainly focused us and heightened our senses to listening and looking out for wildlife.

So, leaving the tents all erected, we drove down into Banff and found a saloon where we had ourselves a few beers. It felt like the 'Klondike 'might have been after a day's prospecting, pretty busy and getting a little wild with lots of whooping and hollering and clearly full of locals rather than tourists.

We headed back to the campsite in a rather jovial mood after a few sherbets. As we reached the place where we had camped, our headlights scoured the clearing, bathing the tents in light, which were pitched on the edge of the tree line. We soon realised that the two-man tent, the smallest one, had someone, or something, in it.

We all tentatively approached the tent, then recognised that it was not a person but a bear cub in the tent, helping itself to some biscuits we had inadvertently left in there. It was making a bit of a mess.

It dawned on us quickly that as per the ranger's talk earlier, if the bear cub is there, the mother bear would not be far away and God forbid we get in between the two!

We retreated! We all sat in the car and waited. It was not long before we saw the entry of the mother into the arena and she enticed the baby bear out, and then much to our relief as we held our breath, they passed by and then away from our car. For what seemed an age the mother tried to get into the bear-proof garbage bin barely four metres from us, then, unsuccessful, they departed into the night.

It was a good half-hour before we were prepared to get out of the car. Cautiously we surveyed the nearby trees, then decided to snuggle into our sleeping bags and turn in for the night. Our previous light spirits from our evening's carousing were now a little tense and pensive. We were all in the large six-berth and two-berth tents, all, that is, except Dave Mc, who decided he was sleeping in the car with the doors locked. He didn't realise the bear would just see him as a meal in a tin.

I must admit I heard every twig crack and every branch sway in the breeze that night, but it was a very exciting experience. We high-tailed out of there pretty early next morning and had a substantial breakfast in Banff before hitting the road and heading the short distance to Lake Louise.

It's a fabulous vista looking up towards the glacier at Lake Louise and we could see why they built the grand dame of the Chateau Lake Louise hotel there. Chateau Lake Louise was built by the railroads in their heyday and this location is a real jewel in the Rockies.

In fairness there were loads of lovely lakes and sights as we travelled through the Rockies. Nevertheless it was a truly awesome spectacle.

We decided to head through Golden and Revelstoke in Glacier National Park, old logging and mining communities, rather than take the attractive northerly route to Jasper as time was of the essence.

We eventually reached a place called Kamloops where the terrain changes in appearance and in truth is quite an underwhelming and arid place. The town at the time felt run down and not a place we fancied dallying.

We headed on into the west coast city of Vancouver, a city we all enjoyed. I recall strolling around Stanley Park and taking photographs of the First Nations Totem Poles. Guess what, there was yet another tower in evidence. The Harbour Centre, what they today call the 'Vancouver Lookout', at over 550 feet high was opened 13 August 1977 by American astronaut Neil Armstrong, the first man on the moon.

Vancouver has plenty to offer, with regular sea planes and ferries heading over to Vancouver Island. We crossed the Lions Gate Bridge to Vancouver's scenic North Shore and the Capilano Suspension Bridge Park. We enjoyed the swaying walkway across the Capilano Suspension Bridge and drove on to Grouse mountain to take a chairlift to get a panoramic view of Greater Vancouver.

We returned to English Bay and carried on through Chinatown to historic Gastown, the heart of Vancouver, where the city first began. There is a steam-driven clock here and apparently named after a guy called Gassy Jack who basically got his name because he talked a lot!

So back into the USA we went and to the very interesting coastal city of Seattle. Expo had been held there in 1962 and there was still a fair bit of that world exhibition left over, such as various international food stalls, which introduced us to some Asian foods. Well, something had to challenge the peanut butter and jelly sandwich that our new-found American cousins had subjected us to for the last few weeks. The Expo site had just been taken over by the city to use as a tourist magnet and it all revolved around … yes, you have it in one, another tower, this time called the Space Needle, which we duly ascended.

We had been introduced to Yesler Road, a steep hill down which apparently timber was sent to Yesler's timber yard during the timber boom. It was known as Skid Road and when a period of financial instability hit and a recession was in evidence lots of folk were laid off work and they all congregated here. It became known as 'Skid Row' and this term became a national reference for unemployment and then used worldwide.

Seattle is the place where the first Starbucks appeared in 1971, the home of Boeing and the location for a number of films like *Sleepless in Seattle*, of course. We enjoyed exploring the vibrant Pike marketplace, but it was the underground tour that was truly fascinating. This subterranean labyrinth was apparently created after the 'Great Fire of Seattle' in 1889 when the city fathers decided to build brick retaining walls and raise the city to a higher level, one to two storeys above the original. This gave us the novelty almost a century later of the underground tour, where you can witness the streets of nineteenth-century Seattle, many still with shopfronts intact. Pretty neat, as the Americans say.

We learned that the Seattle we know today was raised higher off the tidelands to help with the water pressure for public use, which had badly failed the fire service when trying to put out the Great Fire. The improved water pressure also aided the new flushing toilet to thrive, thanks to the designs of inventors like Thomas J Crapper. Guess where a certain slang word comes from!

Ready for more wilderness after our city stay, we soon headed for the mountains, Mount Rainier the destination. We drove so far up the winding mountain road to a visitor centre. It was a gorgeous sunny summers day, so, ascending to 14,410 feet above sea level without a hint of altitude sickness, we walked the trail in flip-flops up onto the snowline and threw snowballs in August. Mount Rainier stands as an iconic peak in the Washington landscape. An active volcano, Mount Rainier is the most glaciated peak in the contiguous 48 mainland states of the USA supplying six major rivers. Sub-alpine wildflower meadows ring this icy volcano while ancient forest cloaks Mount Rainier's lower slopes.

After drinking in some fabulous scenery and meeting some noteworthy characters we headed on the long drive down US Route 1 to San Francisco, sharing the driving through the night. We arrived at the Golden Gate Bridge, somehow expecting something grand and glistening in gold, but all we saw was a dull-looking bridge adorned with what seemed like red lead paint.

But the rest of San Francisco did not disappoint. It was a cornucopia of varied sights to assault your senses. Fisherman's Wharf and the view across to Alcatraz were just as we had seen them in the movies. There was vibrant activity everywhere: restaurants, street artists on land and, in the sea, bobbing seals and ferries dotted around the bay, going in particular to

Alcatraz in San Francisco Bay. The small island was developed with facilities for a lighthouse, a military fortification, a military prison and a federal prison from 1934 until 21 March 1963. The water currents around the island are considered so strong no prisoner could ever escape by swimming from the island, especially when you add in the threat of being shark bait., which presumably decreased the chances or indeed desire of any inmate trying to escape. Although we didn't have a lot of money Alcatraz beckoned across the water, so we took a trip out there. It was fascinating and well worth the visit, and fortunately they didn't keep us there.

Back in the city, we shimmied down the legendary Lombard street in our big American station wagon. This street was billed as the most crooked/bendy in the world. We jumped on the cable car, which was great fun on the ups and downs of the hilly streets. We could just imagine the cop car chases we had seen in the movies and later in seventies television shows like *Starsky and Hutch*.

Forsaking city life, we travelled east inland through Yosemite National Park with terrific sights like El Kapitan and Half Dome, wonderful natural rock formations often targets for free climbers, and the stupendous Ribbon Falls. I have a photograph of Graham and myself stood right on the rocks at the edge of the awesome Bridal Veil Falls. Impressed, we travelled east over Tioga Pass and then through Kings Canyon and Sequoia Canyon National Parks before returning to the coast.

Our first experience of Mexican food was in Los Angeles, in a small restaurant called Pico Pica Rico on Magnolia Boulevard. It was delicious. I had never heard of tacos, burritos or enchiladas and we had to have a few beers to swill down the hotter ones. Of course, we hit Disneyland, the original Disney theme park, followed by Knott's Berry Farm, which was a neat experience and a good taster for the at that time newly opened Disney World in Orlando.

Venturing into the desert we experienced Death Valley, a land of extreme temperatures ranging from a winter low of -2.6°C to 57.6°C in the summer, and small staging posts like Stovepipe Wells and Furnace Creek are like an oasis of sanity and respite on the tough journey where the unforgiving terrain is punctuated with animal skulls and skeletons.

At one picnic area we walked a short way along a trail into the barren and foreboding rocky wilderness and came across a rattle snake, which was in the process of devouring a gopher that it still held in its distended mouth.

We started taking photographs with our basic instamatic cameras typical of the era, the snake's rattler working overtime. Then all of a sudden, the snake spat out the gopher. Boy, did we move fast. My flip-flops shot up in the air and I had to go and get them later, in the hope the snake had slithered off!

Deadwood in the Black Hills of South Dakota was a target destination because of its rich western cowboy history and of course part of that was the 'Deadwood Stage' route and the infamous Wild Bill Hickok. The discovery of gold in the Black Hills in 1874 set off one of the last great gold rushes in the country. In 1876, miners moved into the northern Black Hills. That's where they came across a gulch full of dead trees and a creek full of gold … and Deadwood was born. Practically overnight, the tiny gold camp boomed into a town that played by its own rules, attracting outlaws, gamblers and gunslingers along with the gold seekers.

Wild Bill Hickok was one of those men who came seeking his fortune. But just a few short weeks after arriving, he was gunned down while holding a poker hand of aces and eights – forever after known as the 'Dead Man's Hand'. Calamity Jane also made a name for herself in these parts and is buried next to Hickok in Mount Moriah Cemetery, set above the town. Seeing their final resting places was a highlight for us lads who were brought up watching westerns.

Other legends, like Potato Creek Johnny, Seth Bullock and Al Swearengen, also created their legacies in this tiny Black Hills town. Deadwood has survived three major fires and numerous economic hardships, pushing it to the verge of becoming another Old West ghost town. But in 1989 limited-bet gambling was legalised and Deadwood was reborn and is thriving once again.

It was a town characterised by this proliferation of casinos when we arrived in 1977, and not a very attractive one at that. One of our most vivid memories was the car breaking down here in Deadwood before later keeling over in Alabama. We went to a gas station in Deadwood with a motor repair shop, where we asked for help. At the garage we met an elderly chap who sported a very lined face like a road map. He wanted to charge us a lot of dollars without any guarantee of repair, so we relied on Purky, who was a trainee mechanic back home, to do the job. Fortunately, he managed to get the car going, so we pushed on.

But that wasn't the end of our automobile problems. The darn car also broke down in Las Cruces, New Mexico. It was getting close to being unaffordable. I seem to remember waiting in a cheap motel itching like crazy from mosquito bites, which some pharmacy-bought calamine lotion fortunately made easier.

We listened to the car radio a lot as we travelled around. We soon realised radio stations used to really push some tracks, presumably for reward from the record companies. Tracks like 'Undercover angel' by Alan O Day, 'The joker' by Steve Miller, 'Dreams' by Fleetwood Mac, 'Just want to be your everything' by Andy Gibb, 'Gonna Fly Now', the Rocky 2 theme tune, were played so often we knew all the words and sang along quite merrily. We could have formed a band!

In Arizona's Black Hills, we came to Jerome, named after Eugene Jerome, cousin to Winston Churchill's mother, Jennie. It was here as we turned off the highway that we heard some sad news on the radio. Love him or loathe him, a true musical giant of our era had passed away. Elvis had died on 16 August 1977.

In Las Vegas we visited Mr Sy's casino, which was across the road from the Stardust casino and hotel. We were given some tokens to use in the machines, large metal discs that won us nothing, of course, the plan was to get us tempted and gambling our own money but we had too little to risk it.

All casinos had offers on to attract you in, some gave a free drink or in the case of Mr Sy's two drinks for the price of one. Others gave free food like steak and eggs so we cruised from casino to casino picking up the freebies as we didn't have a lot of money in the kitty.

We were amused to see wedding chapels and right next door, premises where you could get divorced. We lads were trying to get Lydia to marry one of us, then go next door and get divorced. We almost did for a laugh, but luckily even our feckless young minds thought better of it!

Grand Canyon stood out as a magical place, but conditions were very hazy because of the heat, so we couldn't appreciate the true magnitude of it all. The Grand Canyon is a steep-sided canyon carved by the Colorado River. It's 277 miles long, reaches 18 miles wide in places and is over a mile deep in places. The north rim is closed in winter because of its elevation but the south rim is open all year round. It gets up to around 100°F (37.8°C) in summer and -27°C in winter. Wow!

Eventually we made our way through Arizona and witnessed amazing Indian sites like Montezuma's, which was a community built into a hillside forty-five minutes south of Flagstaff. Built into the limestone rock, these thousand-year-old ruins were once home to ancient farmers known as the 'Sinagua' Indians. We walked up to them and through them, climbing ladders from tier to tier. Constant tourism was ruining them, so access is now denied.

San Antonio, Texas was high on our Texan city list. The Paseo de Grande walkway beside the Rio Grande river was atmospheric at night with all the cantinas lit up and buzzing and Mexican mariachi tunes filtering into the night air from the lively bars, inviting you in. Best of all is to visit the Alamo, where John Wayne and the ragged assortment of resisters made their last stand at the Alamo against the Mexicans. Here was where about 250 men were besieged by a superior size and armed force of 1,800 regular Mexican army troops and resisted for 13 days. Famous men like Davy Crockett and the founder of the Bowie knife, Jim Bowie, died fighting for Texan freedom. This was a story we were brought up on so to be exploring this old Mission building was just goose bump territory.

In Houston we visited the NASA Space Center and Mission Control, famous for the words we all knew when younger from the space missions: 'Houston, this is Houston, do you read me?' or the more frequently quoted: 'Houston, we have a problem.' It was fun having a go in the simulator and touching a rock that had been brought from the moon.

At the other end of the history scale off Highway 90 is Langtry, Texas. Famous for the legendary Judge Roy Bean today, the Jersey Lilly and the opera house Bean built in honour of his long-distance, unrequited (and one-sided) romance with English singer Lillian Langtry, combine to create the Judge Roy Bean Visitor Center – I had no idea Lily Langtry was so favoured by an English King Edward VII and was also a mistress of the Earl of Shrewsbury. What a gal!

We next headed for Tombstone in Cochise county, Arizona and the infamous scene of the gunfight at the OK Corral. On main street at the Birdcage saloon, the barman amused us by pouring a beer and sliding it the length of the bar to us just like in the westerns.

Needless to say, they did a re-enactment of the gunfight at the OK Corral too. I was beguiled by the Boot Hill graveyard and seeing graves with names we grew up watching westerns and reading books about, the likes of the gang who went up against Wyatt Earp. We were amazed to see Billy the Kid's grave, a name that we grew up with and synonymous with the Wild West. The Earp brothers Wyatt, Virgil and Morgan along with Doc Holliday arrived in December 1879 and mid-1880. The Earps had ongoing conflicts with cowboys consisting of the Clanton brothers Ike and Billy, Frank and Tom McLaury assisted by Billy Claiborne. The cowboys repeatedly taunted the Earp brothers over many months before the shoot-out on 26 October 1881. The historic gunfight is often portrayed as occurring at the OK Corral, but history says it actually happened nearby in an empty lot on Fremont Street. One epitaph in the cemetery at Boot Hill tickled us: 'Here lies Lester More killed by slugs from an A44, no less no more'.

Now on to El Paso. I had always wanted to go there, after listening to the Marty Robbins song, which my Auntie Doris and Uncle Harry used to play all the time. But believe me, it's not quite as romantic or atmospheric when you arrive! It's very close to the Mexican border and was a very Spanish-speaking town and felt down at heel.

Carlsbad Cavern, one of over 100 caves below the hills and ridges set in the Chihuahan desert, New Mexico, was a neat surprise. We waited until dusk and with others crowded with anticipation around the 'natural' entrance to the cave. As the sun set, thousands of bats swirled erratically like a plague, as they flew out into the night sky and over the ridge, emerging almost as one, like a genie out of a bottle. Such an awesome sight and well worth the visit.

We reached Birmingham, the capital of Alabama in the deep south, and camped one night on a campsite. Next morning very early, we were startled awake by the attention of two police officers brandishing guns. It appeared that a gas station has been robbed the evening before and somebody had pointed the law enforcement in our direction, but after a brief chat they realised we did not fit the description, nor had we been near this gas station. Phew, saved!

Our donated automobile finally met its inevitable demise when, after running maintenance, it eventually keeled over and died in Alabama needing repairs beyond our budget. We were left with our only option. We split into pairs and hitch-hiked out of Alabama.

Gra and I decided to head south to Florida. We picked up a couple of rides, quite quickly, one of them with a chap called Mike. Our new travelling companion was a tall, thin, long-haired, unkempt hippy type, a real character. We chatted with him as we cruised the highways heading east. Before long Mike pulled into a gas station and promptly asked us to put some gas in.

This was not quite what we had planned, but we did reluctantly stump up the dollars first time around, thinking it wasn't too expensive and it saw us on our way. We settled in for the long journey from Alabama to Florida.

Every time we got low on fuel, though, the same thing happened. We had made a rod for our own backs and Mike was not slow at coming forward. Gra and I decided we would do it just a couple more times, then seek a new lift. On the upside Mike was an interesting character. His goal was to reach Florida, in his words to find a 'sugar mama!'

I remember one evening stopping in the heat of the evening and Mike getting some eggs from the trunk of the car and cracking them on the bonnet. It was so hot they fried almost instantly and we ate them off the bonnet (or should we say the hood).

If we found we were going the wrong way (Gra and I were navigating) Mike would simply do a U-turn, often down the huge highway divide to a concrete drainage area and back up the other side. Gra and I would be bouncing around in the back, our heads hitting the roof of the car as we tried to avoid injury negotiating the mound of Mike's junk we were sitting on. I do not recall seat belts and even if there were some a dime to a dollar they would not have worked in this jalopy. We had a number of reasons to worry about Mike. Something wasn't quite right and we had to be on our mettle. He avoided questions on why he was heading to Florida and indeed was very general about where he had come from.

Our journey had been fairly uneventful until just before we reached the Alabama/Florida state line, when we were pulled over by two beefy state troopers. 'Get out of the vehicle.' As we did so the officers shouted a further command: 'Put your hands on top of the car and spread your legs.' They kept watch from a distance and trained their huge firearms on us.

We did as they said, they approached us with caution and then searched us. Satisfied we had nothing criminal on our persons they asked Mike for his driving licence and other documents, which he could not produce.

We had our driving licences and passports which they were happy with, even making some friendly comments about being limeys and some distant relative one of them had in Nor-wich(as they pronounced it). In spite of their demeanour softening a little we feared we were going to be arrested, as it was clear to us now that this car was stolen.

Both officers returned to their car, made a radio call, had a chat, then returned. Amazingly, the officers told us to get in the car and told me to drive, having seen our British licence documents. They told us to hightail it over the state line which was very close, but if they saw us again, we would be arrested and put in jail!

We crossed the state line, then Mike pulled into a rest area and started emptying the trunk of the car. He began throwing things away.

'Why don't you want these? I asked.

Mike replied: 'They belong to my girlfriend, we've split, so she won't be needing these any more. At the next opportunity, we felt enough was enough. We made our excuses and hitched again on our own

Florida was a very different state to the ones we had travelled through up to this point. It was very flat and featureless inland but this was contrasted by the wonderful white sand beaches. On the east coast at a campsite, we met a young girl called Karen, who was camping with her parents. It was a very wet night so we gave up on our not-so-waterproof tent to shelter under the covered restroom area and by chance met Karen there.

I know it sounds dubious, two dudes hanging around the restrooms. We said hello and got chatting and were introduced to her family. As a throwaway statement she said to pop in if we passed her home in New Jersey.

We carried on hitch-hiking and met some great people. We were then picked up by a lovely chap called Butch Campbell and his wife, who took us north up the coast to Vero beach. There was a tropical storm coming in and he and his family invited us to stay the night and were very welcoming and hospitable. I will always remember standing with Butch and his dog on the seafront close to their house and feeling the tropical storm make landfall. It just about blew us over and across the road, scary but exhilarating.

As we continued our path north, aware we had to be back in New York for a return flight in early September, we actually hitch-hiked into Disney World, put our backpacks in left luggage and had a day at the park. Back in 1977 Disney World was the new kid on the block, the original having been opened as Disneyland in California in 1955. Disney World was opened the year I started in the travel industry, 1971. We loved the day, it truly was magical, from 'It's a Small World' to Pirates of the Caribbean and Space Mountain.

We hitch-hiked out and were picked up by a fun couple called Joe and Eileen Karch, who were heading north to Jacksonville back to their naval base accommodation. We spent a pretty wild night with them.

Still a little intoxicated from the shenanigans with our US Navy friends, my heart was beating like it wanted to leave my chest. Goodness knows what those guys laid on us the previous evening. We were thumbing a ride next morning and we were picked up by a patrol car with a very friendly and larger-than-life police officer at the wheel. He saw our Union Jacks on our backpacks and told us we were not allowed to hitch there, so told us to hop in and he gave us a lift to a better spot. En route he told us quite frankly that he was gay and said in the UK he believed that was frowned upon and then, before we could even make a comment, he proceeded to tell us about a terrible woman who was very anti-gays called Anita Bryan. Having been educated royally in sexual orientation and the persecution of gays and their rights we were dropped off and he bid us farewell!

We now hitched up through Georgia and South Carolina. Our target was a rendezvous with Wendy, she of the chance meeting on the French channel ferry from the previous December, so we headed for North Carolina and the capital Raleigh.

It was late when we finally reached North Carolina after hitching up the Interstate 95 and being dropped off on the road to Raleigh I 70. We were tired and dishevelled so we were not surprised when a police cruiser slid alongside us and asked where we were going and what we were up to. The police officer was very understanding and welcomed us to Johnson County. He said we were unlikely to get a ride at this time of night and asked us to jump in and he took us to Johnson County Courthouse in the little town of Smithfield. We wondered if we were actually going be arrested but the officer said, look. it's early hours of the morning now so get your bedroll out and get a few hours' sleep before the cleaning staff and workers arrive and he allowed us to sleep on the marble floor of this imposing building.

We were amazed and grateful in equal measure. Would it happen over forty years later? I doubt it. We were told that Ava Gardner, the famous film actress, was born not far away in a little place called Grabtown, so you live and learn. They were very proud of this association and claim to fame and rightly so.

Next day we were back on the road again before going on to see Wend in Raleigh. I think in many ways Wendy, who was sharing an apartment with a friend Geri Neal at the time, a lovely lady, was very surprised to see us, even though we'd said we hoped to come and visit. But we were made welcome and set up camp on the lounge floor and couch.

It was very hot and humid and it was so good to take a shower and relax by the apartment complex swimming pool. Wend was working at the Angus

Barn steak house to earn some money. She said they were not really paid a salary, they lived on the generous tips.

So began a really enjoyable week or so exploring downtown Raleigh and the bars and clubs and sports venues it had to offer. It was so chilled and laid back compared to back home. We were invited around to meet Wendy's father, Doc Holden, and some of her family, which was a real pleasure. Little did I know this was going to be my first father-in-law, what a lovely chap and a gentleman. His house, 3313 Felton Place, was a large green wooden house on the corner with a pond behind it amongst the trees. It looked small from the roadside but it was like the Tardis with five bedrooms and a sunken living room. Wendy's sisters, Rosemary, Daphne and Julie were all living with Dr Holden there.

The house itself was three storeys, but you walked into the top level straight into the kitchen, which had a lounge just off it. At the bottom of the house was a fabulous den. This was an upside-down house built into a little hillside amongst the trees. It was massive compared to my little council house home back on the Meole Estate.

I have this terrifying memory of Doc asking me to help him get the leaves out of the guttering, which was okay at the front of the house but the scary height on the other side where the land fell away sharply was pretty tough for someone afraid of heights. I scaled the ladder and tried to stay cool.

Wend had arranged for a girlfriend of hers, Sheri, to meet up with us, and all four of us got together in downtown Raleigh. Her friend was a real livewire, her first words being: 'Goddamn! I've forgotten my purse, I just cannot get my s**t together.' Well, back in the seventies you rarely heard a young lady speak like this, so we were quite shocked, but soon got used to it, especially after our jaunt around the southern states with 'hippy' Mike, who was fond of calling everyone a 'mother f****r' sometimes even as a term of endearment. Yep, as they say in the States, 'go figure'. That said, over the years I have found Americans always seem to be 'getting their shit together' like some mobile linguistic sewage service.

To start our evening off well we had a drink at the Doc's and then he gave us a ride downtown and gave us a bottle of 'crème de menthe', another first. We loved the taste and passed it around. We actually finished the bottle and couldn't understand why people were smiling and laughing at us until it was pointed out how green our tongues were.

We hit a few bars and one in particular stood out for its character, a bar/restaurant called 'Darryls' on Hillsboro Street. Apparently two of the owners were collectors of a disparate assortment of unique items such as elevators, Ferris wheel seats, gas lights, jail cells, cast iron spiral staircases, barrels, musical instruments, abandoned cars and double-decker buses, and chandeliers that were so big they dropped the height of two storeys. So, booths were set out like jail cells .and a spiral staircase was in evidence.

We had a bite to eat and we were all getting on quite well together, more as pals than partners until Sheri got up and started dancing on the tiny dance floor. Then she went over to the jukebox and put on a tune she loved and came back to the table and asked us all if we wanted to 'shag'!

Well, as you can imagine our faces were a picture, and following a rather pregnant pause Wendy said, 'Yes, come on, lets shag,' and we were encouraged up on to the dance floor where we jigged around to the music with great alacrity. Returning to our table the girls ordered another pitcher of beer and we couldn't help but ask almost in a whisper, leaning into the table, 'When you say, "do you want to shag?"' We could hardly find the words to finish our question, but before we could Wend said, 'Yes! We love that dance' and everything became clear to us. We explained what the term meant back home and everyone fell about laughing.

We had a great night. We returned back to Wend's apartment and got a good night's rest. We had a fabulous time in North Carolina, it was great to not be on the road and live a little like a local. It was the first time a group of us went 'skinny dipping', which again was a new phrase to me at the time. Not everyone joined in with this swim late at night in a lake (everyone seems to call this 'wild swimming' now as if it's a new pursuit). Perhaps it was the hot evening or the Carolina moon but to me it was magical, like something out of a movie. However, when people we met said 'Oh, I wouldn't do that, not with snakes in the lakes,' it sort of took away something from the whole experience.

We drove around in Wend's car, which had 4WD a/c that's four windows down as the air conditioner didn't work and the Carolina air was hotter outside than inside the car but it was like a fan effect. We listened to country music on the radio and pop chart tunes of the time and heard advertisements inviting everyone down to the local shopping centre for 'shagging on the mall', which a few days ago would have left us speechless but now just left us with a smirk!

We went out one night and called at a liquor store to get a bottle of spirit before going to a night club. This was because clubs were not allowed to serve liquor in 1977, only mixers (which they charged you a lot for). When we got to the club we had to have the liquor in the brown paper bag from the store where we bought it. We could fill our glasses up with it and drink but we were not allowed to display what we were drinking by taking it out of the bag, even though everyone knew. It was called the 'brown bag law' … we were just tickled by this ridiculous palaver. Apparently, a year later the law started to change.

On another night we went to a bar-cum-club with a dance floor in a more rural area and when some music Gra and I liked came on we got up and danced to it. When we returned to the table Wendy asked if we'd noticed that we were getting some odd looks. We said no, we were totally oblivious. She explained that people would think we were gay as two guys dancing together was just not done in the USA at that time. We asked for our tab in our deepest voice and left rapido!

On our final night the Doc took us all out to the Angus Barn for one of their famous steaks. Unfortunately Wendy was waitressing, but that was a benefit too, as we got great service, we were given an introductory daiquiri or two on arrival, all free because it was some special night. It was the first time we had ever had daiquiris and I loved the lime ones. We were hooked!

Time to move on! We said our goodbyes and promised to keep in touch, we had enjoyed a fab time. We had barely a dollar or two left now and decided to head north as quickly as possible to New York City. With our dwindling resources it made sense to call by our new-found friend Karen, the nurse from New Jersey. We called the number she had given us and told her we had hitched to her town. When we arrived in New Jersey at Karen's apartment we met her friend, who was Filipino and kept by a sugar daddy. Karen was out working when we arrived.

We were welcomed and fed. When Karen arrived home, she was really pleased to see us.

Next night she said she was working again, but not at her day job as a nurse but at a bar down the road. Her flat mate dropped us at the bar, and we figured Karen would make sure we were okay for a free drink or two.

When we got in, there was no sign of her, then the lights went down a little and some music started and two women appeared on stage and gyrated around a pole. We sat at the bar open mouthed. Karen was a pole dancer and very good she was, too.

Begging Rides ticket from police officer in New Jersey

We had a great time with Karen and her friend for a few days but knew we had to get back to NYC for our flights back home. We were going to meet Lydia and stay at Larry and Becky's again. Karen dropped us off on the ramp road to the freeway and we said our goodbyes. It was a slow day to get rides, but within thirty minutes a state trooper turned up. He was jovial and asked us what we were doing and where we were going.

He saw the Union Jacks on our rucksacks and talked about the Beatles and other British connections he was familiar with. Then his demeanour hardened and he informed us we were not allowed to hitch on the ramp and asked us what was in our packs. We said just clothes and personal belongings and he asked to have a look.

It suddenly dawned on us that on one of the rides whilst we had been travelling up the east coast, we had met a Vietnam vet who we got on well with. He had given us a bag of Colombian Red Bud, which was a type of marijuana. Our rationale for retaining it was we thought that, as we had no money, we could give this to Larry and Becky as some sort of payment and thanks for staying. As I recall, it was in my bag. Now at that time, in New York State, possession of narcotics was a mandatory custodial offence and we were not sure if New Jersey was the same.

Anyway, we both started to loosen the ties on the flaps that covered the tops of our backpacks. Graham's was tied tightly and he was wrestling with it

September 1977

a little. All of a sudden the flap restraint gave way and snapped back to reveal a camera with its lens sticking up on the top. Immediately the police officer recoiled and drew his pistol, thinking we were armed, and when he realised his mistake he started to laugh uncontrollably, probably out of relief.

He reholstered his weapon and said, 'Look, guys. I'm going to have to write you a ticket but as you are leaving the state just keep it as a souvenir.' I still have this ticket today, our crime being 'Begging rides'. Luckily, he seemed to have forgotten entirely about searching our bags.

After catching up in New York with Larry and Becky we flew home. We had changed, we were more worldly, confident to the point of cocky, and we knew we wanted more out of life!

CHAPTER SEVEN

———

Searching for the end of the rainbow

Returning from the North American odyssey in 1977 I rejoined my old firm Gold Case Travel in Shrewsbury. This time I was appointed as a relief manager. It allowed me to work all over the country, as the company had a national chain of travel agencies. My remit was covering for vacation and sick leave.

Gold Case Travel also owned Shearings Holidays, or Salopia Coaches as they were known back then, and I would often go up to Whitchurch to manage the Gold Case branch there, if the manager Dave Skerritt was on leave. Dave later ran Wem Travel with his wife Dianne. He loved his rail travel and his knowledge was excellent. I really enjoyed this time working in different communities and meeting some interesting people and new staff in new locations. Occasionally it was locally at places like Ludlow in South Shropshire or Newcastle-under-Lyme in Staffordshire.

Working in the Cumberland Hotel at Marble Arch, I covered the travel agency manager Derek Day for a good six months, as he was off on long-term sickness leave. This hotel was a four star hotel in a fabulous position on Oxford Street with views over Marble Arch and Hyde Park. It was renowned as a hangout for the amazing sixties guitarist, Jimi Hendrix. Hendrix was known for bringing his female conquests back to the Cumberland, and the hotel was allegedly listed on his death certificate as his last-known place of residence.

My office was only a little kiosk in the lobby. Most of the work was revalidations of airline tickets to reconfirm flights for business travellers and other clients. Occasionally we changed itineraries for them or sold them theatre tickets. We used a theatre company called G S Lashmar at the time. Tips were often quite good. The best one I had was from a Japanese businessman, £20, which was a lot considering even including my London weighting allowance I only earned £30 a week.

When time allowed, I had lunch at a discounted rate in the hotel's staff cafeteria, which saved me money. Going out at night was too expensive, but I occasionally had a Friday evening drink before going home with the other travel agency clerk and some hotel staff.

I was always intrigued by the very smart ladies meeting men in the busy hotel bar at the Cumberland. I asked my older colleague if they were film stars or debutantes I might have heard of. She just smiled benevolently at the country boy she saw before her and said, 'They may have been in some films but not the sort you see at the Odeon, they're top-class call girls.'

I was staying in a little hotel called the Strutton Park in Notting Hill. After the first week I soon worked out it was quicker and more pleasant to walk through Hyde Park to work than taking the tube, naive as I was, I had been boarding the train for just two stops wedged into packed carriages, some people trying to read newspapers.

The fellow who managed the Strutton Park Hotel was an American. It transpired he was a school friend of Patrick Duffy who played Bobby Ewing in the seventies soap hit, *Dallas*.

I ate breakfast in the hotel, which was not up to much but filled a corner, and I enjoyed the banter I had with the waiting staff, who all appeared to be European. I had just a small en-suite hotel room as my temporary abode, which after a few weeks became a bit claustrophobic. I often walked around London in the evening checking out various central locations, familiarising myself with the beautiful architecture, the history and the bright lights, such a contrast to the sleepy market town of Shrewsbury I grew up in, but not spending any money as sightseeing was a cheap pastime.

In 1978 everything went downhill again. I am always a believer in fate, even though at the time I was devastated to be made redundant. Gold Case Travel were taken over by Ellerman Sunflight. Ironically we used to enjoy weekend trips to Majorca at a hugely discounted rate with a tour company called Mato Jetway, which was now part of this company. We had to call last minute on a Friday, on occasion speaking to the owner Doug Ellis himself, for a Saturday flight, coming back Sunday, as I recall. Doug later became the owner of Aston Villa.

Being made redundant at Christmas was a bit of a blow, particularly so because, in spite of my initial optimism and belief in my own abilities, it was hard to find a suitable managerial opportunity. There were travel vacancies but mostly for senior consultants and I was looking for a management post.

Eventually I took a job, opening a new branch in Cambridge for Pickfords, who were better known for furniture removals. I went down to their head office in Enfield for an interview and was offered the job. My first week in Cambridge was in a bed and breakfast where I met a chap called Ron, who was a professor at Tenpe University, Phoenix, Arizona. We shared a dinner and brew or two. He was a very interesting and pleasant fellow and invited me over to Arizona. I did go at a later date but by then he had moved on from Tenpe University. In Cambridge, I lived in a large bedsit and was enjoying exploring a new city with university life, dodging the bicycles and watching folk punting along the River Cam.

It soon became clear that Pickfords was not very organised, as you might expect from a relatively new entrant into the travel industry. I had no brochures for my brochure racks and despite constant phone calls and requests to head office, this remained the case. I eventually went down to Enfield to see the Managing Director at my own expense. He was very reassuring and things improved quickly but then disintegrated again just as quickly.

After a month I handed in my notice, because in the meantime I had found a job as General Manager of a small provincial company called Gwynedd Travel at Porthmadog in Wales. The beauty of this appointment was that it came with accommodation in the form of a one-bedroomed flat above the office and was on the coast with access to the beautiful national park of Snowdonia. I saw it more as a lifestyle move than a career move at the time.

So here I was on the west coast of Wales, on £30 a week plus a free apartment and also my boss helped me buy a car from the auctions at Prees in Shropshire. My new motor was a lovely white MG 1300, which was great when it went okay, but it kept getting dirt in the twin carburettors and eventually was the bane of my life as it kept breaking down on my constant forays back and forth to Salop.

But what a fab time I had here on the coast of Mid-Wales. Often in the lunch period I would go for a run to the village of Borth-y-Gest and along the beach beyond or just sit eating my sandwich by the harbour with its sailboats bobbing in the water and I'd watch the world go by. It felt like being paid for being on holiday.

I kept fit by running to Morfa Bychan along the road and then back along the beaches and coastal paths to Porthmadog. I played in the local very successful five-a-side team league at the local sports hall. I had made good friends like Bryn the Post through playing on the same team. The main downside, though, was the guy running our team and ironically it was him who had invited me to join. Mike was the husband of Diana Jones, the daughter of Gwynedd Travel's owner Glyn Jones. Mike was working at the inland revenue department in Porthmadog. 'Mike the Tax' was very much a Welsh nationalist, and very anti-English, so I was surprised he invited me to play.

We qualified for the national finals in Cardiff after a herculean effort following a disaster in the qualifying event of having a man sent off and going down to four men in the regional finals at Wrexham, but somehow, we still managed to win. But as soon as we were through, Mike dropped a bombshell on me, saying quite blatantly and unapologetically that he just wanted Welshmen in his team. Having a Welsh father did not count.

Satisfyingly, perhaps, they were knocked out without winning another game in Cardiff. I felt sorry for my fellow players but felt Mike got what he deserved … nothing!

I took a Football Association of Wales coaching award to allow me to become a qualified football coach. This was interesting and very testing, as it was taught in Welsh. We would line up on the field in Portmadog and do drills and our coach would tell us what to do in Welsh with the occasional English instruction, when he remembered, for my benefit. All I could do was observe and follow. Fortunately the written exam was in English and I managed to pass it.

During that first summer in Portmadog, Wendy, who I hadn't seen since the previous summer in the States, came over to see her grandmother in Tavistock, Devon and decided to come up to see me in Porthmadog. We hit it off straight away and carried on where we left off in North Carolina the August before and instead of coming for a week she decided to stay and was with me for the rest of my time in Wales until October 1979, almost eighteen months.

We managed to get her some puppet shows as she was a fabulous ventriloquist entertainer and she also had steady employment with Porthmadog Potteries at their factory. Her job was initially to do tours of the factory each day in the summer and soon to add a little puppet show as she escorted the Porthmadog Pottery tour guests. In the winter season when the tours did not operate, she designed and painted pottery for them.

One of the travel tours that Glyn Jones had set up was a trip to Paris for the France vs Wales rugby union game. We even arranged a special air charter from RAF Anglesey, which involved having to charter a jet plane, a Bac-111 as I remember, and requesting special customs/immigration facilities, all of which made the cost prohibitive for the average supporter.

Now Glyn was not at all happy about Wendy moving in with me, as we were not married, of course, and this was chapel country. He was even less happy when I said I was not going to Paris to assist him unless Wendy could come. He dug his heels in and I had to go without Wendy.

We travelled by coach up to the north Wales coast and everyone was very jolly and excited. What John Glyn Jones had failed to tell me until we boarded the rail special from Llandudno Junction was that we didn't have enough match tickets for everyone booked on the tour. He then casually announced to me that it was my job to arrange a lottery on the train to see who got the tickets we did have and then tell everyone the bad news!

This backfired on him as everyone knew it was his fault without me saying a word. Fortunately, they thought more of me trying to sort it out than of him giving me the poisoned chalice. Gareth the Fish and Bryn the Post and others supported and backed me for sorting it out (by now I was known as Morgan the Travel). We did manage to get a few more tickets from the French Rugby Union in Paris but about seven of us were without tickets. John Glyn gave me some money to give them a good time in Montmartre and watch the rugby on the television in a bar. In the end, a good time was had by all. Well, perhaps tempered by the thought that John Glyn and the others had got to go to the game.

Unfortunately, that evening John Glyn fell off a pavement, fracturing his ankle, and was hospitalised, but not before those with him had a whip-round to pay the ambulance driver, as the French medical team would not take the distressed fellow away until they knew who was paying. We should be thankful for our National Health Service, which we so often take for granted. Poor old JGJ, though!

Needless to say, it was left to me to take the train back with the guys to North Wales and then the coach connection from Llandudno Junction to Porthmadog. As I recall, we did leave at least one guy in Paris who had never returned to the hotel the previous evening. Unfortunately, I neglected to do a coach collection for the driver, so got into hot water for that from J G Jones when he got back from France. This was a learning curve for a young chap like me, a mistake made purely out of inexperience with that protocol and on the assumption JGJ had already taken care of it in paying for the coach. Despite me arriving early morning into Porthmadog and walking two miles to tell John Glyn's wife of the reason why he was not returning with us, JGJ took a while to forgive me, in hindsight I think it was a way of deflecting any criticism of him for his poor match ticket management.

A bit of snow on the hills towards Snowdonia in the winter immediately made you want to go and explore. It saw me making a sledge from old pieces of wood and nails and Wendy and myself taking our dog Lulu (a delightful Papillon) for a spot of sledging. This adventure was rather short-lived as I bounced down one bumpy slope and felt a huge surge of pain in my backside. I was now attached to the sledge by a six-inch nail. OUCH! It was a rather embarrassing experience going to the local doctor for a tetanus shot and explaining the problem. At least by then I had removed the sledge!

I used to love running along the road to Morfa Bychan and Black Rock Sands and returning along the beach and coastal paths. It was beautiful. I especially enjoyed the walks up Moel y Gest, a small peak (peak baggers call them Marilyns over 500 feet, like this one, and Munros if they are over 3,000 feet like Snowdon) that looks down over the coastline and the village of Tremadog the other side.

One of the other hats I wore with Gwynedd Travel was as a taxi driver using JGJ's own canary yellow Mercedes to pick up folk and transport them around the local area. Most guests were from the Portmeirion Hotel. I always gave them a tour of the scenic route via Tremadog, showing them where Lawrence of Arabia was born and where Percy Bysshe Shelley used to live and visit there.

During the summer we organised day coach excursions from Porthmadog to Llandudno, which connected with the ferry over to the Isle of Man. These were long days and full of primarily 'Brummie' holidaymakers who were staying in the Porthmadog area, many of whom came back a little worse for wear as they drank cheap booze in Douglas while others brought kippers back, of course. One man in particular was so drunk he struggled to walk along the pier in Llandudno back to the coach with my cajoling him along. All the other passengers wanted me to leave him, but I persevered and got him on the coach. Fortunately, he fell asleep and didn't cause any more trouble.

*

We had managed to get Wendy an excellent agent, a lovely chap called Tom Layton and ex-para, who took part in the Battle to take Arnhem. We used a local gig she had been booked into at Pwllheli, Butlins, to springboard more bookings.

Through Tom we harnessed support from the Reuben brothers, who became a sort of sponsor-cum-performer angel. It quickly led to a much higher profile, obtaining appearances on TV shows like Sooty and Sweep, Tiswas, Lucky Numbers and others. Wend sang though her ventriloquial skills without moving her lips and Cliff Richard just moved his mouth and mimed and to all intents and purposes it was as if he was singing! Wendy worked with Chris Tarrant and Sally James and many other artists like comedian Frank Carson.

Chester the crow & Wendy
Ventriloquist Extraordinaire

Wendy's mum Ursula was the god-daughter of Rudyard Kipling, the well-known poet and writer, and she had lots of books with his signatures on, some signed 'Uncle Rud'. Wendy's grandparents, Ernest and Frances Stanley, had an ocean-going yacht and Rudyard Kipling and his wife Caroline were often guests and indeed the invitation became reciprocal as the Stanleys used to visit the Kiplings at Batemans, their East Sussex home.

Whilst Wendy had all this English heritage and even though she was doing well, she was pining for home. Her mum and all her family except her grandmother were back in the USA, so I could understand her homesickness. I don't think the approaching lousy winter weather was helping her to settle, despite having our adopted dog Lulu and the friendships we had built, notably with Matilda, whose mother had been nanny to Lawrence of Arabia.

Alas, it was time to leave, and to Wendy's surprise I said I would like to come with her. I don't think she realised I was serious at first. Before we made final plans to depart, we had a great late summer holiday. We travelled for two weeks around the Greek islands with our good friends Gra and Jan, visiting Halki, Rhodes, Crete, Santorini and many other islands. We started the trip by exploring Athens and the Plaka district and, of course, visiting the Acropolis and Parthenon. We then took a ferry from Piraeus and went to explore the islands, travelling deck class as it was cheaper and infinitely airier, especially when folk became seasick!

Santorini was a fabulously romantic island although rather devoid of sandy beaches, but Thira will stay in my mind if only for the ascent by donkey on a steep winding path from the sea to the cliff-top town and the moonlit night looking over to the caldera of a nearby volcanic island. It was truly a vista straight out of a romantic film. We took a service bus across the island to explore a bit further. We stopped in Messaria and I remember tasting olives for the first time. I loved them!

Eventually on our island-hopping trip we reached an island called Halki. Only two people disembarked and we made a spontaneous decision to join them, rushing off before the boat left and we were so glad we did. We loved the life here. The six of us were greeted by the Mayor, who owned one of the two restaurants on the island. It faced the landing stage of the main village on the island named Emborio, where the boat had docked.

He offered our party of four an apartment, which was basic but delightful, set on rocks on the water's edge, lapped by the Ionian Sea, just a short stroll from the waterfront promenade. The architecture reminded me of Italy and the church of St Nicholas stood out as its tower rose against the background of a clear blue sky, as did the clock tower, which was a fine stone building with a clock face that told the correct time twice a day!

A group of local men always sat smoking, their faces craggy and weather-beaten, as they watched the women folk, who were clothed in funereal black, repairing nets ready for the night's fishing exploits.

There was no ferry for six days, and there were no vehicles on the island either, which made it a real exclusive haven for us. Nowadays things have changed but in a very controlled way. There is a taxi (just one) and a minibus and that's it, apart from the ubiquitous donkeys carrying panniers of olives and other produce when needed.

We used to go to a sandy beach called Potamos just over a small rise from the port of Halki, no more than 400 metres, I would say. We had it to ourselves, a turquoise blue clear sea, with a gradual slope of sand. It was magical as we swam and sunbathed on what seemed like our own private beach. We had found a little slice of heaven and felt very covetous of our discovery

One day we climbed the steep, winding road past locals collecting olives from the craggy trees which populate most Greek Islands. Some passed us going to Emborio with their loads, the donkeys swinging from side to side under the weight of their cargo. It was so hot even though late September that we were wilting just walking. Those poor donkeys must have been exhausted, but they were uncomplaining as the occasional swipe with a switch from their masters chivied them on their way if they dared to stop and rest.

Eventually we reached a small track to the left which took us to a promontory with fabulous views from an old disused church, its once grand and vivid wall mosaics faded by the ravages of time. It was very like the little chapel in the later film *Mamma Mia.*

During our stay we had a tab at the restaurant owned or run by the Mayor's family. My recollection is that all four of us wanted to pay as we went so we could see how much money we had left, but the restaurant owner insisted on us not worrying and just pay when we leave. We ate and drank very well on plenty of fish and freshly caught lobster and we were fearing we might end up marooned, washing dishes until we had paid off our debt, but we needn't have worried as the bill came to just £31 at the end of our stay for all of us. We were flabbergasted and delighted in equal measure! We complained they must have undercharged us but they diligently checked the account fearing we were complaining it was too much! We left very happy travellers!

On one overnight ferry when I was very tired this young Greek kept playing the bouzouki, which was very nice for a short while, but it went on and on and started to grate. My hunger and weariness had made me tetchy and the persistent din took me over the top. Quite out of character, I told the guy rather forthrightly where he may want to stow his musical instrument. Aah! the impetuosity of youth.

We eventually visited Rhodes and enjoyed the wildlife in the form of lizards in our room. We loved the old walled town. This is where I was first introduced to kalamari and even today I do not get why people see this rubbery delicacy as tasty.

CHAPTER EIGHT

The North American dream

Summer season finished at the Porthmadog pottery where Wendy had been working and, missing home, Wendy could wait no longer, she was going back home to the USA.

Although she was initially surprised that I really did want to go with her, in time we both embraced the idea. For me I saw it as a great adventure to live in another country, especially the home of the 'Wild West' and JFK. North America had left an indelible impression on me when I visited in 1977 and I was excited to face a new challenge.

In October 1979 we flew out and stayed with Wendy's family in Raleigh, North Carolina. They were most welcoming and of course intrigued to meet the man in Wendy's life for the last eighteen months.

Trying to help me to find work Wendy's father, Doc Holden, introduced me to a fellow called Bruce Dollar who owned Meridian Travel. In spite of a very encouraging meeting, it was all to no avail as it transpired that there was no chance of me getting a Green Card. I soon found out I had to be outside the USA to apply for this 'alien residency' permit known as the Green Card. Of course, the term 'alien' does not exactly have the attraction you might want to line up for, but in reality, it is a much sought-after term by many in the world!

We decided to move to a country as close by as possible while I was applying for this prized document. The nearest obvious place was Ottawa, the capital of Canada, where Wendy had once lived and her dad still had friends who could help us.

For just a heartbeat before we emigrated yet again, there was even some consideràtion given to my marrying an American girl who was an American citizen, in a marriage of convenience as a friend of Wendy's needed some money. It was all tongue in cheek as I advised that we would need to consummate the marriage to make it legal, so needless to say the idea was not pursued. Wendy was a permanent resident and a Canadian citizen and British passport holder so marrying Wendy so I could stay and work in the USA was not an option.

While waiting to move to Canada, I enjoyed my time in Raleigh. I would visit the malls like close-by Crabtree Valley Mall, all a revelation to me after shopping in the traditional high street in England, which at the time had yet to embrace the idea of out-of-town malls with their mix of small shops, large stores, restaurants and movie theatres.

One afternoon in downtown Raleigh we witnessed a Ku Klux Klan rally, which was surreal, all these hooded Klansmen and people up on grassy knolls where we were standing barracking them. It was scary that this sort of activity was still prevalent. Shortly afterwards on 3 November 1979 a 'Death to the Klan' march in Greensboro resulted in a shoot-out between members of the Communist Workers Party, the Ku Klux Klan and a neo-Nazi group, leaving five dead. Sixteen people were arrested and six were brought to trial. Despite video evidence showing quite clearly the perpetrators getting out of a vehicle, going around to the trunk (boot), extracting weapons and gunning down people armed only with clubs, no one was convicted.

Wendy confirmed to me that racial tensions were still very high and in spite of desegregation in restaurants, on buses and in schools there were still constant assaults on both sides. She recalled that when the schools were desegregated she was cornered in a restroom by three black girls and verbally abused and shoved for no reason, in no uncertain terms threatening her, she never went alone again. The worm was turning!

Things had come a long way from the Rosa Parks incident, when she refused to give up a seat to a white person on a Montgomery, Alabama bus and which led to a boycott of the buses and then a year later a ban on segregation in Montgomery. But there were some rocky times ahead, in spite of all the civil rights acts passed in the sixties, law was one thing, but implementing it was another. It was going to be slow progress in the south.

While I was in Raleigh, I started helping out as a local soccer league coach and enjoyed coaching the boys' team called the Vikings. A lot of emphasis was put on attack in many teams, but if you didn't have the personnel to carry that through some severe losses were in order. When I joined the team we regularly lost 6- or 7-nil, but as we got better organised and changed formation to make it harder for the other team to attack the losses were reducing to 3- or 4-nil and just before the end of the season we even scored! I know that doesn't sound much, but, believe me, the lads played to the plan and achieved a lot. I was sad I

First ever road race

missed the end-of-season awards as they were a lovely group of young boys and fellow coaches.

Meanwhile Sally (Wendy's eldest sister) had encouraged me to enter a 10k road running race. I had never run this sort of distance, the most being

Ron running a 10K in Virginia

a bit of cross country at school but I was fit and young at twenty-five, so decided to give it a go.

After picking up my Pepsi Cola-sponsored T-shirt as part of the registration and race fee deal, I was a little too enthusiastic and ran the route just the day before the main event (what an idiot). At eight o'clock on the Saturday morning I was out toeing the line with a 1,000-plus other runners. I was keen and inexperienced and shot off with the leading pack, but after just 400 metres I realised they were far too quick for me. My lungs were burning. Also, my body was still tired from the previous day's efforts, my quadriceps and hamstrings sore. I tried to keep going but I suffered from severe stitch. I battled on and eventually finished in about 43 minutes but was 177th out of more than 1,000.

My experience taught me a lot. I knew with the right training and effort I might be able to do quite well at this running malarkey. So, a seed was sown in my mind and this seed started to germinate in the time I would spend in Canada.

We used some personal savings and bought an old car called an Oldsmobile 98. Ours was a dullish bottle green, a grand old limousine-type sedan that only cost us US$400. Big cars were not in vogue at that time because of the cost of fuel due to the world oil crisis. This car we bought only did about 17 miles to the US gallon. We said our good byes and drove to Canada, staying in motels en-route. Thus, our Canadian adventure began.

It was an unseasonably warm autumn when we arrived, but the bitter cold of winter and the snow and ice were soon upon us. We spent our first week in the furnished and serviced Algonquin apartments in Downtown Ottawa. Initially I headed to a cashpoint every day to draw money out from our savings for our living expenses as without jobs we had no income. We then had to look for a more permanent base within a budget we could afford.

Our first rented apartment was on Carling Avenue, where the ceiling used to keep us awake at night making alarming noises as the steel frame expanded and contracted, as by now winter had arrived with a vengeance. The structure would startle us as it resounded with a large crack as if the roof above our head was going to cave in.

Wend could work because she still held Canadian citizenship from when she emigrated to Canada as a child with her family. Following a number of interviews. Wendy eventually found a receptionist job at a real estate company at Bells Corners, a suburb of Ottawa. Her boss was Peter

Assaly of the building company Thomas Assaly and Sons. I think they were of Lebanese descent and Wendy thought it was a slightly shady 'Mafioso' type of operation, but we could not be choosy. After all, this opportunity came after a failed attempt to land a job at the Big Bud discount store, the equivalent of Poundland today.

We needed somewhere to live that was a little cheaper and more comfortable than Carling Avenue. We wanted underground parking so that our car would be out of the bitterly cold winter temperatures that affected its starting capabilities. After some searching, we moved to some modern apartments at Bayshore not far from Bayshore Shopping Centre with a heated underground car park and a fairly low rent. The address was 2395 Richmond Ave, a place called Aspen Towers. We were almost at the top in our 16th floor apartment and looked down over the Outaouais river towards Quebec.

Our kitchenette was the scene of many a preparation of the 22-cent packets of Kraft macaroni cheese, we could not afford much more. All our income was going on the rent and keeping the car on the road to allow Wendy to get to work. Furnishing was nil, that's right, nothing! We had no money so could not afford anything to sit on, so we initially slept and sat on the floor.

One evening I picked up Wend from work and on my drive in I spotted the discarded base of a broken box bed on a skip at Bells Corners, presumably from a house clearance, where Wend worked. We somehow crammed it into the car and that's what we slept on from then on.

We used to go to a St Vincent's thrift shop to get odds and ends. In due course, we added furniture such as a rocking chair and a bean bag and we hired a television for a few dollars a month.

We bought ourselves some ice skates from the thrift shop, too. I learned to skate on the Rideau Canal, reputedly the longest skating rink in the world, all serviced by the local council and kept smooth and in good shape for the skating public. Most neighbourhoods had little ice hockey-cum-skating rinks too.

I remember one cold winter evening queuing outside a cinema to see the hit movie of the day *Kramer vs. Kramer* featuring Meryl Streep and Dustin Hoffman. Boy it was cold, temperatures close on -15°C, but worth the wait.

One winter's day on the way back from work we were broadsided by a car who clipped our car tail end as he crossed carriageways with a totally reckless move. I got out to see if the other driver was alright, but could immediately smell alcohol on him, and he started swearing aggressively. I went to a nearby gas station and called the police.

When the officer came, the other driver sat in the front of the patrol car with the officer and Wend and I sat in the back as it was early evening and freezing cold, whilst the officer had to fill in forms to report the incident. Whilst in there, the other driver who had hit us, lit up a cigarette, and the officer admonished him for that and told him to put it out, which he duly did, but unfortunately, he did so on the leg of the officer, needless to say he was arrested and we all went to the station to file the report.

Eventually we put in an insurance claim. We were quite surprised when the insurance company, instead of offering to fix the vehicle, offered us US$400, which we accepted. The bottom line was we received the same amount of money which we had bought the car for. Our Oldsmobile was still drive-able with minor damage to the driver's rear wing. A win-win outcome to our little escapade!

I remember on some late nights going to Dunkin Donuts as a treat, just because we could, and it was so different to the UK, where everything, even the pubs, were closed as you approached midnight. It was notable the car park was usually full of police cruisers and the doughy franchise full of police officers having doughnuts and coffee and their physiques often reflected this diet.

North American television was full of advertisements and was tedious to watch but the commercials did make me smile being a lot different to back home. One that stuck in my mind was 'Do you suffer from haemorrhoids? Then you need Preparation H'. I just couldn't imagine this on UK television at the time.

Beer advertisements were all about extolling the virtues of socialising with a cold beer, yet strangely if you analysed the advertisements you never saw anyone actually drinking a beer. There was a good reason for this as most channels we watched were governed by US advertising standards laws, which said you couldn't be seen to be drinking beer in advertisements. How bizarre! It underlines the strong religious lobby in the USA from the Baptist and other churches.

We made some really good friends whilst we were in Ottawa. The McKercher family in particular were great. Dr Bob McKercher was a doctor when Peter Holden (Wendy's father) practiced there. His daughter Phyllis and son Bob Mckercher still lived there and they welcomed us and helped greatly with our integration into Ottawan life. Bob McKercher introduced me to a guy called Ron Duncan when we were running on the indoor track at Carleton university. Ron and Bob would suggest an occasional winter outside run, perhaps along the canal, and were amazed at me running in shorts in Arctic weather conditions. I earned the nickname 'Crazy Brit'. Little did they know I could not afford tracksuit bottoms. We needed to eat!

Ron ran with a running group called the Sunset Striders, so before long I joined them. Eventually in the spring and summer, Ron and I used to go running up on the scenic Gananoque Parkway in the Laurentian mountains of Quebec, which was great resistance training and improved our stamina no end. This was the time when Quebec wanted separation from Canada and independence, led by the revered René Lévesque. We would sometimes come back to our car to find the tyres and bodywork would have been damaged and the vehicle broken into, purely because the perpetrators saw Ontario plates.

I also joined the National Capital Road Runners, now part of Run Ottawa, and used to run every weekend in staged, timed race events and get newsletters with our results.

I was looking at various ways to get citizenship and work and spoke to a lawyer called Sol Wiseman who I got on well with. He recognised my ability as a travel man and came up with a proposal out of left field, to use a baseball Americanism.

He wanted to set up a travel agency in the Turks and Caicos Islands and asked me if I would be prepared to go down there and manage it. It soon became clear this might have also been a way of getting money off shore for him but that apart I was quite taken by the idea, as I was enjoying my running and especially training for marathons.

I contacted the Turks and Caicos Athletic Association and asked what their view would be if I moved there and enquired if I qualified to represent the Turks and Caicos at the marathon in the Olympics. I figured I would continue improving and loved the idea of being in the Olympics, given back then there was no minimum time needed to enter. Before long I had a reply saying I would qualify after a period of residency there and they were happy for me to compete for them as they had no marathon runners at the time.

For just a short while I fantasised about being at the Olympics. I even figured I might be able to play for their football team in the World Cup qualifiers … I was certainly full of myself. Wendy was not as happy about the plan and my enthusiasm waned as the uncertainty of income and the fear we may become involved in something not quite above board made us wary.

Meanwhile a fellow called Terry Fox, who had been diagnosed with cancer, was running coast to coast across Canada to raise money and awareness of cancer on what he christened the 'Marathon of Hope'. Tough for anyone but this young fellow with cancer and a prosthetic leg certainly had the odds stacked against him!

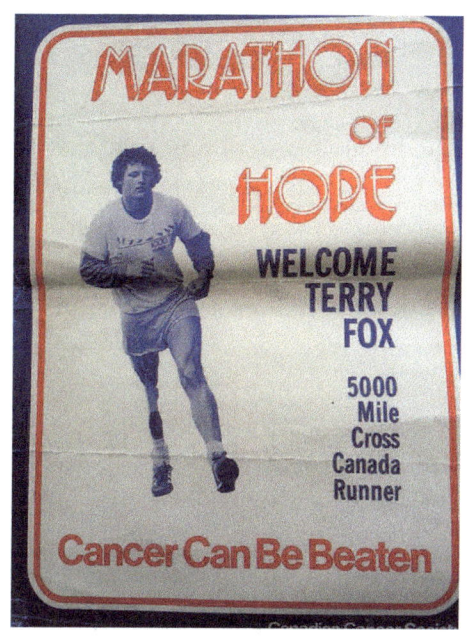

Marathon of Hope running poster

Whilst he was fundraising and running through Ottawa I had the great honour of meeting him and running some distance with him in and around Ottawa on 30 June 1980. Almost a year later to the day, 28 June 1981 Terry lost his battle against cancer, aged just twenty-two. He was an amazing young man and has inspired many others with his 'Marathon of Hope' and there are now statues of him at Victoria, British Columbia and in Ottawa. His legacy lives on through the Terry Fox Run, the world's largest one-day fundraiser for cancer research each year and the Terry Fox Foundation, which organises the runs and has helped Canadian scientists make numerous advances in the fight against cancer.

Terry's drive and sheer determination got him to just outside Thunder Bay, the 3,339-miles (5,373km) mark on his 5,000-mile adventure from near St John's in Newfoundland to Victoria in British Columbia. The cancer had reached his lungs and it became increasingly difficult for him to breathe. He had been running a marathon a day for 143 days. Yes, you are reading that correctly, yes, there were injuries that slowed him up and the odd rest day but you can only be in awe of this young man's grit and determination.

When I went for a run outside in the winter, any moisture on my hair or my beard would freeze immediately, and I was transformed into a Father Christmas figure within seconds. It was often difficult running in these conditions but every time I faced adversity in my running, be it marathons or ultramarathons, I would think of Terry and what he endured and it would immediately put my pain into perspective. Ironically, thirty-five years after meeting him I would have my own run-in with cancer.

In order to work. I needed a Canadian Green Card and with the assistance of Jamie White, an apartment neighbour, we set about obtaining one. Jamie was formerly in the Rhodesian army and a very forthright character. He shared an apartment with his Yorkshire girlfriend, Sharon Fuller. As an aside one day when we were down in his apartment he said, 'Can you hold my pet while I clean its little house out?'

I said, 'Sure'. Jamie went into another room and returned with a tarantula which he promptly put in my hands. I stood very still for about fifteen minutes not moving a muscle as this huge spider's abdomen palpated on my open palms. I think he thought I would freak out and I was surprised I didn't. Quarter of an hour later I stoically handed it back to him and we went for lunch!

Our nemesis in immigration was a very suspicious and unfriendly chap called Mr Peter Lovekin. He was very pedantic and we cannot help feeling he was against us from the start, everything took so long and he nitpicked whatever you said or provided. Unfortunately, over the course of time Jamie's well-meaning bullishness may have added to Lovekin's stubbornness, but Jamie was well meaning.

Still with no work permit, I started work on a voluntary basis at the Ontario Motor League Worldwide Travel, which was similar to AA travel in the UK. I couldn't bring in any money, but the experience gave me a good insight into the Canadian travel operation. At work in winter, having left the sanctuary of my underground heated apartment complex car park, I used to have to plug the engine block in to a power point in the shopping centre car park to keep it warm, otherwise it would freeze solid in the minus 20°C temperatures plus wind chill.

Whilst in Ottawa we did have a good time on little money. Mum and my brother Chris visited in the early summer and we showed them around, including having a couple of days driving down to Toronto and Niagara Falls with a little cruise on the St Lawrence Seaway around the Thousand Islands.

We actually hired some furniture for the two weeks Chris and Mum were with us. We hired a bed and sofa, so they didn't realise how destitute we were.

I was continuing my running and as the spring and summer came I had been training for the Ottawa Capital marathon. I often ran alongside the beautiful Rideau Canal and sometimes the river. In the winter I used to run a long rectangular route from Aspen Towers into Nepean using Baseline Road.

We got some ventriloquist gigs for Wendy, which earned us extra money, by advertising in local free newspapers. We were watching our rented cable television one night and saw some of the commercials, which we thought were amateurish and lack-lustre. I suggested approaching one of the advertisers we saw, Di Lawri, a local car dealer in a place called Stittsville in the suburbs of Ottawa.

Mr Di Lawri advertised a lot and we felt we could help him to improve his ads. Following an interview in which Wendy showed off her fantastic ventriloquial skills, to our astonishment he decided to take us on, beguiled by Wendy's superb ventriloquism and the novelty of our approach. We wrote the script, Wendy delivered the message with her wonderful characters and ventriloquism in the advertisement and made a fair bit of money.

Di Lawri, 'Just beyond the fringe', had radio commercials too, so we took their strapline and wrote a few new scripts, which were recorded as television commercials at the Canadian Broadcasting Company extolling the virtues of their Chevrolet dealership at Stittsville, Ontario.

Meanwhile my residency permit application process was dragging on and I still could not work for remuneration. Eventually Jamie lobbied an MP called Lloyd Francis and we went to see him at the Houses of Parliament by the Chateau Laurier on the Rideau canal and Outouais River. What a wonderful experience to enter these hallowed halls of power and governance. It was a fabulous old building, very similar to our own Houses of Parliament on the Thames We were encouraged by our meeting.

Amazingly, rather than my visa application progressing positively, an astounding chain of events unfolded.

One day I was working in the office at Ontario Motor League, Lincoln Plaza, still unpaid, when I was invited into the office of the agency manager, a pleasant and compliant Asian guy called Dick.

Dick was a permanent resident Green Card holder himself; also in attendance were the accountant and franchise owner. They told me they were pleased with my efforts and would cut me a check at the end of the week as they had spoken to immigration and my permit was being issued.

Well, I was pretty happy with this, but just a few minutes after I'd returned to my desk, as if tipped off or waiting outside, two RCMP officers breezed in and told me to come to the back office and I was joined by all the office and department managers, before being told I was being arrested for working illegally. One of the officers was the husband of a fellow female employee at the Motor League, who seemed to be quite aware of what was happening. I had been set up!

So began a nightmare of fingerprinting, mug shots, court cases and eventual trial at which I was found guilty and given a two-year sentence suspended for two years.

I was made to feel like a criminal, yet I had done everything above board. It was clear they were making a scapegoat of me. I decided I had had enough and wanted to go home. Ironically, the week after we left the visa for me to stay and work in Canada was issued. I learned a lot about life and opportunity and also about racism and dishonourable racist fellow workers.

One other major development took place in Ottawa! Wendy and I were married at the registry office in Ottawa on St Valentine's Day, 14 February 1980. We had two witnesses Patti and her husband who both worked at Bells Corners with Wend. It was a 9am ceremony because that was the only appointment time they had left. We had a couple of photographs taken, then went for breakfast next door in a cafe. We had bought wedding rings at Mappin and Webb.

Phyllis Mckercher and her apartment mate Brenda and a few other friends threw a party at Chateau Laurier, the fabulous hotel on the Rideau Canal, and then Wend and I went for dinner at Crock and Block, a steak house nearby to celebrate.

Next day, Wend went back to work. Theoretically this should have helped my work visa plans as I was now married to a Canadian citizen, but it was viewed by the court as an arranged marriage in spite of the fact that we had lived together in the UK and North America for more than eighteen months ... crazy or what?

A sad postscript to our time in Ottawa was our good friend Phyllis McKercher, who loved her sub-aqua diving and was an instructor actually died on a quarry dive whilst tutoring. She was a lovely lady.

CHAPTER NINE

———

Back home fired up!

Adventure over for the time being we returned to Blighty.

Once back in Shrewsbury I dropped by on an old friend of mine from Luxitours days, Terry Jones, who was Manager of Yeoman Millers Travel in Roushill, Shrewsbury and he told me of his plans to open a new travel agency in Oswestry, with two silent partners from that area, Rick Carr, who owned the Queens Hotel (the corner of which was soon to be the new travel agency), and Parker Clare.

I was excited about this for him, but he explained he could not open it for three months because of the notice he was serving with Yeoman Millers. I said in a throw-away comment, 'I'll open it for you and if a job becomes available later, maybe you would consider me.'

It came to pass that Fourwinds Travel was born in 1980, and we had a fruitful four-year relationship through a few ups and downs business-wise. We ended with a number of branches in Shropshire and Mid-Wales.

Whilst at Fourwinds I developed an interest in tour operation and tour managing, almost by chance. These meant creating itineraries, putting groups together and then escorting the groups on holidays rather than just buying a packaged tour from a tour operator such as Cosmos or Thomsons and adding a local transfer to the airports.

The main catalyst for this development really stemmed from Criddle Billington Feeds (CBF), an animal feed company. One of their regional salesmen Ken Morris, who lived in Mid-Wales, wanted to change to a local agent that could organise annual tours for CBF, rather than a company they used to use in Stoke on Trent.

He approached us for a quote and we quickly built a good rapport which lasted for many decades. These tours were part of an incentive for farmers to buy feed from CBF as they had discounts on the tour price; the more feed they bought, the more discount they received off the tour.

However, the precursor and stimulus for doing our own tours was that in the first year we arranged a ready-made tour with Cosmos on the Rhine, a coach tour with coach transfers from the Shropshire and Mid Wales area. I did not escort this tour because of the costs involved. On his return Ken had not been happy with various aspects of this tour including the guide/tour manager. It was this experience that prompted my first venture into tailor-making holidays.

Our first tour together used a local coach company called Minsterley Motors and a driver I had travelled with a few times before called Trevor, who was a real character and a great driver. Although he knew the ropes, it transpired that what he did not know on this first trip was the route. I had foolishly (albeit naturally) expected him to know the way and the journey stage times etc. In fairness, having collected the forty-nine people on board and travelled down across the channel to Rheims for the first night, all was tickety-boo and going according to plan. Well, perhaps all except the timing. Arriving a lot later than Trevor had forecast should have been an early warning to me.

When we pulled up at the characterful small hotel in the shadow of Rheims cathedral I went to the reception desk, whilst Trevor brought in all the bags with the help of passengers and hotel staff. The tired and hungry farmers filed into the small reception area, I handed out the keys and folk made their way to the rooms.

I waited in reception to make sure all was okay before going to my room, which was how I always operated. Soon Ken and two other couples returned to reception. Ken was red faced and very uptight. His fingers rubbing together, as they always did when he was anxious or agitated. 'All three couples have been allocated one big room with connecting doors. This is not on,' he exclaimed.

I spoke to the desk and they said 'Oui, monsieur, this is a suite of chambres for the family.' I then realised what had happened. All three couples had the surname Morris and as they were on the manifest with each other, the hotel had presumed it was one family, a fair assumption. Our hotel was full. There was no chance of reallocating rooms, so we had to insist that the manager go and lock each of the interconnecting doors, to keep all the couples happy.

I could just imagine them arriving in their rooms and exploring, opening doors and coming face to face with their fellow travellers. Good job they noticed straight away, rather than during the night.

All resolved, I was going to head to my room. However, another couple came and said, 'Our room is no more than a cubby hole under the stairs.'

I followed them and agreed, it was rather small. I offered to swap with them but they said no and that was that, there was no other solution.

We all went for something to eat and drink. I made sure the wine flowed freely and everyone was happy.

Trevor insisted on an early start the next day, as we had a long journey and wanted to arrive around teatime. This second stage of our journey, however, can only be described as calamitous and comedic.

We had some comfort stops to break up the journey, of course. On one, we asked the attendant to put some fuel in the coach and fortunately I noticed just in time that he was about to put the diesel in the water tank. We would so nearly have been going nowhere.

As the day progressed, I'd look at the map from time to time. I came to the conclusion as 5pm, then 6pm passed that we were nowhere near our ultimate goal. I checked with Trevor and he said it was taking longer than he thought. He reassessed. We were not going to be there until about 8pm, 7.30pm if we were lucky. I decided to call through to our destination, the Hotel Excelsior Splendide on Lake Como.

I advised our progress had not been what we had hoped for. The manager there, a wonderful chap called Emilio Gilardoni, when I stated our current location and our calculated arrival time, said he would be surprised if we were that early. Before I could respond, I ran out of money. No mobile phones in those days, or phone cards, just a telephone box and change!

We ploughed on, reaching the southern end of Lake Como at 9pm. By now darkness had fallen and we could see the lights of properties twinkling around the lake. We now had another dilemma: which side of the lake to take to get to the northern shore. We looked at the map and Trevor suggested taking the eastern side as it looked the shortest. I pointed out that the road looked better on the western side, but in honesty I did not do it with enough conviction. We took the road on the eastern side. Within an hour we were at walking pace, the road became very narrow and I could see the lights of the homes along the lake, which by now were way below us. There was not much stopping us from a rather rapid descent into the lake, the coach being at times as wide as the road. Suddenly we came to a tunnel with a roof of rocks protruding downwards in a jagged, fearsome form of natural danger that would have damaged any high vehicle if you did not judge your passage carefully..

Ken volunteered to get out and walk-in front of the coach, to guide Trevor through. Everyone became rather quiet as we inched our way along, but we were eventually through and soon saw the Bellagio oasis at the end of the lake. With great relief and very weary, we safely arrived at the hotel at 11pm.

Amazingly all the hotel and restaurant staff had stayed and served us a full dinner. All were tired and I was very relieved. We were served antipasti and everyone wolfed it down. We thought that was the main meal but dishes kept coming until we were stuffed full. What a wonderful reception.

Next morning, I opened my floor-to-ceiling bedroom windows to gaze across the sun-kissed lake. What a wonderful vista. As the sun poured in the frustrations of yesterday were now a memory and all was right with the world.

After breakfast the hotel staff made it clear that we should have come across on the car ferry at the northern end of the lake, having taken the western road. We would have saved an hour at least, and avoided all the angst. The added financial burden, which had not been costed in, continued each day, as we had to pay for the coach and passengers on the ferry every time we left the hotel to go anywhere. That said it was a far better option than travelling the road we had arrived on. Never again did I make the mistake of letting the coach company or indeed any other ground handler take care of the route and any tour content without me checking it thoroughly in advance. A lesson learned!

We had some lovely tours from our base at Bellagio. One day we headed to St Moritz, which proved a real highlight. It was made even more memorable by the screams of the passengers as driver Trevor negotiated some horrific hairpin bends in and out of the lake area.

As we ascended then descended, sometimes having to shunt back and forth to make our way around the bends with the rear of the coach hanging over the roadside, one lady moved up to the front of the bus, a bit like the old story of a first-class passenger on an aircraft thinking if the plane crashed she would be safer in first class.

Whilst the clients were mooching around St Moritz I went with Trevor to refuel the coach. Now that was an education, as we could not understand how to operate the diesel pumps and before we knew it were filling up from a pump which took cash in the form of coins only, not notes. Trust me, it took a very long time and a lot of Swiss Franc coins to fill the huge fuel tank.

We included a day at leisure as part of the holiday and a number of the holiday guests were relaxed in the Italian sunshine, lounging around the lovely little swimming pool at the rear of the hotel, which was positioned below a tiered garden.

I came back from exploring around the quaint but steep village of Bellagio just in time to witness a feat of derring-do by our driver Trevor, who had been sharing copious beers with a number of his fellow guests.

He was holding a picnic table umbrella on the lawn above the pool. I arrived in time to see him run across the grass and hang-glide into the pool on the level below. He just about made it.

I have to say it was a sight that made you wince and simultaneously laugh. It was clear to me he was pretty squiffy. I said, 'Trev, you've been drinking!' He laughed and said, 'Too true.'

'We're going on an evening excursion to a restaurant for a meal and folklore show and will be driving up the side of the mountains above the lake,' I reminded him.

Trevor looked bemused and then was quite adamant he had never been told, insisting he knew nothing about it. I produced the itinerary, of which he had a copy. Reading it, he struggled for an excuse, before shaking his head and admitting he must have been looking at the wrong day.

All afternoon we plied him with black coffee. I was so worried about the excursion I wanted to cancel it. We had a council pow-wow between myself, tour leader Ken and Trevor, and decided that, as long as we took it steady, we would be okay, and indeed we were. Trevor seemed on the ball but I sat by him talking to him all the way to keep him focused and alert. However, that was the longest evening and I was so glad when we all returned safe and sound.

I had learned yet another important lesson! Control and check every facet of the tour, do not delegate to others such as coach companies you think (or hope) should know better! Fortunately, it did not put me off tour operation and learning from this steep curve I went on to take many groups to numerous countries, perfecting my tour choreography as I gained knowledge and experience. More about that later.

We had some lovely clients and some interesting ones, too. One of our regular clients was Per Lindstrand, the hot-air balloon manufacturer and partner in the balloon transatlantic crossing attempt with Richard Branson.

Two lovely clients among many were Molly and Frank Evans, who farmed at West Felton near Oswestry. Frank had been a bachelor and dedicated farmer until late in his life when he met Molly, a funny and straight-talking northern lass. They had married and lived together on the farm. I had been booking flights to Europe for them when they decided to buy a villa on the Algarve and so started a very long friendship. We even used to go and stay with them in their villa above Faro in a place called Pechao.

My good friend David and myself recall one such visit where we were trying to develop our business, finding villas for sale and introducing buyers to the villa owners with a view to our travel agency, of which we were partners at the time, getting more flights, car hire and rental business, and receiving commission for selling villas and apartments in the sun.

David and I would drive around the Algarve, listening to our only music tape, one of Kenny Rogers' greatest hits (RIP Kenny), singing along word for word, windows down, with the warm breeze carrying our lyrics to the passing villages and their inhabitants as we passed by looking for properties for sale in good locations. Sight of a 'Vende casa' sign and the car would make a sharp stop and we would check out the property, take some photographs if we liked it, although no mobile phone in those days, just good old-fashioned Truprint prints when we got home.

Dave and I almost bought a villa above Molly and Frank's place, but the asking price seemed a bit heavy for a property that needed a little redeveloping. In hindsight its location and the amount of land available should have been a definite buy. It was the one that got away!

So many memories. Molly had been a school cook, which belied her talents as a chef. The food she prepared was wonderful. She would prepare breakfast and then pour us a Maciera brandy. It seemed a little early but a habit was developing for these ex-pats.

One evening meal Molly asked David and I to whip some cream with a hand whisk. Well, we were young guys so had a competition taking it in turns to whip the cream to see who could whip the fastest. Molly reappeared and said, 'What's this? It's been whipped so much you've made butter!'

David and I met a few of the local football team in one of the two restaurant bars one evening and were invited along to training, with Frank (once a good footballer himself) coming along to watch. It was a fast-paced technical game on a dustbowl of a pitch but very enjoyable. Frank's observations and conclusion were that David was a very good, technically astute footballer. As for me, he said I didn't let much get past me, which being polite meant I was a bit of a clogger with not much footballing prowess. Ha ha! You have to know your place!

We enjoyed expeditions in the evenings to a restaurant called Chicken Georges in the Monchique mountains, which was always full, mostly with Portuguese, but Molly and Frank were regulars and greeted as such. The chicken meals were excellent and that's why people came from miles around.

We had a farewell meal there and insisted on paying for Frank, Molly and one or two others. We had enjoyed a fabulous meal. Even though English wasn't widely spoken back then we used to play charades by flapping our arms and clucking to indicate we wanted chicken and if our actions didn't work for what we wanted we followed the staff into the kitchen and pointed to things. The whole meal and drinks for the evening came to just over £30. Amazing, we thought they had got it wrong for sure but they insisted that was the cost. Happy times!

CHAPTER TEN

———

Off and running

Whilst I had been in Ottawa, I had enjoyed my running but on return to England in 1980 I found there was no similar running/jogging groups, only clubs for more serious athletes.

It just so happened that the first London Marathon had been advertised for March 1981 and I applied to take part, thoroughly motivated and swept along by the promotional videos of the forthcoming event and the thought of completing the ultimate running challenge. When I received the entry form in the post, it asked what club I belonged to. Of course, I wasn't a member of any club, so I pondered what club name I could come up with, in effect starting my own joggers group.

I read copiously the missives of running gurus and experienced distance runners. I had picked up a book called *The Complete Book of Marathon Running* by a chap called Jim Fixx, which was excellent in guiding me in my preparation. Another influential guide was Arthur Lydiard's five principles of training, which are as sound today as they were in the sixties. Lydiard coached and influenced great runners like Peter Snell and John Walker (the famous Kiwi miler). I consumed articles and books by him, Ron Hill, Joe Henderson and so on, all trying to improve my running performances. I even followed Nathan Pritikin's diet book of the day to try and optimise my body's effort. Running magazines also brought to my attention stories of Kathy Switzer being the first woman to finish a marathon in spite of the organisers trying to stop her, even mid-race.

My hero and inspiration was a softly spoken Lancastrian from Accrington called Ron Hill. He was renowned for his record of running every day and in fact not many years ago he missed one day due to ill health following running consecutively for 19,032 days. WOW! He was a superb runner, running sub-2 hours 10 mins many years ago, before it was the norm. I had the great pleasure of chatting with him on the telephone for about forty minutes a couple of years ago. He is struggling with Parkinson's Disease now but told me a few little stories about his career and it was clear he still has a passion for running even though his body is letting him down. We vowed to meet at his local park run one day. I hope we can do this!

I still try and run with the Shropshire Shufflers and I have been most impressed with the way the stewardship of the club has developed without losing its ethos of inclusion and social interaction whilst exercising as well as achieving goals from couch to 5K for those just needing to exercise and get fitter, to top runners.

It's also great to see the Parkrun going from strength to strength around the country, but especially in the Quarry Park, Shrewsbury thanks to all

Ron and Di Park Run Shufflers group

the great volunteers and the inception and development by Paul and Susan Bowes in Shrewsbury.

Like many, I was never keen on running at school, and cross country was something to be avoided! So, it was a great surprise to me when I moved to North Carolina in the late seventies to be persuaded to run a local 10K road race. I think it was the fact I was getting a free t shirt for entering that swayed me.

A few months later I moved to Ottawa in Canada and as I couldn't work because of permits I pursued the gentle art of jogging with a newly formed club called the 'Sunset Striders', based at Carleton University and but for them perhaps my interest in road running would have waned.

What I liked most about the Striders was the camaraderie and support each member gave to one another. An experienced sports medicine guy from the University led the group and we started with warm-up exercises not too dissimilar to those we use today and then set off on a steady run together mostly along the Rideau canal which was very scenic and for the most part flat.

We had social events where we could share tips and ideas and talk about other things in our lives at BBQs or fundraisers. All this time I didn't realise the seed of the Shropshire Shufflers was slowly germinating.

Winters in Canada were tough to run in, as the snowfall and extremely low temperatures with wind chill factor slowed us up dramatically. I had a beard at the time and as soon as I ventured outside the moisture on my face would freeze immediately and sometimes my eyelids would freeze shut. Now that is scary!!

As I was not working Wendy and I didn't have a lot of money, I had a pair of shorts and T-shirts but no tracksuit so initially ran with a few T-shirts on and a pair of shorts in this cold weather, which earned me the name from local runners as the 'Crazy Brit' and at one function I was presented with a T-shirt inscribed with this monicker. Little did they know it was not that I was tough but just broke!

I had also joined the National Road Runners Association that ran monthly time trials at a local air base around Ottawa and this was a great way of measuring my improvement As the spring came 10Ks and other events were many and road running was becoming popular.

Indeed in 1980 I ran my first marathon along a beautiful, flat course in the National Capital Marathon in Ottawa. I had read many running magazines and books by running gurus Jim Fixx and of course Bill Rodgers (a prominent USA marathoner of the era), but nothing really prepared me for that warm May day. I was a bit of a purist in those days, running a marathon meant just that, no walking at all, but about the 25-mile mark my legs, which had become very heavy, just stopped. (Do you know that feeling?) Somehow I managed to lean forward and get going again, finishing in a fairly respectable time of 3 hours 24 minutes 55 seconds.

Not long after I returned to England and continued my jogging, but missed the friendship and social aspect of the Sunset Striders. All that seemed to be on offer locally were Athletics Clubs which gave the impression that you had to be a swift runner to join and were considered elitist.

I plodded along and then saw that the first London Marathon was being run in the Spring of 1981 so decided to enter. Before I filled the form in, I wanted to have a club name to run for. I was tempted to use the Sunset Striders.

We began trying to think of a name for the club: Shrewsbury Strollers, Shrewsbury Striders … and then Wendy and I initially came up with the Shrewsbury Shufflers and then amended this to the Shropshire Shufflers as we had a vision of having small groups in each area of Shropshire, a little like a franchise, and so the Shropshire Shufflers were born. I was on the landing at 171 Mount Pleasant Road at the time and can see me uttering the words right now: 'What about the Shropshire Shufflers!!!'

I entered the first London Marathon as a Shropshire Shuffler, as I didn't understand about affiliated clubs and unattached runners at this stage.

My thoughts in the twelve-week training period in which I intensified my training were also on forming a club like the Sunset Striders to offer friendship and support to folk who would like to run but felt self-conscious about running on their own. You must remember that you rarely saw anyone out road running at the time. You were in fact an oddity if you did so!

Leading up to the London Marathon, my training had been quite good. Through the winter I noted that once I ran over thirty miles a week, I started to pick up injuries and colds, so pared back my efforts. The day before the big event, I was one of over 7,000 runners who registered at the Strand Hotel, London. I was on the start line of the first ever London Marathon.

On the day of the 1981 London Marathon, you could not help but be moved by the throng of humanity, buzzing with excitement despite the miserable wet Sunday morning that 29 March 1981 had presented us with.

Placards in the crowds of 'You can do it' and 'Come on Mum' set the scene, celebrities like Jimmy Saville in a gold lamé tracksuit added to the occasion. At 9am, the 25-pounder naval cannon sounded and over seven thousand runners headed off on the journey that would launch the real birth of fun running in the United Kingdom!! I loved it!

It was a great occasion, well supported around the course by cheering public and musical entertainment as we excitedly made our way past iconic London sights like the Cutty Sark and Tower Bridge. I managed a time of 3 hours 15 minutes, finishing along the side of Buckingham Palace in Birdcage Walk. What a wonderful experience! Nothing can ever prepare you for this spectacle in our capital, the support and camaraderie is wonderful. Following this inaugural event and its television coverage, lots of neighbours and friends talked about it in the weeks and months to come. Many stories could be told just about that first marathon but I will regale you with just one!

Feeling surprisingly good I started to run with a group of folks around the 20-mile mark and it soon became clear amongst these was a well-known athletics character of the time, David Bedford, who I believe may have been involved with the organisation of the marathon too. This made me feel

London Marathon 1981, hanky on head!

great running with the former 10,000-metre world record holder, though unexpectedly at about 22 miles he just veered off into the crowd while we just continued.

For a few weeks I lived off the story I had been running with David Bedford and had actually beaten him, until I read this article in a newspaper:

On the eve of the first London Marathon David Bedford accepted a £100 bet that he couldn't get around the course. The night before the race, he ate a chicken vindaloo and drank 12 pints of beer. He got to bed at 4.30am and, after one hour's sleep, set off like a train. After 22 miles he was seen leaning over a manhole cover being sick. He did finish and collected his £100'

Following the first London Marathon in 1981 great interest developed and I bumped into a good friend of mine Keith Ivison who was most interested in starting to run and train for the next year's London Marathon.

Keith, myself, Steve Wellington and Robin Jones ran the Shrewsbury Carnival race on the 25th May 1981, as the first unofficial Shuffler Race together.

As the weeks progressed, we decided to form a club together with a few like-minded people and had our first committee meeting in the living room of Keith and Chris Ivison's home. It therefore transpired that on 27 July 1981 the inaugural meeting of the Shropshire Shufflers took place with Ron and Wendy Morgan, Keith and Chris Ivison, and Janice (sister to Chris). At later meetings Robin Jones and Steve Wellington were in attendance. The Shropshire Shufflers were now officially on the road!

Nowadays I can't tell you how good it makes me feel to see so many Shufflers meeting and running together and how wonderful it was to see

the many Shuffler vests at the recent Lake Vyrnwy Half Marathon.

I tried to improve my times and targeted my second London Marathon in 1982 which I was fortunate

Robin Jones, Keith Ivison, Ron Morgan Steve Wellington, 25 May 1981, Shrewsbury Carnival 5 mile race

to be accepted for. Bob Parker, a former Shrewsbury School master, who is still about now in his early eighties, had joined the Shufflers, and he began coaching and training with me for the 1982 marathon. In spite of his efforts to get me sub-3 hours (I failed by 15 mins), Bob ran with me, aged I guess 44 years old at the time. He was full of worldly, sound advice. As I recall he had run 2 hours 40 minutes in the past. In London in spite of our planned pace we reached the half-way stage and inexplicably I started to struggle, by 15 miles I was off pace and by 17 I was struggling. It became a long haul to the finish.

I was slowing Bob up, he stuck with me for a while but I implored him to go on, so he trotted ahead finishing in 2 hours 58 minutes 28 seconds. That was the clock time, so he was probably nearer 2 hours 50 minutes allowing for crossing the start line.

To this day I do not know what went wrong, my training and schedule of increasing distance events had gone well, running 20 miles at least twice in training.

We had started the Shufflers but many problems lay ahead before we were accepted by the athletic community. We had to get ourselves organised and Keith Ivison was elected the first Chairman, and he went on to give sterling service to the club and help the Shufflers develop.

Chris Ivison's father, Ron Barrett of Barretts Bakery, accepted our nomination and became our first President on 10 August 1981 and was a great source of support in many ways in enabling this fledgling club to flourish.

At the same August meeting we discussed which companies we would approach for our shirts and how we could become officially recognised as a running club by joining the Shropshire Amateur Athletic Association and the Midland Amateur Athletic Association. It was also at this meeting that we set our first annual subscription at £2.00 per person and agreed our club colours were to be the very patriotic red, white and blue.

A car-load of us headed Stoke-on-Trent way and Bourne Sports to source the kit, where we found some red, white and blue horizontal banded shorts and bought some of those off the rack. Wendy had designed the Shuffler logo and we wanted this printed on a T-shirt and vest, so decided to have these made bespoke, which initially John Bourne arranged for us and we later bought others from other suppliers to gain a better price advantage as member numbers grew. Thus, the fantastic and unique logo-branded shirt became well known on the road race circuit.

We applied for membership of the Shropshire Amateur Athletic Association but were turned down without reason. On 28 September 1981 we met for the first time with Shrewsbury Athletic Club in an effort to understand why they had been a staunch opponent of us becoming affiliated to the Shropshire AAA. Shrewsbury essentially felt that the town did not need another club; we could see their point but they could not see that our club was not aimed at the more elite runners but at those who wanted to run for fun and to enjoy camaraderie. This committee could also not see that the perceived elitism of an established bastion such as Shrewsbury AC was a deterrent to someone who lacked confidence in their ability and who perhaps did not want to race but just run with people of similar ability from the beginner to the veteran and to make the whole process as much social as getting or keeping a level of fitness. This was a great chance to put our case.

We were taking small but assured steps forward. On 30 November that year our committee meeting notes showed that we now had eight members and our first batch of ten club shirts had arrived, Ron Barrett having generously donated £60 to the Shufflers to facilitate this.

We started to produce our own publication, aptly called 'The Shuffler', where we included future race fixtures and social functions as well as race results and articles aimed at improving people's appreciation of running, including training and diet advice and even a column for injury advice by 'Dr Shuff', news of our efforts to raise monies for local charities as well as letters from Shuffler members.

By 18 January 1982 our membership had grown considerably and we were now regularly meeting at the Port Hill Bridge by the Quarry at 11am every Sunday morning. On this date fourteen members ran together.

In this month we also asked June Tudor (Hon Secretary) of the Shropshire AAA why we were turned down, but nothing concrete was proffered other than the aforementioned observations. We decided we would reapply until we were accepted or turned down again with a good reason.

We also sent Jimmy Saville a club T-shirt and honorary membership. I am sure we would not have done so if we knew what we know now but his character was larger than life and we saw him at the London Marathon each year so we felt his notoriety and indeed popularity could enhance our standing, but fortunately in hindsight he never responded or acknowledged us so nothing transpired.

1st February 1982 Evan Williams from Shrewsbury Athletic Club and a second claim member of the Shufflers gave the club members an excellent insight into running preparation, everything from choosing running shoes to running marathons. Second claim means his first loyalty is to Shrewsbury Athletic club but he can run for the Shufflers as they had second claim on him, however If you compete in a race in which your First and Second Claim Clubs are in contention then you must represent your First Claim Club. If the competition does not offer a team prize then you may compete for either club

With the advent of spring and the jogging phenomenon that was taking the nation by storm we were being joined by more and more Shufflers and on 10 March 1982, we were sent a very important letter. It read:

I am writing to inform you that the affiliation of the Shropshire Shufflers to the Shropshire AAA and Shropshire Women's AAA has been accepted. I have sent your club rules to the Midland Counties AAA. The affiliation is for your club to enter road running and cross-country races held under AAA and WAAA rules only (underlined).'
Yours sincerely
   June A Tudor Hon Secretary
      Shropshire AAA & Shropshire Women's AAA

Yes! We had finally been accepted, I cannot tell you how excited we were, I also cannot impress on you how difficult it was to be accepted by the running community as we were seen as devaluing the sport.

By our second anniversary we had 121 members, a massive growth, including a children's section. We went on to enter and perform well in numbers for many events, not least of which was the 1982 London Marathon. June Tudor and her husband George, stalwarts of the athletic scene, went on to be very good friends of the Shufflers and indeed a number of Shrewsbury Athletic Club members joined us as second claim runners. Community spirit and understanding had prevailed.

It's fair to say we took on the establishment and in my view, everyone won!

I think time has shown we as the 'Shufflers' have made a great contribution to the sport and I feel it is important to acknowledge the efforts of those initial members and officers of the Shufflers, who through their persistence and great efforts made sure the Shufflers were not just here as a passing fad but by working hard for AAA affiliation gave respectability and permanence to the club and most importantly gave an organisation to the average person in the street to be part of, an opportunity to exercise and socialise with fellow runners irrespective of ability.

Today's members and officers are a testament to those efforts all those years ago, as they have continued and improved upon the structure and operation of the club, increased the number of members and developed a great coaching and support system that is flourishing, and you as an individual, by participating and encouraging one another, can be proud to be a Shuffler!

Now we were an established running club and AAA member we looked at how we could benefit our members and support them with training and other ideas. We could now flourish and expand, but we never wanted to lose sight of our social and fun running ethos. We looked for varied running events and also increased our training days together from just the 11am meet on a Sunday morning at the Port Hill Bridge to meeting midweek at St Peters Church, Monkmoor. Some of the more experienced of us took on the role as coach and mentor but this was way before the official coaching courses the Shuffler coaches undertake now.

One of our favourite runs used to be the 'Turkey Trot', which was held in Wolverhampton just before Christmas each year. I think it was about 5 miles but remember running one year with Kathy Smallwood, who was a pretty good GB international track runner. At the end you queue up behind a refrigerated truck and collect a frozen turkey. Trust me, it was tricky to carry what amounted to a pretty big piece of frozen meat whilst perspiring like crazy from your race efforts. Still, Tiny Tim did not go hungry that Christmas!

It might be said that this unusual reward as a method of motivation may end up in finishers getting the bird! We also used to enter a race called the 'Christmas Pudding' race. Guess what you received at the end of that one. Talk about singing for your supper, this was running for your dinner!

On 16 October 1981 Shrewsbury Athletic Club and the Shropshire Shufflers got together and formed a team to take on other teams, notably Telford, our main rivals, on a 14-mile course as part of the Shropshire Mini Olympics. We were billed as the Shrewsbury area team with ten runners, eight from Shrewsbury Athletic Club and Keith Ivison and myself representing the Shufflers. Each of us ran a short leg of the relay, which was from the Column in Shrewsbury to Oakengates in Telford. The relay exchange points were set along the route, and legs varied from half a mile to two miles. We entered Wellington along Haygate Road towards the finish at Oakengates. Shrewsbury area won in a new course record of 67 minutes.

By now we had achieved quite a lot. Our ladies team had won the team award at the Shropshire Cross Country Championship, and some of our runners were posting great times and high positions in races, but even more importantly races were witnessing a sea of red, white and blue vests and shirts with the Shropshire Shufflers emblazoned on them.

I came across a list of club records dated 1985 from one of the Shufflers newsletters. Here are a few of them for your interest:

Club Record for 10 miles was held by Peter Farebrother who ran 57 minutes 30 secs at Telford in 1985 and Jan Navas who ran 66 minutes at Tipton in 1982.

Best 10K times were Steve Holgate in 1985 at Bridgnorth in 35 minutes 32 seconds and first lady was Christine Ivison at Coal and Clay (Potteries) 1984 who ran 44 minutes 41 seconds.

Half Marathon was again Peter Farebrother, 75 minutes 6 seconds at Winsford in 1984; first lady Janice Navas Birmingham 1983 ran 89 minutes 42 seconds.

Marathon, unsurprisingly, Peter Farebrother 2 hours 41 minutes 28 seconds at Telford in 1984. Fastest lady was Jackie Tooze at Wolverhampton in 1984 3 hours 40 minutes.

Peter Farebrother was a veteran when posting these times and I wouldn't dream of suggesting that one or two of the above ladies were too. (Well, that's up to you to guess, but bear in mind certainly then you only had to be 35 as a lady to be classed as a veteran.)

As stated, many times before, whilst we gave due accolade to these great efforts it was all about running together and having fun. We really had some great days out and the cameraderie and banter was all important much as it is today, the last one in was just as important as the first one and we all stayed and supported and cheered each other in all the events.

Today, when we see all at the 'Park Runs' up and down the country and see runners all the time, it is hard to believe runners on the street were rare in 1981 and always heckled by motorists and passers-by and that was just the bloke's, ladies had to be escorted in the early days because that's how life was! How times have changed!

On a personal note, I went on to run a total of six marathons, three of them the first three London marathons 1981/82/83, the Paris Marathon in 1985 and in between the British American Marathon in Manchester. Here I ran against/with one of my heroes Ron Hill.

We all lined up on what we knew to be an undulating course. I had trained well and felt very fit. I took off with the group of front runners which, of course, included Ron Hill.

It soon became clear after about 8 miles I could not keep up that pace so dropped into the chasing runners, which included Lesley Watson, the top British lady runner. We were moving swiftly, then at about 11 miles I had a very painful stitch which would not go away and I dropped further back, eventually having to stop, which I had never done in a race. I struggled to get going as the lactic acid built up in my leg muscles and I eventually finished in about 3 hours 25 minutes, by now so out of it I had to go to the medical tent as Plodders Lane, the steep hill at the end really took its toll. I was so exhausted they could not get a thermometer in my mouth so had to use another orifice. I recovered quickly, though, thank goodness. I had learned an important lesson to run within your training abilities and pace yourself.

In March 1984 a very well-known local lady was abducted and found murdered five miles from her home in Shrewsbury, her name was Hilda Murrell we knew her as the daughter of the Murrells Rose Nurseries family. There was a huge police investigation and the public were asked to come forward with any information especially if they had been in the vicinity of Haughmond Hill on a number of days in late March.

Witnesses stated that a running man had been seen, and when checking my running diary I realised I had been in that area on the dates the police were interested in, as it was one of my running routes along the river from Castlefields following a footpath then the old canal path before hitting the road through Uffington before taking a left up to a stile on the Upton Magna road and crossing a field onto the back of Haughmond Hill making a circular journey retracing my steps to head back home, about 10 miles in all as I recall.

I quickly contacted the police thinking anything I could offer in the way of observation might help. Days turned into weeks in which I heard

Shufflers runners

nothing. Eventually I received a call that the police would be calling at my house to take a statement.

Two plain clothes officers arrived at 3, Benyon Street in Castlefields. I showed them into our small front room where they proceeded to ask a few basic questions about my whereabouts on the days in question. Where had I run to exactly and at what time? What had I been wearing? What do I recall seeing?

I was quite precise about times as of course I was timing my training effort on my reliable low budget Casio wrist watch and said I was quite conspicuous wearing my royal blue Adidas tracksuit with the well-known logo and vertical white stripes and as it was a little chilly wearing a black woolly hat which sported a colourful green, red and white bobble.

As far as my observations were concerned I did recall a car parked at the stile on the Uffington to Upton Magna road, but nothing new there, as there were often walkers or runners who parked there. I gave them my vague recollections. I had recorded quite a fast 10 mile time but I decided not to share that with them.

With a sigh and apparent lack of interest they shut their notebooks, nodded and left without so much as a 'thank you so much for coming forward'. My impression was they seemed exhausted and a little disenchanted with the whole investigation.

Eventually a local man was convicted but to this day many, including Hilda Murrell's nephew Commander Robert Green, Royal Navy now retired and living in New Zealand where he continues to research his aunt's demise, believes there were more clandestine parties involved in this rather conspiracy inflamed murder investigation.

Paris in 85 was a similar debacle but of my own making, as I was not entered and took a group to Paris as a tour manager. A member of our party dropped out and I could not resist taking their number and starting, even though I had not run any more than 10 miles for almost a year. By the 16th mile I was exhausted. I had managed to get that far driven along by the camaraderie and atmosphere. Around this time, I made the error of having a banana from a feed station, something I found crazy to have on there anyway, but I was desperate for energy. I soon developed a stitch and felt unwell. Yet Chris Ivison remembers eating a gingerbread man and loving it. All sorts of things to give us energy were on offer, but not an isotonic drink in sight!

I eventually managed to finish in just over 4 hours, ironically, I seem to recall the first Shuffler to get over the line. Then unsurprisingly my body rebelled and broke down. Trying to get warm, I headed for a subterranean toilet at the end of the race, where I started to hyperventilate and felt very weak and disorientated. No one knew where I was, so it was a very disturbing experience. Fortunately, I was able to recover and re-join the group. The moral of the story is always train properly and prepare for the event.

My best time ended up at 3 hours 15 minutes at one of the London Marathons. The support all around was magical but the first London in 1981 was special with only about 7,000 entrants in those days and I gather we are nearer 42,000 now, so six times more. Marathons were challenging and hard work doing the training when you are trying to raise a young family and run a business. Gradually the running fell by the wayside with living overseas and working.

CHAPTER ELEVEN

————

Cricket, football and ultramarathon trekking

I have always loved my sport and one of my life's highlights was playing cricket for the Under 19 Wakeman team and winning the Shropshire trophy against Ellesmere College in the final. I was barely 17 and the team's wicketkeeper and fortunately my games master saw some ability in me and encouraged me.

From this achievement I was selected to play a game for Shropshire Under 19s against Shrewsbury Schoolboys at Shrewsbury School. For a lad who lacked confidence and felt I should not be there, it was rather daunting.

We batted first, and I was one of the last in. No sooner was I at the crease than the batsman facing the bowler at the other end hit the ball into the covers and shouted to run. Well, I was already backing up and set off quickly, then half-way down the wicket he said 'No'. I tried to get back, but I was run out. What an ignominious end to my Shropshire cricket career, run out without facing a ball.

I then went out to keep wicket. Boy, this was tough, the fast bowlers were so fast I could barely see the ball and I was way behind the wicket, so what the batsman were feeling or seeing I have no idea. I did okay but was never brought into the mainstream squad.

I do have to point out I learned a lot and improved with school first XI

Wakeman School team, winners of the under-19 trophy — front (left to right): S. Price, R. Morgan, M. Bould (capt.), B. Davies and D. Clarke. Back row (left to right): M. Smith, J. Burn, T. Gennoe, M. Stephenson, K. Minton and B. Groom.

Cricket U 19 team picture

games but I was in very talented company, vying for positions in the county team. A young fellow called Andy Lloyd from Oswestry, eighteen months younger than me, went on to play for England. He played in one Test match and three one-day matches for England in 1984.

His only Test was against the West Indies in June 1984. After making ten runs, and batting for thirty-three minutes, he was hit on the head by the West Indian fast bowler Malcolm Marshall. Luckily he was wearing a helmet, but he spent several days in hospital and did not play for the remainder of 1984. He never played for England again and is the only Test Match opening batsman never to have been dismissed from the national team.

Lloyd was captain of Warwickshire. He made a total of 17,211 runs in all first-class cricket with twenty-nine centuries, and he took twenty-three wickets.

I was up directly against Geoff Humpage from Bridgnorth School for the wicketkeeper position. Geoff was a couple of months older than me, and played three one-day internationals in 1981 as well as playing as a middle-order batsman and wicketkeeper for Warwickshire.

I, on the other hand, carried on playing for local side Shelton and enjoyed my games with them, travelling all over the county and indeed further afield with games in the West Midlands and, most prominent in my memory, to Greenfields in Lancashire one bank holiday weekend.

Greenfields near Oldham was a typical mill town ground, with a rather grey, polluted watercourse called the River Chew running alongside. Lancashire was a hotbed of cricket and the number of fans attending was amazing. I could not believe the turn-out. Of course, many sides in those days up in Lancashire and Yorkshire had overseas professionals, players I remember seeing on the international stage like Wes Hall, Viv Richards and so on. It was like a big day out for these fans, who actually paid an entrance fee and I am told that if someone performed really well, they had a whip-round for them. It wasn't just blokes there, but families too, all loving it, with their sandwiches, flasks and beers and dressed up as if they were going to church.

We bowled first and their first two batsmen retired 50 not out. We figured it was going to be a long day. One of their guys hit a drive past me and I chased as hard as I could towards the boundary. Just as I thought I had lost the chase to stop a certain four runs, I tripped on the undulating outfield and fell forward and with outstretched hand stopped the ball crossing the boundary. The crowd went into raptures and I heard shouts of 'Brilliant fielding' and 'Eee, well done, lad'. I was chuffed, even if my throw in was a bit wayward. We managed to restrict them to about 179 and amazingly we beat them, even I got a run or two at the end. The best thing for me, though, was that my dear old dad had travelled up on the coach and was there to watch.

We used to play a variety of sides around Shropshire and usually on bank holiday weekends we played a team from further afield, one from the Goodyear factory Wolverhampton/Birmingham way. When we played the home fixture their coach arrived and all these rather big West Indian chaps loped off the coach all looking pretty fit until the last one off was a diminutive white guy. Sure-enough that was the team make-up and whilst we were expecting a really tough game in this era of fantastic West Indian Test match teams with exceptionally fast bowling, we managed to weather the opening bowlers. Fortunately for us there were no Wes Hall-type fast bowlers, just competent, keen fellows and we went on to win the game.

When we played the return game at their place it was the year of the World cup in Mexico 1970 and we were watching England play West Germany in their splendid sports and social club in the quarter finals. England took a two-nil lead and we thought we were on the way to victory.

It was our turn to field and by the time we came in for tea it was extra time, the Germans having drawn level, and to our misery the Germans won 3-2, Muller scoring the winning goal. To add insult to injury we lost the game of cricket so not a very good June day.

I also played for Bugle in the Cornish league whilst I was at college in St Austell in 1977. All associated with the club were so friendly and I was blessed to play on many Cornish grounds and see a lot of Cornwall. I was back to playing wicketkeeper and I did reasonably well.

One day we had batted first and I had opened the innings. I only scored about 10 runs but I stuck around a fair while, a bit of a Geoff Boycott without the runs!

Then the opposition went in to bat. I had a cracking game, stumping seven of them, they were so rattled they started complaining to the umpire that I was gaining an unfair advantage by knocking the bails off before receiving the ball or taking the ball in front of the wicket. The reality was our team's spin and slower bowling was good and I was young and quick. Anyway, they were all out for something ridiculous like 39. We decided to play the away fixture right then and there as hardly any time had elapsed. Wouldn't you know it, we lost!

I have always enjoyed the outdoors and reckon it's in my genes as Dad was born in a mountain village on the edge of the Brecon Beacons and also my mum lived on smallholdings in rural locations.

No doubt I developed the yearning for the coast from the fond memories I had of the seaside holidays in Pembrokeshire in my childhood, but I seem to be drawn to the Shropshire Hills, indeed any hill or mountain. If I see one, I just feel I have to climb it. As one of my heroes, George Mallory, said when asked why he wanted to climb Everest, 'Because it's there!'

I have always enjoyed trekking and running challenges right from the first sponsored walk I did in my teens around 1970 for the Save the Children Fund. It sparked a desire to be on the hills and mountains but also to challenge my body and mind to achieve feats of endurance. With 'Brakey', my mate John Brake, I caught the transfer bus from Shirehall by Lord Hill's Column out to Church Stretton School. People had already started, as you just checked in and walked at your leisure. Many were raising sponsorship, we just paid the entry fee as it was a last-minute decision.

It was apparently about 10 miles to Devil's Chair on the Stiperstones and back. We set off in pursuit of all these walkers passing them one by one until we figured we were in the lead. We couldn't help it, we were and still are competitive. Then as we approached the rocky path to the Devil's Chair, we were surprised to see an older chap walking with a stick. He was moving quite quickly so we increased our pace with a view to passing him on the way back. But we just could not make any inroads into his lead. He kept a steady but very fast pace and we were getting more and more tired. We had not brought any drinks or snacks with us, so we weakened by the mile, and were very miffed to come in second, even though it was not even a race.

Worse was to come, because we were back so early, the coach back to Shrewsbury was not going to be for about another four hours. We decided to hitch-hike back to town, but we failed to get a ride so we walked all the way home to Shrewsbury, probably about another thirteen miles. We were singing songs, especially marching songs like 'Mabel … we love you Mabel' to keep up our morale.

On reflection, this was the start of our trekking adventures. We started doing lots of walks and knew the Long Mynd and the Shropshire Hills like the backs of our hands. We completed three Long Mynd Hikes, and three Across Wales walks. But prior to the trekking ultramarathons, I personally completed six marathons, starting with Ottawa in 1980.

It was not so much the events as the training I enjoyed, especially with the trekking. With my great pal Gra (Graham 'Wobble' or 'Wob' Croft) we

Three Amigos on Tour (Across Wales)

used to go up to the Church Stretton area after work. About 7pm we would walk the hills usually Ragleth and Caradoc, and occasionally Lawley too, for three hours and more, but still be back in the pub before closing time for a beer. Whatever problems we had in life be it work and a bad day or whatever, these were all left in the valley below whilst we enjoyed the banter, vistas, fresh air and exercise. Absolutely priceless!

The Long Mynd Hike, organised by the Second Long Mynd Scout Group, takes place over the first weekend in October every year and you never know quite what weather is going to greet you. It is listed as a competitive 50-mile distance hike and you have to be over 18 to enter, but it always felt a lot more than 50 miles, maybe because we got lost now and then, having to navigate as well between checkpoints and many of the checkpoints were at the top of hills. Brian Faulkner with fellow pals George Davies and Mike Allen started the Long Mynd Hike over fifty years ago and we were often calling on their house in Church Stretton as one of my pals was fond of Brian's sister Tina. Joe Faulkner, the well-known trail runner and outdoors guy, was another brother who popped in and out when we visited.

The following is an official overview of the rules published by the Long Mynd Hike:

The route: *The hike now follows a set figure-of-eight route over the undulating and hilly countryside of South Shropshire and the Welsh Marches border area, with about 8,000 feet of climbing covering eight summits. The start is at 13:00 hours on the first Saturday in October from Church Stretton often referred to as 'Little Switzerland' and the object is to complete the course in under 24 hours. Church Stretton School is used as the hike HQ for the weekend and is where registration and kit check takes place, car parking is available and all hikers must report here first. The hike now starts from a location to the East of the A49.*

There are rules for your own safety such as: *Suitable footwear must be worn, waterproofs and a specified amount (or more) of safety equipment must be carried, which is fully detailed in the rules.* [Most runners would cram the bare minimum into a bumbag or sling. Nowadays I see some pretty fancy hiking/runners' vest with great storage of minimal folded kit.]

All checkpoints are linked by radio to the hike HQ at the school, first aid and rescue/sweep teams provide backup and transport is provided from retirement points back to the school for competitors who are unable to complete the route.

What is provided: *Drinks are provided for competitors at roadside checkpoints and the school. The school also provides breakfast for competitors from early on the* Sunday [that was always an incentive to get back within the 24 hours], *has showers available and classrooms for sleeping. Everyone who starts the hike receives a certificate of their achievement showing the distance completed and time taken, the souvenir tally carried en-route is returned with the certificate.* [Nowadays your friends and family can track your progress around the course on the FellTrack.com website. But years ago, you had a map and compass, no GPS, and unless you physically came into a checkpoint tent you were not sure if your friend or kin had reached that far or passed through. You need to check with the officials who had lists of those who had arrived and departed or indeed dropped out.

Trophies: *12 individual trophies and two team trophies are awarded.* [There was always a prizegiving but we never troubled those who awarded them.]

The first year three of us did the hike together but Wob hurt his ankle and his feet were badly blistered, so he had to drop out somewhere around Black

The Three Amigos

Rhadley. Brakey was in a bit of pain too but was drug dependent, sponsored by Nurofen, so he got through and we just managed to run down to the finish a few minutes before the 24 hours was up. (Well, we didn't actually run, we hobbled.)

Next year the Three Amigos were determined to improve, so we trained diligently and, better than that, we learned every single short cut. We knew the course backwards and finished nearer 18 hours

At our third and final attempt before we moved on to other ultra-projects, we steamed in under 16 hours, which is all very well but the first guys in did it in around 8 hours, so let's put this in perspective. This was the era of Dave Newsom, Wagstaffe, Cartwright and Greaves. I believe the record is 7 hours 51 minutes by Andy Davies, achieved in 2007. I also believe The Shropshire Shufflers won the team trophy one year in the nineties.

The course has changed from the original started in 1967, once in 2002 due to land access issues and then again in later years, so it is difficult to compare like for like. We used to finish with Ragleth then Caradoc and back to Stretton; now I believe Caradoc is at the beginning and Ragleth is the last hill checkpoint.

Whatever the route, it is a fair undertaking and not for the faint-hearted, especially when you realise that starting at 1pm it is dark by about 6pm, so most of the event is at night using head torches and maps. Plenty of opportunity for mishaps and navigational disasters!

Naturally there were plenty of tales from the treks and marathons, lots of them on one particular event, Across Wales. There are a growing number of ultramarathons worldwide now, including the well-known British one, the Bob Graham Round, about 64 miles, the Snowdownian

Frankie Goes to Habberley
Team walking

Paddy Buckley Round and the Scottish one, the Charlie Ramsey Round, all of which were a lot tougher than the ones we competed in as there is far more ascent and descent. A number of events go for longer than two days too, like the infamous Dragon's Back. This incredible five-day slog is 315 kilometres long with 15,500 metres of ascent across wild, trackless, remote and mountainous terrain in Wales.

On the Across Wales Ultramarathon we used to stay the night in the Memorial Hall in Clun before being bussed to the start at dawn the following morning. On one occasion Brakey, Wob and myself went for a few beers at the Sun Inn and other local hostelries and returned to find everyone had turned in. We tried to be quiet, a few people muttered and turned over on their roll mats. We were soon asleep, but not for long. The whole hall was awake, head torches were going on, people were getting up and puffing and blowing, expressing disgust and discontent at our late arrival and the sleep-apnea king Wob as they passed us on the way to the bathroom, and all because Wob snored so loudly no one could sleep … er, except him!

Across Wales is an approximately 45-mile (as the crow flies) challenge that has to be completed in under 18 hours from the Anchor Inn, Newcastle-under-Clun on the Kerry Ridgeway to Clarach Bay. In reality you could easily do 50 to 60 miles dependent on your navigation. So about 3.45am we were all awoken as the hall lights went on, time for quick ablutions and a kit check before a cup of tea and toast and jam and into the coach at 5am on a fresh Saturday September morning.

Again, you needed OS maps and compass and it was all about hitting six checkpoints, which included the finish line, to get your tally punched to prove you had done the course. To gain an advantage we waded across the Severn near its source to save going down to a bridge. We then started to climb onto the inhospitable and windswept landscape to the summit of Plynlimmon at 2,647 feet, source of the River Severn, we were neck and neck with a few guys at the front, and then edged ahead of them.

Looking back at our pursuers we spotted one was in distress, purely because there was a drop off the high land with a bit of scrambling to continue the descent and journey towards the next check point and ultimately our eventual goal of the seaside. This chap had a fear of heights, so we went back up and guided him down. It didn't affect our time a whole lot and we trundled off to checkpoint 4 at Nant-y-Moch dam. In those days there were only about sixty teams allowed to enter. We finished as a group in ninth place, well inside the 18-hour deadline and were so chuffed with ourselves arriving at Clarach Bay, we of course had to have a paddle … well, we were at the seaside and salt water works wonders on blistered and tired feet.

As a change from hiking and running, I went to some trials for a new baseball team starting in Shrewsbury with the wonderfully named founder, Danny Crockett, but I don't think it managed to get going beyond a few friendlies.

I enjoy all sports but especially my football and beloved local team Shrewsbury Town. Like all kids we had a first division (now Premier league) team we supported and mine was Tottenham Hotspur.

In 61/62 Spurs were brilliant, with players like Danny Blanchflower and Jimmy Greaves, John White, Terry Dyson, and winning the FA Cup against Burnley and winning the league too. But when all is said and done Shrewsbury is my team, it's the one that can make my week brilliant or miserable, dependent on its results.

My Mum was a keen Town fan, having been taken by her dad when she was just five years old. Her first game was in 1930, mine was the 1962/63 season. We were then in the league proper, in the third division (division one today).

My first game was on 13 October 1962 and the opposition was Bristol Rovers. Town won 7-2 and that was it, I was hooked. I thought all the games were going to be like that! Fifty-five-plus years on, I am wiser but still daft enough to go as often as I can. A first and yet to be matched was two hat tricks in the same game by Jimmy McLaughlin and Frank Clarke. It was only Clarke's second game, whilst Pat Wright made his debut and became a firm Town favourite.

Mum was like the Pied Piper as she would often take more than ten of us on the bus down to the Swan and then to the old Meadow. We used to stand on the Tech End (Cecil Duffill Terrace or Wakeman End). My dad only took me once to a Shropshire Senior Cup game against Sankeys (later GKN), when we won 4-0. Dad took me in what was 'C' Stand in those days, because as Mum got older, she started to take me in the stand so we could sit. It wasn't the same atmosphere as the terrace but I was just happy to be there.

As I grew older the Station End was the place to be for all us young bucks and this later transferred allegiance to the Riverside as the popular standing area for the singing fraternity! You could go to the Nags Head or Crown public houses and rush to the match a few minutes before the start.

There was never much of a queue as you entered the creaking old solid iron turnstile and paid your money. It was a wonderful feeling I don't quite have with the new ground.

We saw lots of brilliant games and terrific footballers over the years. My Mum's favourite was a tricky little winger called Trevor Meredith, and of course we were all in awe of Arthur Rowley, still football's leading goal scorer with 434 goals. I remember going to his testimonial games, yes, two on the same evening, on 27 October 1965. First, an International XI vs ex Wolves Stars. Tom Finney played for the International team, Bert Williams was in goal for the Wolves team. Then it was Shrewsbury vs Wolves, a strong side including Ron Flowers. The highlight for me was seeing Sir Stanley Matthews play at the age of fifty. He played a cameo for Shrewsbury and still looked very fit! However, the player that sticks in my mind for Shrewsbury from that era was Peter Broadbent. He was a class above everyone else, having dropped down the leagues. He was transferred back up when he left us.

I also got to see Bobby Moore play in a game at the Gay Meadow when West Ham came to play Shrewsbury. In a friendly in aid of the Shropshire World Cup Fund, Moore scored twice for West Ham. Martin Peters and Harry Redknapp were also playing in a strong Hammers team.

Cup games stick in your mind. I remember Mum taking me to Manchester City where we drew 1-1 having taken the lead but Derek Kevan equalised late on. We went on to win the replay at the Gay Meadow.

Going to Leeds and losing 2-0 to a top team of the day with players like Billy Bremner, Jack Charlton, Johnny Giles and a tricky winger called Albert Johannsen was no shame. We went on the train and I can remember 500 of us parading from the station. Mum and I had made a banner saying 'Up the Town' on a piece of bedsheet. Needless to say, we left it in the ground!

Chelsea away, again for a 5th round tie, was a magnificent performance. We lost 3-2 but one of their goals was a very dubious decision, offside in my book, so again we were not disgraced

One season myself and some friends took the coach to Wrexham with other Shrewsbury fans to take in the league game with Wrexham football club a hotly contested derby with the added spice of a Welsh vs English club rivalry! We thoroughly enjoyed the game especially as Shrewsbury were victors by a 2-1 margin. Unfortunately, after the game there was a lot of trouble between fans which stopped us getting back to our coaches, we tried to go to the train station but had the same hindrances. So, hemmed in, we reluctantly returned to the ground which was now like a ghost town, we spotted the Shrewsbury Town coach was still there.

We knocked on the changing room door and to our surprise Shrewsbury Manager Alan Durban greeted us, we explained our dilemma that we had no way of getting home. He was firm and a little dismissive at first before saying wait outside and we will see. About thirty minutes later the players emerged and boarded the coach Alan Durban and the others walked past us and we thought that was it, we were going to be stranded, but as Alan Durban stepped onto the team coach he turned around and said 'well come on then'!

Thus, we returned home with our Shrewsbury Town heroes and were pleased as punch!

Over the years there have been many great games, not least of which was the final game of the 1978/79 season when we beat Exeter 4-1, Jake King scoring twice to secure promotion to the second division (now the Championship) for the first time in our lives. Magical!

We went on to play several seasons in the second division, playing big teams like Chelsea, West Ham, Newcastle, Sunderland and Fulham.

Of course, there have been awful days, like being relegated into non-league, but let's not dwell on those! Sad days also include the last game at the Gay Meadow in 2007 against Grimsby Town. Yes, it's wonderful to have a new stadium with better facilities, but the character of the Gay Meadow will live long in my mind and heart. The evening games where from the Tech End you could see Laura's Tower in the Castle grounds. Looking the other way between the floodlights and beyond the ground you could see the school gardens and statue behind the Tech End.

You could walk around the ground and chat to people you knew, and you could change ends at half time. There was so much flexibility, which added to the enjoyment. Of course, there were the negatives, like the awful pungent toilet facilities, but, heh, I did say the ground was characterful!

As Wakeman School pupils, we used to stand on the staircase of the school, which was situated behind the Tech End (so called because the school was once the Technical College). We would open the cantilevered windows to watch the game, sometimes staying behind for an evening game as the building was open for the art school on D Floor.

My cousin Trevor Boucher played for the Town in the late sixties/early seventies, but did not manage to break into the first team. When Arthur Rowley became manager at Southend he went down there. Trev used to get me autographs. I used to see him when we had breaks in class. We'd go down and play footy at lunchtime with a tennis ball.

My good old mum, knowing I was a keen Spurs fan and an enthusiastic if not talented young footballer, wrote off on my behalf to Spurs, her letter extolled my virtues as only a mother can and, lo and behold, I received a letter from Spurs signed by Bill Nicholson asking me to attend their training ground at Cheshunt for a trial. I was chuffed to bits and kept the letter for many years, though sadly it was lost in one of the many house moves I embarked on over the years. I never attended as I guess I knew I was not good enough but the offer and letter itself was a thrill in itself! I should have gone just to meet Jimmy Greaves, my hero, but I did get to see him around 1968 playing for Spurs at Wolves.

Since I've had cancer Mauricio Pochettino when Manager/head coach of Tottenham Hotspur kindly wrote to me thanking me for almost six decades of support and wishing me well on behalf of the club, himself, staff and players, which brightened me up no end. What a kind gesture.

Over the years I went to a few cup finals at the old Wembley in crowds of 100,000, like League Cup Final Manchester city vs Newcastle Utd in 1976, which city won 2-1, with players such as Joe Royle, Peter Barnes, Malcolm Macdonald and Asa Hartford, who was eventually manager at Shrewsbury. The match was refereed by the excellent FIFA and British referee Jack Taylor

As a Spurs fan I was lucky enough to see them in the 1980-81 cup final, again with a crowd of 100,000 when they played against Manchester City once again, who took the lead through Tommy Hutchinson but the match ended 1-1 thanks to an own goal from City. I was also at the replay again at Wembley five days later when Ricky Villa the Argentinian gave Spurs the lead. City bounced back and not only equalised but scored a second to lead 2-1 but after Garth Crooks equalised it set the game up for a fantastic Spurs winner when Ricky Villa dribbled past four defenders to score the winner and his second goal.

The next season we were back at Wembley again when Spurs drew 1-1 with QPR, Hoddle scoring the Spurs goal. So in a repeat of the previous season, we went to a replay five days later to witness a lacklustre game that Spurs won 1-0 with a Hoddle penalty.

I also saw Spurs lift he UEFA cup after a final against Anderlecht in 1984 when they drew the second leg 1-1 at White Hart Lane after a late Spurs equaliser by Graham Roberts, which took the game to extra time and then penalties, which Spurs fortunately won 4-3. A real nail biter.

Sadly, I didn't have the same luck with my beloved Shrewsbury Town, having now been to Wembley five times and witnessed them lose every time, twice to Rotherham! Still, it was wonderful to have watched Shrewsbury at the old Wembley in 1996, and then in 2007 at the new Wembley in a record crowd for Shrewsbury of 61,589 fans against Bristol Rovers. Alas we lost 3-1 after taking the lead, and have since lost there in two other finals. In 2018 we were at Wembley twice and disappointed twice.

CHAPTER TWELVE

——————

Life constantly morphs

Just taking you back a bit in time. In 1974 I had started dating a young lady called Evelyn and when she returned from being an au pair in Germany our relationship progressed and we became engaged. After a brief hiatus in our relationship Rhian was born in February 1975, but sadly I was deprived of bringing up my first child as the beautiful girl's maternal grandmother believed her daughter Evelyn could do better than me and needed a man with prospects. At the end of the day as per any mother she had her daughter's best interests at heart. I am now blessed many years on to have had a loving father–daughter relationship with Rhian for the last twenty-plus years and witnessed her children, my grandchildren, growing up. So, making new memories with Rhian, Owen and Drew has been a wonderful addition to my life.

Wendy and I had started our own family in 1981. On 30 October, Wendy and I were blessed with our first son, Peter. What a wonderful day that was. I loved looking after him and willingly was up every night changing him, feeding him and singing him back to sleep. He must have a Freudian complex about Rod Stewart as I was keen on Rod at the time and Peter heard a lot of Maggie May in the early hours of the morning.

Fourwinds Travel was also burgeoning, and apart from the general trade and group tours which we had been developing we were also looking at pop concerts to develop our income. Our personalised tours were selling well, too well, and it was a steep learning curve for me. We booked far too many people on a four-day mini cruise from Newcastle to Esbjerg using the rather smart DFDS ferries. On board we were spoilt with a fabulous smorgasbord buffet and entertainment in the evening.

We docked for one night in Esbjerg and took an afternoon excursion to the twelfth-century former capital of Denmark, Ribe. These tours grew in popularity from an initial tour taking 40 people to the final tour of 153 people. Such a large number of people proved to be a nightmare. I was the escorting tour manager with three full coaches under my 'control' and no help! It was very, very arduous and stressful and proved quite a challenge in spite of my best efforts.

By the end this was a well-oiled tour operation that had scaled up drastically in numbers. We managed to have a wonderful time with no hitches until we arrived back at Newcastle docks. I was one of the last off the boat making sure we had everyone with us.

I had instructed everyone to look out for the three coaches we had waiting for us ready to board and told the coach company to instruct their drivers to wait for me to disembark. Assuming my instructions would be followed to the letter I thought all would be good as gold.

However, things rapidly went awry. A rogue driver in number one coach boarded about forty passengers and then just shouted down the coach 'Is everyone here?' a passenger said 'Yes' and the driver took off with about thirteen spare seats, which of course meant we had thirteen passengers with no seats on the remaining coaches.

Now this was in the days before mobile phones. After disembarkation I was presented with this catastrophe. I hotfooted it to a dock office and called the police. I asked them to stop the coach on the motorway and wait for us at a service station.

Meanwhile the remaining passengers crammed on to the two remaining coaches. Sure enough, we caught up with coach number one who had been waylaid by a motorway police patrol and pulled into a service station. We rearranged the luggage and redistributed the passengers and we all got home safely!

In hindsight it was amusing but I was very upset indeed with the driver. I refused to use that company again if they used that driver.

Travel agency commissions were meagre and ever under pressure, so we had to diversify away from just selling the normal travel products. Terry and I both loved our music and indeed for the most part the same artists too so we expanded into pop concerts. Our plan was simple; if we could bulk-buy pop concert tickets and add coaches to them, we could make some serious money.

Our problem was getting the promoters to sell in bulk to us. For our first concert we focused on a Rod Stewart 12,000-capacity concert at the National Exhibition Centre in Birmingham and decided we would try and buy as many tickets as possible. We contacted the promoter Harvey Goldsmith and spoke to his personal assistant Juliet who screened his calls, but the upshot was they would not accept a cheque, nor were they happy in selling a large amount of tickets to us. Understandable, as we were unknown to the promoter and lacked credibility or form.

However, after a passionate plea through Juliet, then a brief telephone conversation with Harvey we came to a compromise or rather a proposition. Harvey said, 'If you can bring £20,000 in cash down to London, I will give you the tickets.' So, game on.

We asked for 1,200 tickets, which translated to twenty-four coach-loads. We were very excited, until it dawned upon us … Where were we going to get £20,000 from quickly?

To cut a long story short, after begging and borrowing and being creative the money was raised and I set off on our behalf heading down to Euston on the train with a plastic carrier bag full of cash. I made my way to the promoter's address in London and down into a small but well-appointed professional basement office and finally met Juliet in person.

After a short while I was invited through to Harvey Goldsmith's office. A short amicable conversation took place and we exchanged the tickets for the cash.

From that day on, we were made. He trusted us and we bought many tickets, just by sending him a cheque, because we had proved our honesty and credibility. Harvey Goldsmith went on to greater things, organising Live Aid with Bob Geldof amongst other achievements. He still operates today and has been awarded the CBE.

It was a huge logistical operation getting everyone to these concerts, picking up all over Shropshire with Oswestry, Shrewsbury and Telford being our main pick-up points.

We went on to do many more concerts, such as Elton John, The Who and the Rolling Stones. We often did less popular, more niche artists, for which we bought around 200 tickets.

Peter, Rhian Ron and Tom

We had a very successful era. Before too long, other local companies like Owens jumped on the bandwagon and our numbers diminished, but what a time we had!

Fourwinds developed offices in Market Drayton and Whitchurch in addition to our Oswestry head office, plus Welshpool and Newtown in Powys. However, I'm afraid that despite my efforts, I felt the rewards and appreciation from the partners of Fourwinds Travel were not forthcoming as promised, and it was time to move on.

In 1984 I headed for pastures new. I returned to my home town and a company called Travelon. It was owned by Bob and Dorothy Freeman, but the business was struggling having moved from under the Market Hall building at Shoplatch to the old Royal Salop Infirmary building which Bob Freeman's company was redeveloping with apartments and shops.

I was made a director and within eighteen months a freshly galvanised and motivated staff supported me to improve the company's fortunes into a multi-million-pound business. Awards came flowing in: best Thomsons sales agent in Shrewsbury despite there being two Lunn Poly offices who were owned by Thomsons; best Neilsons ski agent in the West Midlands. Turnover rose from £543,000 to £1.2 million.

It was while developing the ski side of the business that I was invited to a ski educational in Austria. I joined a group of travel agents travelling out to Saalbach with Neilsons ski company.

On this trip I met some interesting characters and shared a room with John Powell from Wolverhampton whose firm owned Powell's Cottages, a company renting out holiday cottages in the UK.

Saalbach is a pretty little Alpine village with chocolate-box vistas complete with onion-domed church tower at the end of the valley. We stayed in a small comfortable guest house not far from the nursery slopes where we started to develop our ski skills. We were accompanied by a ski instructor as well as ski representatives from Neilsons.

We had all over-imbibed the night before our first day of skiing at the welcome party and a few sore heads took to the gradual incline of the learners' slope which we descended and returned up a T-bar lift, which was an experience in itself. We had all tried out skiing on a dry ski slope prior to our Austrian arrival, which gave us a little practice.

I soon began to enjoy the freedom and thrill of skiing down the slope, although we all had a few falls. It was a sport I wished I had tried years earlier!

After lunch we were given a free afternoon to explore the resort. Our tour rep and the ski guide said they were going to head onto the main slopes and do some red and black runs. Without thinking, John and I, egging each other on and as impetuous as ever, agreed to go with them, encouraged by the fact that there was a beginners' slope unusually situated at the top of the main slope which you reached by cable car.

I have always loved mountains, which is strange considering I am not very happy if I am at the edge of a tall building always saying I have a fear of heights, but put me high on a mountain, even on a cliff edge, and for some reason it does not faze me … How does that work?

Encouraged by the guides that we would ski down to a little mountain restaurant and have some lunch and a beverage or two we were rather taken by the thought, so were easily persuaded.

Taking the gondola up to the top of the mountain we put on our skis

Learning to ski in Saalbach

and wondered at the panorama of Alpine peaks before us. It was a truly exhilarating experience. We set off on a fairly easy run down the Bernkogel peak under a little bridge and then we were at the top of another slope, which seemed to disappear beneath us. As we pushed off, being encouraged to traverse the slope from side to side, our new-found skills, which were still primarily snow-plough based, were found to be wanting.

This was far too steep a slope to snow plough on, we needed to parallel ski and schuss our way down, a skill we had yet to learn, let alone perfect. Before I knew it I was falling and then hurtling downhill on my back with my skis in the air (don't ask) before I tried to roll and one of the skis caught the slope and jackknifed, separating from my boot before heading off downhill, whilst I eventually came to a not very dignified stop.

John didn't fare much better and thanks to our fellow skiers and the guides with us they rescued my ski and we managed to limp slowly to the

Meribel Chalet ski pals!

restaurant half-way down the slope to regroup and for our guides to have their fun at our expense.

An hour later, cheese toastie on board and fuelled with a number of schnapps we set off anew and this time fear was not on the menu. Relaxed, we attacked the downward slope with glee and somehow managed to get to the bottom without further mishap, our reputation restored and a good pat on the back from our guides confirmed the same!

John and I were so into skiing now we just skied every minute we could. Much to my surprise I won the group slalom competition, supporting the fact of how much I had improved, albeit on a (literal) crash course!

On the last day we wondered why we were the only two going up on the early gondola to the top of the mountain until we realised we were skiing down in a whiteout! Our goggles were covered in snow so we had to ski without them and ski very gingerly using our new-found skills of a sort of parallel skiing to get down the mountain safely. When we reached the mid-station the cable cars had stopped and we had no option but to sit tight or

Dianne and Ron in Meribel

ski down the rest. Somehow, we skied down to the village to the amazement of the lift operators and other witnesses we encountered, having survived the folly of naivety and youth.

Years later I returned to skiing when Dianne and I decided to join a Chalet house party of friends in Meribel in France, and we had a fantastic time we followed this with a wonderful adventure in the United States Rockies at a place called Breckenridge at a higher

Ron, Gra, and John 'The Three Amigos' in Meribel

altitude again and this was another time it felt strange to simply breathe as the lungs searched for the scarcity of oxygen in the rarefied climate even when we were sitting or sleeping!

All through these holidays various escapades, near misses and inexplicable actions tested our agility and luck and it surprises me even today looking back as to how we managed to get home unscathed by injury, if not by fright!

One time I was going up on the ski lift and one of my skis fell off so I had to exit the lift on one ski and trust me it was not as elegant as a swan gliding on a lake as you may witness in the ice galas, I just ended up in a crumpled heap whilst following skiers tried to avoid me as they disembarked. On another occasion we reached a turn-around, where there was a very small window of opportunity to alight the lift, I missed mine thorough lack of timing and concentration but still decided to depart the lift by now I was going back up in the air about three metres from the ground I descended and to my surprise although landing with a jolt I stayed on my skis it could have ended in tears for sure!

Lots of opportunities come your way in life and one was when I was helping make some flight arrangements for the former Bridgnorth Town Football Club manager, Ken Jones, who was setting up a soccer training camp in the States. We got talking and he said he was looking for some coaches who had a coaching award to satisfy the American stipulations on qualification to work in the States. As it happened my F.A. of Wales coaching award just about sufficed and I was sorely tempted, it would have been wonderful to coach in the sunshine and be paid for it, but I ummed and aaahed and decided against it, and recommended some other guys to the chap.

Little did I know not many months later I would be in the States for other reasons.

CHAPTER THIRTEEN

Life has its ways

All was going really well in the beautiful county of Shropshire, then we heard my father-in-law in North Carolina had been diagnosed with cancer and there was serious concern about his longevity.

Wendy and I and our small son Peter aged about five went to see him.

Needless to say, Wendy was very upset at the prospect of losing her Dad and wanted to travel and stay with her father, whom she worshipped. He was a great role model and father and she wanted to spend with him what we perceived might be his last days.

I was now in a dilemma, having returned on my own to Travelon. I felt we had built a sound financial future for our family. Nevertheless, Wendy was quickly ensconced back into Carolinian life. Before too long I returned to the States again, and we made the joint momentous decision whilst I was out there. I would stay too.

I had to inform Bob and Dorothy Freeman, my fellow and senior directors from across the water, that I would be relinquishing my leadership of Travelon and subsequently lost my directorship and share benefit, but family is all important and takes precedence and whilst I was sad not to return to Travelon there is more to life than money.

I was sad to let Bob and Dorothy down and indeed the team I worked with. I am sure they felt let down too, even though their business was doing really well and improved dramatically in the time I had been there and was on an upward trajectory thanks to the hard work of all the team.

Fortunately, in the past we had already done some groundwork on applying for a Green Card which would give us permanent residency, so we reignited this application before it expired and were ultimately successful and able to work in the USA too. A whole new adventure had started, I was disappointed to lose everything from Travelon but felt fate had dealt us a hand.

In 1986 I had visited a local company, Travel Agents International at Hillandale Road in Durham to revalidate some flight tickets back to England and Jean the owner had said as a throw-away comment that if I was ever looking for a job to let her know.

So that's what I did. Amazingly, not only did Jean give me a job, she made me Manager! I was getting a darn good salary and we found a one-bedroomed apartment down the road at 2702H Camellia Street.

Travel agencies in the States were a different experience altogether from working in the UK. We used computers, which we had not seen in a British travel agency outside the London area, and we used a reservation system called SODA, which was the Eastern airlines online system for booking flights. Unlike British agencies, very few travel brochures were on display.

There were obvious cultural differences when dealing with clients, not least of which was the language. I used to mystify people when I said, 'I will pop your tickets in the post,' before listening to the deafening silence on the other end of the phone and correcting myself with: 'I will mail your tickets to you!'

One business account we held was with Duke University, a fantastic educational establishment and renowned for one particular sport, basketball. We were privileged to service the team account, which was a very prestigious and lucrative honour. I never let on that on the sporting front I favoured NC State 'the Wolfpack', which my brother-in-law Geoff had introduced me to at the end of the seventies.

Some basketball players came and picked their tickets up from the office. It was like seeing a giraffe loping through the parking lot towards us as they were so tall and they had to duck to get through the door because they were invariably anywhere between about 6 feet 10 inches and 7 feet plus.

One coordinator at the university obviously had a problem with me being British. One day she rang and complained. Apparently, her grouse was that I had written to her giving some information, and in that correspondence, I had used the word 'colour' and had spelt it the English way. Her complaint was that my English was poor as I could not even spell the word 'color' like Americans spell it.

In fairness to the boss of my company she said to this troubled lady 'Mam, at the end of the day it is us, the Americans, who have corrupted the language, not the other way round.' The lady was sent away with her proverbial tail between her legs and we still kept the account, a win-win situation. Well done, Jean!

One day I asked one of the female staff for a rubber without thinking about the connotations of that word in America. She thought I was intimating I might need help with contraception. I soon changed my enquiry, asking instead for an eraser!

I needed to try and keep fit so thought I would join a local soccer team to train, even if I didn't feature in the team. I called the local soccer league and they said they would let me know. I was obviously having trouble speaking American because I started receiving calls from people who wanted to join 'my' soccer team.

Within weeks Durham Hotspur had been formed with me as manager, coach, administrator and general dogs-body! At least my Football Association of Wales coaching award came in handy, and I had also coached local youngsters, in-particular a team called the Vikings in the past, which helped me understand the culture and attitude towards soccer in the USA.

Durham Hotspur was a motley assortment of individuals, and the beauty of the local league we joined was that it was for male and female players. I will not forget our first training session. As I waited to see who, if anybody, would actually show up, a pick-up truck cruised onto the playing field area, and two guys alighted. The truck had the Confederate flag displayed in the back of the cab and two rifles on a rack in front of the flag.

These guys stepped out of the truck dressed in T-shirt and shorts, cowboy hat and cowboy boots and announced in a southern drawl 'Hey, dude, you the Durham Hotspur coach? We're here to play some soccer … let's get it on'. This set the tone for the quality of my new teammates who turned out to be, shall we say, a motley crew but a delightful bunch of people. In fairness, the country cowboys were great guys, one a lot better ability-wise than the other but both keen. We had a decent winger and that was about it. One woman in the team wore a leg brace, underlining our diversity. That first season we managed to finish bottom of the league, losing all our games bar one at the end, which we drew. Taking the positives, it showed we were improving.

One day we only had seven players turn up and we managed to restrict the final score to a 5-0 loss! I don't think we got out of our half for the full ninety minutes after kick-off!

For our second season we signed a couple of Nigerian lads from the local university and one or two other players who were more athletic. We won a cup game and a few other league games and ended up third from bottom and I even scored a goal as player manager. All in all, it was a great experience. I played against some players who went on to play for the USA National Women's soccer team. They were good players and just as physical as the fellas!

In one game the opposition were taking a corner. I was marking one of their women players, and, before I knew it, she had pushed me and put me in the dirt, in the meantime scoring from a header. I had not expected her to challenge quite so firmly. It didn't happen again. Next attack and she enjoyed eating grass, so to speak!

One sad event happened when I was out running. I entered a cross-country event over forest trails called 'Race for Life', which was also a charity event and common in the UK now too. As the trails were through woods there was no one around the course except the checkpoint teams at kilometre markers. As we entered into the last kilometre and came up a steep hill, I was racing a guy to the top. When we reached the top and the track flattened out, we spotted a runner in the ditch in front of us. He was obviously in distress and in spite of us challenging for a top ten position we didn't even think about it and stopped to help him.

As we rolled him over, he was blue in the face, other guys stopped and we got him out of the ditch and took it in turns to give him CPR. A lifeguard who was running in the event suggested we lift his legs to get the blood flowing back towards his heart. One other chap ran back to the last checkpoint to get help.

Soon a flatbed truck arrived and took him to the medical services at the end of the race.

We walked to the finish line rather shaken by the whole experience to be told he had died. He was only a lad in his late teens and it was ironic this event was called the 'Race for Life'. Apparently, he had a heart defect. His face and the futility of our efforts have stayed with me for life!

At Travel Agents International I had improved the business significantly with the help of the great staff there, especially Angela and Sheri and support from the owner Jean. Sheri was a character and a fabulous softball player. I enjoyed the novelty of flying to meetings in Chicago and other cities and to our headquarters in Seminole, Florida, where they put us up in the Clearwater Beach hotel. On one of these trips, we worked hard all day and went out and partied in local bars, ending with a late evening in the jacuzzi, an evening extended mainly because of an interloper who appeared off the beach with a cooler full of beer and wine.

Our new-found friend, referred to his drink's cooler as a 'Clearwater briefcase'. Well, we helped him empty his briefcase, and eventually we were asked to leave the pool area as no bottles were allowed there for safety reasons. I think it was more the fact that the noise level had increased, as it does with the imbibing of copious amounts of alcohol.

On this trip I was introduced to tequila slammers and other exotic drinks, as well as the 'beer to go' concept, where we would have an unfinished beer transferred to a paper cup and then get in the car and drive to another bar. Crazy when I think of the driving under the influence laws and penalties, let alone the safety issues. Looking back, it was truly reckless to drink and drive, but the culture was much different thirty or forty years ago.

One morning I awoke and realised I was a little early for class. I showered and went onto my balcony overlooking the beach and took in the view. I watched the Sandpipers hopping on the beach and listened to the surf. I was at one with this beautiful world we live in until I tried to go back inside. The handle of my patio door wouldn't budge. It dawned on me too late that when I had closed the door it had locked. Now this is okay in a motel, but I was eight floors up.

I shouted, but my cries for help were lost on the sea breeze at such an elevation. What should I do? In hindsight I should have waited for the room cleaner to come, but fearful of being late for class and the embarrassment of it all, I tried to lean over and knock on the patio doors of the adjoining room, but couldn't quite reach.

Foolishly I managed to climb over to the next balcony without looking down. Don't ask me how or why, as I have a fear of heights. I was sweating up a storm but miraculously made it without falling to my death, and banged on the window of the adjacent bedroom..

After a minute or so a very bemused chap opened the doors with his wife/lady-friend still in bed. In my best British accent, I explained briefly, thanked him and walked through his bedroom, saying good morning to his partner who was by now sat up in bed open-mouthed with the bedclothes clutched to her chest.

By now I had missed breakfast but made it to the seminar on time!

I met some interesting characters on these seminars. I got to know one guy from the Southern states quite well, and he chatted a lot, especially in our evening drinks sessions. It turned out Don had been in the CIA or, as he said, still was, as you never retire from the CIA. As I recall he had been an overseas teacher in Cambodia at one time, telling me a lot about 'black ops' and clandestine activity in Vietnam, Cambodia and South-East Asia. He was generally a very affable and interesting bloke.

He was also a character. One evening we were in the bar-cum-dance floor at the top of the hotel having a night cap. I wanted to get some shut-eye but he said to me, 'Before you go, Ron, on your way out can you ask that pretty little thing on the table by the window if she would like to jump on the bones of this sweet old man from Georgia.'

I laughed, got up and out of devilment did exactly as he asked, repeating his message verbatim, pointing in his direction as I delivered his message. On my way out I saw her get up and go over to his table. I never did find out whether he received a dressing down. I wonder where Don is now?

I went on a familiarisation trip to Mexico, too, which was fabulous. We stayed at a beautifully positioned property called Las Brisas set up on cliffs overlooking the Pacific Riviera. The resort was Ixtapa and at the time was just developing on an array of unspoilt beaches.

As we travelled from the airport, we could see lots of shacks and could weigh up it was not a very affluent area. It amused me, though, that no matter how poor the people were, one or two shacks had satellite discs on their roofs.

After checking in, we relaxed and next day went to a lovely little fishing village called Zihuatenejo. Little children dressed naturally in their loose calico clothes and hats for shade tried to sell us some souvenirs, others were begging a little as we sat at a bar. We found these small, brown-eyed children so cute you could not resist them, they knew we had the mighty dollar, which was of so much value to them. This was the first time I was given a beer with a lime squeezed into the neck of the bottle. Now it's a given in gastro pubs and bars the world over.

It was a pretty little place but I suspect the innocence of the area has disappeared altogether now. The fishing boats turned over on the beach and the 'frontiers town' feel to the place, I am sure are long gone, replaced by resort properties.

We were being shown a local golf course which was nearing completion. We were surveying the verdant fairways whilst standing on one of the tees when suddenly a large alligator came out of one of the water hazards and ambled along the fairway, stopped briefly and looked towards us. We held our breath. I guess in golf for me that would definitely be a handicap.

This hedonistic existence was shattered one late afternoon when we were told a tropical storm was due. In the early evening, we were in the open-fronted restaurant, taking in the fabulous ocean view and having a bite to eat, when all of a sudden a huge gust of wind billowed through the restaurant and blew a chap's hat off on the table next to us. We knew then that the storm was about to hit.

Everything was battened down, drinks cabinets, tables, chairs, anything that moved. As the gusts became stronger, we were all escorted to the basement, where we spent the best part of the evening and into the early hours drinking, eating and dancing, as they brought a mariachi band down there and we had a great time, so much the better as it didn't cost us a dime.

When we surfaced the storm had passed over and we went to bed. The next morning the area was devastated, with trees blown over, cars damaged and debris everywhere. Anything that wasn't fixed down had been moved violently and deposited hundreds of metres away. If this was a tropical storm, I wondered what the heck would a hurricane be like!

*

Wendy's father Doc Holden had been brilliant to me and I loved him dearly. He was a very interesting chap, having served with the Royal Navy in the Korean War. He was now a psychiatrist at a local federal penitentiary at Butner and had been involved in assessing the murderer Mark Chapman, who was convicted of murdering John Lennon on 8 December 1980. Chapman shot Lennon outside the Dakota Apartment Building in New York City.

Dr Holden was also responsible for assessing John Hinckley Jr, the college dropout who attempted to assassinate Ronald Reagan on 30 March 1981. After over thirty years in a psychiatric hospital, John W. Hinckley Jr. was given conditional release into society on 10 September 2016.

While staying in the States we had rented our house in Benyon Street in the Castlefields area of Shrewsbury to a former work colleague, but my father had called there one day to do some maintenance and found the house not very well looked after, so after a little soul-searching (probably because I always felt we might come back), we decided to sell it which we did for the princely sum of £15400 in 1987.

I came home for just a week and with Mum and Dads help cleared the house, did some remedial work, which was minor until I was clearing the attic and stepped off the wooden support putting my feet through the lath and plaster ceiling, covering my dear old mum with debris as she was directly below in the bedroom. Thank goodness my dad was a great builder and plasterer and repaired it as if it hadn't happened. It still makes me smile when I think about the incident though.

While I was home that summer, I played in a five-a-side tournament for a team of friends of my brothers, which we won, and the following day played a game of cricket with these lads and others and scored over 60 not out at Frankwell. It didn't help my journey back to North Carolina as I was obviously going to miss home. My dear old mum wrote about five times a week enclosing newspaper cuttings and it's only now I have my own lads living 12,000 miles away in Australia that I realise how much she missed us.

I returned to the USA and we pushed on with life. Unlike the UK, the States at the time did not have national health care , so you relied on so-called public hospitals if you couldn't afford health insurance. We were fortunate to have enough money to buy our cover from a company called Kaiser Permanente, which meant we could go into a clinic or doctors, see the physician who could request tests to be done in-house and within a short period of time you had the results. Pretty efficient for over thirty years ago.

The other side of the coin is illustrated by a story I was told about a black youth who was dropped outside a private hospital, suffering with knife wounds. Because he had no insurance, he was sent to a public hospital, but by the time he arrived he was dead. You would have thought that the Hippocratic oath physicians take or even just moral decency should have influenced the actions of medical staff, but life was really cheap so a family lost a son and a brother. All very sad and it made us thankful for the NHS in the UK.

Laws have changed over the years, but the democratic move under Obama towards Medicare was eroded by the Trump administration.

Typically, if a person in the USA needs medical care and does not have health insurance, they must usually pay in full for all medical care. We will have to see what the Biden presidency brings.

Most hospitals are required to follow the Emergency Medical Treatment and Active Labour Act of 1986. The law requires most hospitals to provide an appropriate medical screening examination to a person seeking treatment for a medical condition, regardless of the person's citizenship, legal status or ability to pay. These hospitals are not allowed to transfer or discharge patients who need emergency treatment unless the patient requests to be discharged, the patient is stable or another hospital is better equipped to treat the patient's condition. The hospital will still bill the patient for medical services. The hospital just cannot refuse to provide medical services.

A growing number of people have unpaid medical bills that they will never be able to afford to pay. A hospital might sue the person, which could result in a court ordering the person to pay, or order their employer to give some of their pay to the hospital. A person may file for bankruptcy because of medical bills, which affects their credit rating and ability to rent and get anything done financially.

It has been known for people go to other countries for medical care, or to medical clinics run by charities that cost less. Others just do not buy important medication and then die of medical conditions that could have been treated.

Whilst in North Carolina, our son Peter was growing as a character. He was full of energy and very headstrong. I recall him doing back flips into the community swimming pool even though he could not swim. He would come up for air and somehow reach the side as we watched amazed. He taught himself to swim. He was a survivor, that's for sure.

One day I took a call from his nursery telling me he was in trouble. He was only about four years old. When I arrived, the police had been called because Peter was being disruptive and threatening. It transpired he did not want to take a midday nap as insisted upon by the carers. He had gone outside and, in a temper, kicked off part of the aluminium downspout, which he was using to keep two of the carers at bay.

Why adults have to involve the police I have no idea, he was a small four-year-old, albeit one with an attitude! Needless to say, this commotion did not end well, and he was expelled from nursery.

I was called by his elementary school one day to say come and collect him as the school was in lockdown. I feared he had been up to something dastardly again. In reality he had already walked out because one of his classmates had brought his father's gun to school. He was not threatening anyone with it, but the fear was a shooting, of course.

Willie Priest was a neighbour in the apartment below us. He was a huge African American and a lovely guy, an ex-Green Beret who certainly had a presence about him. We often passed the time of day for a few minutes and from time to time he would ask me for a lift to a shopping centre a mile or two down the road. I was always happy to oblige. He would practice martial arts under the pine trees that surrounded our apartments. I said to him one day that I should learn some martial arts, and without missing a beat he offered to teach me some moves as a thank you for my kindness. He even gave me a martial arts uniform and Peter used to watch as he sat on the apartment steps.

For a big guy Willie moved rapidly and he could have killed me ten times over with his power and speed. But he taught me various moves, how to relax, how to strike like a cobra and target pressure points with the hits, some to immobilise and potentially kill. It was engaging but it was soon clear that I did not have the killer instinct.

One day he invited us downstairs to his apartment for a bite to eat, so Wendy, Peter and I turned up with a small gift and entered his apartment for the first time. We were amazed to find there was no furniture whatsoever. Just a framed picture of his daughter surrounded with a flower garland propped up above the fireplace. He was so proud of his daughter, but rarely got to see her and his wife wanted to move away.

We sat on the floor and shared a mix of snacks and little desserts like 'Moon Pie', two round cookies, with a marshmallow filling in the centre, all dipped in a flavoured coating like an American version of our Wagon Wheels. Willie was a bit down, so we tried to console him. Peter was completely under his spell, he thought he was awesome like some superhero!

Willie had to move out not long afterwards as even though veteran support paid his rent the residents committee had made complaints. The reality was they were frightened and confused by his larger than life, presence. In our opinion he was no threat, just different.

A week or two later Peter had come home from school and was watching early evening television while I prepared dinner, and a local news bulletin broke into the programme with live video from a developing scene about an assailant bank robber who had tried to rob a bank with a barbecue fork.

The robbery had not been successful and the assailant had been pursued to a house in Durham which was now besieged by S.W.A.T. teams. We watched intrigued and listened to broadcast conversation between the police negotiator and the assailant. As we honed in on this amazing turn of events, police fired in tear gas through the windows, the voice on the phone coughed, then recovered as he threw the missiles back out. We would have recognised that voice anywhere.

A little while later after a stand-off, Willie Priest was led out and eventually after trial convicted of armed robbery and sent to the penitentiary for a long sentence. I always felt he had been shunned and not supported enough by society, this in spite of his service for his country.

I travelled to and from the UK a couple of times and plenty of times within the States going to meetings with Travel Agents International. Some of the air crew in the States were quite funny with their public and safety announcements, as I remember. Some I recall are:

There are more than fifty ways to leave your lover but only four ways to leave this aircraft. Check out the emergency exits!

Your seat cushions can be used for flotation in the event of an emergency water landing. Please take them with our compliments.

Over my years in the travel industry, I have had plenty of questions directed at me, especially when tour managing. One that particularly comes to mind is: 'Does the water go all around this island?'

I was at the airport waiting for a flight to Germany once when the airline announced our flight had been cancelled and to go the desk for instructions. I heard the chap in front of me say, 'I've heard the flight is cancelled. Does that apply to business class as well?'

I once heard an American at Windsor Castle say, 'It's such a shame they built this castle so close to Heathrow airport'. You just have to smile!

In spite of the good life in North Carolina, memories of home were stimulated by watching a play of *A Child's Christmas in Wales* one evening at a local theatre. It evoked such a flood of homesickness I decided I was going to go home. One of the popular tracks at that time was 'Sailing' by Rod Stewart and I used to sing that all the time, knowing I wanted to be 'sailing' home. Wendy's father had got over his cancer treatment and been given the all-clear on his latest scans and, heartbreaking, as it was, I broke the news that I wanted to go home.

Wendy's father Doc Holden, was not very happy, which I understand totally and I fully expected Wendy and Peter to stay for as long as they wanted to. I just felt I needed to go home to the UK, and in the end we all returned back to the UK together.

CHAPTER FOURTEEN

Back to the future

We arrived back at Gatwick airport with World Airways, but our hold luggage didn't, which was very worrying as it contained all our earthly belongings. There were some big trunks with household goods such as a large, heavy dining set of Portmeirion pottery that was sentimental to us.

Eventually they found the luggage in Frankfurt and it was all delivered to our address in Shrewsbury a week later. These trunks were so darn heavy it was a blessing they were temporarily lost in transit and finally delivered to our door. Some things are meant to be!

Later that year I enquired about some more flights with World Airways and said, tongue in cheek: 'Can I book some return flights London to Washington but send my luggage to Frankfurt on the return segment?' Apologising, they said they could not do that, so, laughing, I pointed out that they had managed to do it last time I travelled with them. Strangely, they didn't see the funny side of it, after all our British humour is a little different to the Americans', like the spelling of some words.

We were blessed to be staying with a good friend, John Brake, in Reigate, Surrey not far from Gatwick airport.

Not long after our arrival, John received a telephone call from a chap in Shropshire who was trying to track me down after hearing I was back in the country. I don't think I ever found out how the chap in question, Peter Fraser, even knew I was in the UK but, there again, he was a former policeman.

Peter had bought a small chain of travel agents in Mid-Wales called Powys Travel from a chap I used to know, Harvey Cartwright. More accurately it had been run by his daughter Cynthia, known as 'Tinty', since Harvey had passed away.

Peter needed some help in consolidating and developing the company and wondered if the challenge of perhaps managing Welshpool and helping improve Newtown appealed. Well, what could I say? I was flattered and broke, so it was, to use an Americanism, of which I had brought back many, a 'no brainer'.

So here I was in April 1988 straight back into a job. I revelled in the opportunity and was fortunate to rent a house from a friend back in Shrewsbury and set about increasing the business generically as well as bringing in a lot of my former clients. Fortunately, now I was back, my former clients were very loyal and thankfully appreciated my travel services and experience and one by one they came back to me.

I set about improving our service, efficiency, awareness of products and marketing. Nor did I overlook the simple things, like listening to what the client wanted and advising accordingly. I enjoyed working here and appreciated the people I worked with and for. However, it was clear after eighteen months, when I asked about a partnership, even given the outstanding results already achieved, I was met by a strange response of inertia and almost distrust of my motives.

Ron Morgan Travel, Princess Street

Peter's pal, Pat, counselled him against offering me anything other than a profit share. Disappointed, knowing I had proved my worth, I decided it was time to move on. I will always be thankful to Peter and his lovely wife for the opportunity.

Around springtime 1989, I teamed up with my old schoolfriend and travelling buddy, David, and set about looking for some premises in our home town of Shrewsbury. It was time to branch out on our own. Never swim against the tide!

We planned to start our fledgling business as a partnership and looked at quite a few premises, almost taking the old Livesey's printer's premises on St John's Hill. But, just as we seriously considered Livesey's, we found a cheaper, albeit smaller, option.

We decided on the old Honychurch Ironmongery premises at 24/25 Princess Street. It was a double-fronted property with a large display window, which suited us down to the ground as a start-up company. Close to the town square, we witnessed good footfall for a secondary street, and for that reason too, we decided this was the one. Only years later did I found out that my future wife Dianne and her mother Margaret wanted it as a café, but we pipped them to the post. Ironically it is today a cafe!

We had been speaking to a former work colleague, Sue Owen, about coming and managing the business for us, as I saw us as a potential miniple, eventually opening new branches around the county and in Mid-Wales. We wanted a good person in charge of the Shrewsbury office, as we figured that, once we were off and running, I could spend some time developing the company, while David would be a silent partner.

We then had to decide on a name. We thought of calling it Rendezvous Travel (Ron, Dave and you (Sue), a play on words). But logic brought us back to trading on my name, experience and contacts of eighteen years in the business. We decided on 'Ron Morgan Travel'. I felt a bit strange having my name up there on the sign but it proved to be the right decision.

It was a proud day when Mum and I carried the shop sign bearing the name of our company from the signwriters in Meadow Place, up Castle Gates and down Pride Hill. It was wonderful in many ways because my dear old Mum became our bookkeeper in the days before computers, and very good she was too despite her advancing years.

Sue Owen had agreed to join us. We had to develop the infrastructure for the business and importantly interview and appoint staff. We decided we would add another experienced team member and a trainee, making us four in total.

We received over ninety applications. I must have been crazy, but I decided to interview them all. I worked on the theory of giving everyone a chance. It would allow me to meet them and do a bit of public relations at the same time. I hoped that even if they didn't get the job, I could influence them enough to send friends and family to us to book their holidays.

It took a while, but I felt we made the best two choices and so it proved. Debbie joined us from Brian Bass Travel (Brian being my first boss), and Tonia joined us as a trainee.

One of my first tasks was to ingrain the company philosophy into my staff. Our mantra was that we do everything we possibly can for the client from the time they enter the office, keeping them informed throughout the time from booking to holiday completion, including giving them a call when they get back from holiday, to see how their travel plans worked out.

The latter made sure we kept up a good client relationship and resolved straight away any issues that might have occurred on their vacation. Plus, we also learned a lot from the clients about their tour, information we used to benefit other clients in the future. As the Americans would say, 'we went the extra nine yards'. We decided to have a smart uniform, so the staff were presentable and didn't have to think about what they needed to wear each day.

I had built up a large personal clientele, especially groups, one of which I escorted from 1980 right up to when the company was sold in 2001. Indeed, I travelled with a good number of them after that date, as they had become friends.

We were likely the first company in this region to recognise the future was not in selling everything, but in specialising. We decided to make long-haul and luxury travel our niche market. We were aware people called themselves cruise and long-haul experts, but we felt they could not really back up that claim.

We truly felt this had to be an accolade that was earned by educating ourselves. All the team were sent to far-off places to learn about those countries and their cultures and what they could offer the tourist, in everything from accommodation to tours and experiences. There is nothing to match the experience of feeling and sensing the places and people in every destination you can.

We all attended lots of company seminars, learning about products and destinations and specialist functions, like the various Australian tour providers and Tourist Board 'corroborees', where we immersed ourselves in Oz culture with everything from vegemite to boomerangs.

Educationals or familiarisation (fam) trips, as they were variously called, became more exotic as the seventies and eighties progressed. One of my favourite early educational journeys with other travel staff was with Pegasus holidays to St Lucia. I can still feel the humid warmth of the air that hit me as the plane door opened and we disembarked. I experienced the sweet smell of frangipani for the first time as it wafted to me on the Caribbean breeze.

It was a new, small airport terminal. I was one of the first to disembark the aircraft walking down the aircraft steps and taking in the vista of the many palm trees bordering the airport. I entered the terminal with my small bag on my shoulder and walked straight out the other side where a taxi or two were waiting. Meandering back in to look for my hold luggage, it was clear security was not, shall we say, on the ball. Soon a couple of big local fellows in a uniform of brilliant white cotton shirt and black trousers arrived, some plastic strips in the side of the flimsy terminal wall fluttered and the luggage started to appear onto the mechanised baggage belt. I took my maroon suitcase up to the luggage channel and the guys armed with a huge ball of chalk duly left a large X on my baggage, a mark that stayed on for quite a long time.

We had a fabulous time exploring this magical island. I particularly enjoyed Marigot Bay where they filmed some of Dr Doolittle with Rex Harrison. It's fair to say the roads were not very suitable for anything other than a jeep back then. Going down to the bay one time we were greeted with a crater in the middle of the road.

We enjoyed a delightful lunch at the Hummingbird Restaurant as we watched the hummingbirds so small and cute hovering around the foliage alongside Anse Chastenet beach, where we also watched an impromptu game of cricket between a group of young locals. This tranquil and laid back beach is much changed now with the evolutionary and impressive Jade Mountain Resort set into the hillside above.

We travelled on a narrow road up into the mountains where we visited some resort accommodation which looked down on to the coast through the two iconic volcanic peaks on St Lucia called the Pitons. We were welcomed with a refreshing and colourful iced fruit punch cocktail to help assuage the effect of the heat and humidity. We were then guided by their manager around this smart resort. We were shown a spacious suite, very well furnished with attractive good quality furniture all in light, airy pastel shades.

On entering the bedroom, we were faced with a large king-sized bed enclosed by a mosquito net. As my eyes focused in the dappled light of the room, I saw something strange on top of the net, then realised the gaggle of travel agents had gone very quiet. As we neared the bed, one of the party let out a scream and it wasn't me. The something strange was a huge tarantula, its abdomen pulsating as it rested on top of the net. Boy, I would not have wanted to wake up under that net, that's for sure!

A year or so later I travelled for the first time to India with British Airways Speedbird Holidays and fell in love with the place. Delhi, Agra and the Taj Mahal, and Jaipur were on the itinerary, a tour generally referred to as the 'Golden Triangle' in north-east India.

It was probably when I first had a hankering to own my own company and specialise in long-haul holidays. This was an exotic and fabulous country, with the colours and smells assaulting your senses, the history and especially the various remnants of the colonial days really made me want to explore more. We met up in Delhi with other agents who had travelled out with British Airways and who had had the disturbing experience of having to evacuate their plane down the emergency chute after a rather bumpy landing and a smell of fumes filling the cabin. It was ironic, really, as we had travelled out with Air India and were bemoaning the fact that we weren't flying BA. As it happened, we had a fabulous flight, terrific service from the Air India crew, the female attendants or stewardesses as we called them back then were dressed beautifully in their saris and were just as passionate about cricket as their male counterparts.

One thing that makes me laugh even today was as I was walking around the aisles to stretch my legs in flight, I was intrigued, or was it shocked, to see a gentleman, presumably of Indian origin, starting to assemble a primus stove in the aisle in order to cook some food. It was not long before an astonished and vexed member of staff pointed out to him in no uncertain terms the error of his ways. You could see the fellow had not considered the safety factors, more the fact he needed to have a hot drink or food.

On my first morning in Delhi I walked out of the hotel gate just after dawn and enjoyed watching a game of cricket taking place on some wasteland between some young lads. The endemic poverty and caste system were so alien to our experiences back home. On one excursion we were at the Red Fort and I was waiting for the rest of the group to finish the tour. I was leaning against a cart when the sheet on the cart blew back to reveal a human foot with a tag on. We were told it was someone who had died on the street and was waiting to be taken away.

To see the maimed beggars was awful, mutilated on purpose by terrible gangs who used their injuries to curry sympathy from tourists and gain big donations for the gang leaders. We were at some traffic lights one day with the minibus windows open when a woman shoved her baby through the window, offering to sell the child to us. It upset the girls in our party badly and concerned us all that life could be so cheap and people so desperate.

We caught the 'Shatadabi' express train to Agra from New Delhi station. I was hoping for a steam train but, alas, it was modern. Before it arrived, there was a throng of people at the station but all of a sudden, a further mass of people appeared, swarming like bees as they clambered onto the platform under which they had been sleeping. The train arrival was imminent and awaited by a heaving amoeba of humanity. It was a comfortable, fast journey, though, and fortunately we had reserved seats which were respected.

With the poverty and chaos outweighed by the history and iconic buildings prime of which was the Taj Mahal, I was hooked! Beguiled by India I have travelled back since to escort other tourists around this exciting country.

In Agra the Taj Mahal, the marble mausoleum to Mumtaz Mahal, makes a wonderful sight and changes with the light, looking different at sunrise, midday and sunset as the sunlight plays on the walls.

On the first evening at Clarks Hotel where we were staying we were introduced to a young man by one of the local tourism officials hosting us, as he was aware after chatting that I loved cricket. This diminutive character, we were told, was a very good cricketer and even though he was only sixteen years old he would shortly be playing for India. He was a very pleasant lad and we wished him all the best for the future. Within two weeks he made his debut for India in Karachi against Pakistan. How I wish I had taken a photograph with him or even got his autograph, but I have the wonderful memory of meeting him all the same. His name was Sachin Tendulkar, 'the little master', currently the highest run maker of all time in international cricket.

I had been feeling a bit queasy and, in fact, at breakfast every morning more and more of us seemed to be feeling the same. I thought I had a D and V bug, or maybe it was something I had eaten, but the local people pointed out it was nearly always drinking too many ice-cold drinks after spending too much time in the sun. When I thought about it, I agreed. I naturally asked our guide for advice and he said, 'I know just the thing'. He whispered to the head waiter, and before long a waiter returned with a large bowl of natural yoghurt and a glass of lassi, a blend of yoghurt, water and spice.

With the road trips we had ahead, my trusty pack of Imodium was needed, plus a copious supply of toilet paper, which can be in short supply in public and private toilets in India, let alone if you have unscheduled toilet stops.

Fortunately, the yoghurt-based products worked very well and within a few days I was a lot better. Needless to say, I was eating little else, just drinking bottled water, so was becoming a little weak. By the time we reached Jaipur as a group we were fed up of the constant spice in our meals and just wanted something bland, so our diamond of a guide found us a restaurant who cooked us all a huge omelette. Brits abroad, eh!

Growing up, one word my mum always used when something was driving her crazy was 'doolally'. It was only when travelling in India I was informed by a guide that the word came from India. 'Doolally', originally 'doollaly tap' (perhaps where we get 'tapped' from when saying someone is a little mental), apparently goes back to a meaning to lose one's mind and was derived from the boredom of being in the Deolali British Army Transit Camp, and the 'tap' bit from the Sanskrit word 'tapa', meaning fever or heat. My Welsh-speaking relatives point out the Welsh word 'twp' means stupid, unable to reason sensibly, so maybe it's a combination of Welshmen being held at the Deolali camp and being affected by heat stroke and/or malaria … who knows?

*

1989 was a momentous year. Not only did we start our own business but my dear father-in-law Peter Holden passed away and our son Tom was born, all within a month. We opened the shop doors six weeks later.

My wife Wendy was bereft, losing her father, the main constant in her life since birth and found it hard to look after Tom. So I used to go running with him in the pushchair as he got older, just to get him to sleep. We got through it, though, and the following year moved house too.

Our business had to ride a tough time or two, firstly, when the Gulf War started in 1991. All of a sudden fear of the unknown stopped people booking holidays and we went a few months with very little business. We took the decision to retain our staff, in the hope confidence would return and the conflict would subside. Bigger companies like Thomas Cook laid off staff. Our loyalty to our staff paid off, as when the Gulf War finished, we were ready to deal with the sudden surge in business when it arrived, whilst competitors had to play catch-up.

CHAPTER FIFTEEN

———

'Epic RTW' – Round the world adventure!

I had not been well for some time through 1993 and 1994, having belatedly been diagnosed with ulcerative colitis, and had become steadily worse, to the extent that I lost weight and could never be far from a toilet. This resulted in an emergency admission to Royal Shrewsbury Hospital and a traumatic and lengthy period of treatment and assessment.

I was discharged after a long stay in hospital, where at one stage they were going to operate to remove the large bowel and give me a colostomy. However, my consultant managed to get it under control with steroids and an enhanced intravenous feed through the heart as I could not take anything by mouth, and this regimen reprieved me.

While I was in hospital my young sons used to come to visit and it was clear to them that I wasn't well. I remember once when they'd come to see me with Wendy and they were getting ready to leave, Tom skipped ahead down the ward and shouted back, 'See you, Dad, hope you don't die'. Even in my weak state I had to smile.

In April 1995, one of the many agricultural tours I arranged over the years was scheduled. Having branched out from the European jaunts, we were now going for the big one, a round-the-world trip via Hong Kong, Sydney, New Zealand, Fiji and Los Angeles. I knew I wasn't really well enough to travel with them. I had only come out of hospital in the January and whilst I had returned to work after a few weeks' rest out of necessity, something you do when it's your own business and livelihood, I was still under par. So, I considered a plan B.

Dianne, who was initially a client and who Wendy and I had met when she nursed Tom in her professional guise as a nursery nurse at Royal Shrewsbury Hospital, had casually stated one day when picking up her tickets for her own holiday that if I ever needed help taking any groups away, she would be happy to help.

Dianne was well travelled and it transpired a group her parents were booked on to Florida needed an escort because a courier I had arranged dropped out at short notice. Dianne did really well and subsequently escorted groups to California for Ron Morgan Travel and had also taken the farmers party to Cyprus, most of whom were in the Round the World group, so I asked Dianne to escort this trip, stating I would possibly come as far as Hong Kong, just to settle the group and hand over to Dianne before returning home.

I put this to the group organiser Ken, who was initially unhappy with me not being on the whole tour and, in fairness, it was an intricate schedule of flights and tours with many hotel changes. That said, I was still not sure I could last the entire duration of over three weeks. So, with lots of steroids and other prescription drugs packed, Dianne and I set off with the group. Our first stop was Hong Kong.

The approach to the old Hong Kong Kai Tak airport was legendary, requiring excellent flying skills by experienced pilots because of the airport's location being very close to skyscrapers and mountains and the runway jutting out into the harbour and therefore surrounded by water. Apparently, pilots relied not on instruments but on their own visual take on the low-level and very short final approach.

As a passenger it was a fabulous experience, but not for the faint-hearted as we seemed to be within touching distances of people's balconies bedecked with washing. It is not deemed the sixth most dangerous airport in the world for nothing. Having touched down safely after a journey of fifteen hours we stayed at the Regal Hotel Kowloon.

Our first guide was a lovely lad called Gain. He had a great sense of humour and fun, and he had actually studied at Concord College, Acton Burnell in our home county of Shropshire.

Jumbo restaurant

Aberdeen Harbour transport, Hong Kong

As we travelled, I became ill because of the long journeys and stress. Exhaustion brought on the ulcerative colitis, which brings with it debilitating blood loss. I took some steroids I had been prescribed to steady the ship and decided to travel on with the group as they stabilised me a little.

What a fabulous holiday we had, Hong Kong with its traditional Star Ferries, a tram up Victoria Peak to literally look down on the conurbation of high-rise blocks and views across the causeway, Stanley Market where you can buy just about anything, lots of it counterfeit but at a good price! In fact, I have a Phil Collins album at home on which Phil sounds very Chinese but still very enjoyable! We took a traditional vessel, a small junk once used as a Chinese fishing boat, the lady of the family cooking a meal for herself as we sat on little benches on either side, barely above the waterline as this lady guided the small craft around Hong Kong's Aberdeen Harbour, rudder in one hand and stirring some rice on a little primus stove with the other. The tour included an up-close and personal look at the huge and iconic 'Jumbo' Floating Restaurant which can take up to 2,000 diners at a time and is quite a sight lit up at night.

I was not feeling well at all as we checked in for our onward leg, Hong Kong to Sydney. I almost made the decision to go home, but opted to make it to Sydney before taking stock and fortunately I managed to pull round.

In Sydney we checked into the Renaissance Hotel on Pitt Street, all a little tired from the journey but excited to be in Sydney and looking forward to seeing the iconic sights like the Harbour Bridge and of course the Opera House. As usual I checked in every one and when the lobby was empty, I picked up my key and noticed I was on a high floor. As I reached my door and opened it I was astonished to find I had been allocated a junior suite. Thinking this was a mistake, I called the desk to be told that as I was the group leader it was a courtesy upgrade. Now normally I always have the most inferior room in the allocation as I want my clients to have the best but the desk insisted I retained this room. I walked over to the lofty window to be greeted with a view of the famous Opera House framed by two high-rise buildings. It was a very emotional moment, probably more so as I felt weak from the colitis and the journey. I had a real frisson of excitement realising we had travelled across the world to see this fabulous building, designed by Danish architect Jørn Utzon. Its main concert hall can hold almost 2,700 people and boasts one of the world's largest organs – it has 10,000 pipes. To witness the sunrise over the Opera House from my suite was even more amazing.

We took a harbour cruise to see the skyline, bridge and Opera House from a different perspective before taking to a coach to a look-out location at McMahons Point on the North Shore, a residential neighbourhood and devoid of tourists unlike popular viewpoints like Luna Park. We gazed at the city skyline and Darling Harbour, then had a bit of down time taking in the iconic beaches, especially Bondi, famous for its surfing. To our farmer friends this was very different from the green fields of home.

I have since been back a few times and never tire of Sydney and love taking the ferries to places like Cremorne or Kurraba Point or to Manly Beach from Circular Quay. We have every reason to go now to see our sons Peter and Tom and their families.

Still feeling a bit under the weather, I escorted the group to Christchurch, staying overnight at the Quality Inn Anzac Hotel close to Cathedral Square (which is dominated by the Cathedral and is actually cross shaped rather than square). In the later 2001 earthquake Christchurch was devastated. A friend of mine in Christchurch at the time witnessed his vehicle bouncing on the drive in the suburbs whilst he visited his mother-in-law, such was the intensity of this natural disaster. It destroyed the Cathedral's spire and part of its tower, and severely damaged the structure of the remaining building. The rest of the tower was demolished in March 2012.

Having enjoyed the city, which felt a little like a quintessential small English town such as Stratford-upon-Avon, we headed west for a splendid little rail journey called the Tranz Alpine Express to Greymouth. It starts along the Canterbury Plains, and the excitement builds as the train slowly ascends through amazing gorges and valleys towards the Southern Alps. At the alpine village of Arthur's Pass, we had a short stop to stretch our legs and take some photographs before dropping down via the Otira Tunnel to the West Coast terminus at Greymouth, the landscape changing to alpine beech rainforests as we headed to the west coast.

Greymouth was a rather featureless and uninspiring destination, not endeared to us as we were being bitten to pieces by sand flies on a short walk on the beach. After a good meal and cultural dancing and singing show at the hotel we were happy to move south down the west coast to the Franz Joseph Glacier the next day. You can take helicopter flights up onto the glacier, which was named after the Austrian Emperor Franz Joseph by the German explorer von Haast as far back as 1865. At Hokitika, a quaint seaside resort which felt like a frontier town, we visited a jade and glass-blowing factory. We stayed at the Westland Hotel, where I experienced my first taste of Orco John Dory fish, which was delicious.

I recently saw some pictures of the Franz Joseph Glacier and it has receded dramatically in the last twenty-five years since our first visit. We have been told by experts that this glacier retreats and advances in cycles. It apparently retreated several kilometres between the 1940s and 1980s, then the glacier started to advance in 1984 and at times has advanced at the amazing pace of 70 cm a day, about ten times that of typical glaciers.

We found an interesting bit of information whilst there regarding the glacier. A postage stamp was issued in 1946 depicting the view from St James Anglican Church. The church was built in 1931, with a panoramic altar window giving a great view of the glacier. By 1954, the glacier had disappeared from view from the church, but it reappeared again in 1997. This is due to the highly variable conditions on the snowfield, which take around five to six years before they result in changes. The glacier was still advancing up until 2008, but since then it has entered a very rapid phase of retreat. As is the case for most other New Zealand glaciers, mainly found on the eastern side of the Southern Alps, the receding is attributed to global warming.

We continued our road-trip down the west coast, then inland towards Lake Wanaka, now the home of a great buddy of ours John Brake and his Kiwi wife Kaye. This route is spectacular as you go through the small town of Haast, take the breathtaking Haast Pass and cross the Haast river into Mount Aspiring National Park into the Southern Alps. Yes, you'll have noticed that Haast has not been forgotten. You travel from the Franz Josef and Fox Glacier region, and see two large lakes, Hawea and Wanaka. The landscape changes from coastal scrub, rainforest, glaciers and native bush, to alpine meadows, taking in fabulous lakes and mountains.

Wanaka, as attractive as the little town is, is just a diversion on the way to the wonderfully situated Queenstown, my idea of heaven, lying resplendent on the shores of Lake Wakitipu and looking out towards the awesome peaks of the Remarkables. We based ourselves on the lakeside at the Gardens Park Royal Hotel for a few days as there was so much to do in this area. Queenstown has a certain karma about it, you just feel so relaxed here. Sure, it has grown a bit since I was first here about thirty years ago but it's an idyllic setting, and still a small town which sprawls over the surrounding hills. It's an ideal base to explore some adrenalin-rush pursuits like the jetboats or bungee jumping.

Wine tasting in nearby vineyards like Carrick in Bannockburn was fun. A little less than one hour's drive from Queenstown, Carrick is a boutique vineyard that concentrates on things organic, not just in their wine production. We liked the bijou Mt Difficulty vineyard, situated on a hill in Bannockburn overlooking the Cromwell Basin. This vineyard is the perfect place to sit back, relax and enjoy a quiet quaff. All the vineyards here were planted in the 1990s and are named after historic local gold mining areas such as Target Gully, Manson's Farm and Long Gully.

There are some pleasant walks around Queenstown. Even a short stroll in the little park is worth a dalliance.

One of the fab little excursions you can do is on the old TSS Earnslaw, which dates back to 1912. You can just take a leisurely pleasure cruise on the lake, but we went on a cruise right down the length of the lake to Walter Peak Sheep Station, which serves up a fabulous evening meal after a little tour of the active farmstead, where we could get up close to sheep, deer and highland cattle. You can also help feed the farm animals, watch a sheep-shearing demonstration and explore the lakeside gardens. You can do this morning or afternoon, but I do recommend the evening cruise, especially with the communal singing supported by a pianist on the way back. It's a great evening, which I've enjoyed on more than one occasion.

A free shuttle bus from downtown Queenstown took us for the ten-minute drive to the gravel shores of the Shotover River. An experienced guide then gave us a briefing for the hair-raising jet boat ride we'd booked.

We sped past rocky outcrops, skimmed around boulders and zipped through the dramatic and narrow canyons as their walls towered above us. We were amused that the chap who had put together the group we were escorting was a keen motorcyclist, but even he had his head down and his eyes closed for most of the journey as the spray and breeze flew by. It was especially exciting when we reached a wider point in the river and the driver would spin his hand in a circle indicating he was going to do a 360-degree turn, which was a bit like a handbrake turn on the road at speed, but cushioned by the water in this case. After a thrilling twenty-five minutes or so, we returned to base and on the way back to town the transfer coach bubbled with excitement and probably relief that we had survived.

Of course, this area is well known for bungy jumping. In November 1988, bungy pioneers AJ Hackett and Henry van Asch jumped off the historic Kawarau Bridge here, launching the world's first commercial bungy jumping site. If you want to be tied up and thrown off with a friend, then this is the bungy site for you, as it offers Queenstown's only tandem bungy jump. Either way I still find it scary and totally unnatural!

From the Kawarau Bungy Centre, set into the rock face above the magnificent Kawarau Gorge, I was just happy sipping drinks and taking in the atmosphere from the specially constructed viewing platform as jumpers took off screaming and often with embarrassing wardrobe malfunctions. They were then unceremoniously lowered upside down like landing a huge fish into a raft and were paddled back to the riverbank where they could climb the steps up to the departure point, by now recovered, full of bravado and pride and buzzing that they had a tale to tell of survival of the 43-metre leap! All have to be weighed beforehand and the length of the rope adjusted to take in the specifications. Some choose to just take a dunk into the water which for me is even more scary. Some are allowed a free jump if they are of a venerable age or choose too bungy naked! Not bad, considering nowadays the jump costs £100 or more. Strangely, not one of my farmers group who ranged from their fifties to their eighties was inspired enough to take the plunge!

The first time I visited I was with a group of fifteen travel agents, only two of whom were brave enough to have a go and it was free for us! One was our tour leader and a Kiwi, who jumped off no trouble. The young woman who was the other gallant participant had her legs tied together with the bungy cord and shuffled forward onto a precarious platform. She was told to look into the distance and jump towards a point she was focused on, but in reality, she fainted and fell backwards, never achieving her goal!

Back to the farmers' group tour and we enjoyed a leisurely evening on the Skyline gondola, riding to the top of Bob's Peak for the incredible panoramic views. We boarded a four-seater chairlift right to the top of the peak before enjoying a fabulous meal at the restaurant, looking out over Queenstown and the lake below, which as sunset passed became a magic carpet of lights illuminating this fabulous mecca for the outdoor enthusiast and tourist alike. Ken, our tour organiser, was seventy the day we arrived. What a place to celebrate your birthday!

One day we headed the short distance to Arrowtown, an old gold-mining settlement. This beautifully preserved town is just 20 miles from Queenstown and shares its dramatic backdrop, but the pace of life slows down considerably in the quiet Victorian streets, lined with deciduous trees full of vibrant autumnal colours, russet, gold, yellow and brown, sheltering wooden churches and colonial buildings with lots of cafés and restaurants. Main Street has a number of restored shops, which meld with small stone miners' cottages at one end of town, dating back to the 1860s. Rather than dwelling on the obvious history of the buildings, our farmers were more taken by some agricultural equipment, made up of a tractor and ploughshare and spent most of their time here standing by this display and chewing the fat, reminiscing about their earlier years and the similar equipment they used.

After a really enjoyable stay we moved on to Te Anau, a gateway and base to visit the fabulous Fjordland National Park and the famous Milford Sound. Te Anau sits on the shore of the South Island's largest lake, also called Te Anau. We stayed at a novel property called the Village Inn, where all the rooms were behind facades representing shops of a bygone era. We stayed in the outfitters store, livery stable, pharmacy and so on. Many of the rooms had a jacuzzi. Wow!

Milford Sound was a highlight we were all looking forward to as we had a boat tour booked along the Sound towards the open sea. The first time I had been there we actually flew over the sound in a small plane, which was quite an experience as it took off down a short runway and had to bank sharply to miss the mountain called Mitre Peak which lay straight ahead. We loved the trip but found out that a year before six Japanese tourists had died as their plane crashed trying to avoid the peak just after take-off and being involved in a mid-air collision.

On the way to Milford Sound, we stopped at Mirror Lakes to photograph the reflections. We then passed through Homer Tunnel. The rocks up to the right of the tunnel are where Sir Edmund Hillary did some of his training for his ascent of Everest in 1953.

The Fjord Cruise, takes about 1 hour 45 minutes. The ship travelled the full length of the fiord, at times getting up close to the cliffs so we could see the forest and waterfalls like Bridal Veil and Fairy Falls cascading into the sound. We also drifted towards Seal Rock to view some seals, of course. We cruised out towards the Tasman Sea at the mouth of the fjord before returning to the wharf and then to Te Anau.

It was time to travel further south, not quite as far as Stewart Island but to the very Scottish city of Dunedin via Lumsden and Gore, both very agricultural areas with a little visit to a Paua shell factory. The jewellery especially seemed to appeal to my fellow travellers, even if one or two of our group had misheard me when I said we were going to a Paua shell factory and assumed we were visiting a power station. Well, the words are pronounced the same.

We arrived in Dunedin, the second-largest city on the South Island after Christchurch. Its name comes from the Gaelic Dùn Èideann, the Scottish name for Edinburgh. You can see the Scottish influence in the buildings. This is a university town, quiet in January when the students go on holiday. It has a lovely botanical garden, loads of graffiti around town especially in the Octagon, the public square with lots of cafes and a statue of Robbie Burns. It is not a particularly attractive town but has some fine buildings including Presbyterian churches and its law court, but the one that sticks in your mind and worth a look is its railway station. A splendid Renaissance-style creation made from Central Otago bluestone and Aberdeen granite, it's impressive turrets, Royal Doulton porcelain, friezes, stone lions, French tiles and stained-glass windows are all cherries on the cake and make it very photogenic.

APT Group picture

Most of the town developed during the gold rush era in Victorian times. We took afternoon tea at the fine Larnach Castle with its 300-square-foot ballroom, but we found Olveston House, the former home of the Thurman family, a very interesting turn-of-the-century home, now museum, where we learned about their daily lives.

We left Dunedin and headed north past Oamaru, then inland north-west towards the settlement of Tekapo with a population of just over 300 and Mount Cook National Park in the Southern Alps.Our destination was Mount Cook Hermitage Hotel, a 65-km drive up the valley from Twizel. We broke the journey by stopping en route at the small Mount Cook Airport where some of us took a tourist plane up on to the glacier on Mount Cook. We could get out and take a walk on the ice and snow, a real buzz and we would highly recommend this experience. It felt surreal as the sun was shining brightly.

While the party settled into the hotel Di and I took a walking track to a lovely spot called Kea Point, named after the very colourful parrot-like birds that unusually frequent the area. About 48 cm long, the Kea, the world's only alpine parrot, is mostly olive green with brilliant orange under its wings and a large, narrow, curved, grey-brown upper beak.

We had a lovely cosy stay at the Hermitage Hotel, nestled in the beautiful Mt Cook National Park. It was the perfect place to relax and explore from with walks and hikes and even mountain climbing. We had some super food and there's a homely log fire in the lounge for when it gets chilly. The lack of light pollution means some of the world's darkest skies, perfect for stargazing.

Lake Tekapo is about three hours drive south-west of Christchurch in the Mackenzie Basin. The small township faces north across the beautiful glacial lake to the mountainous peaks of the Southern Alps. The lake gets its milky-turquoise colour from the fine rock-flour grated by glaciers and suspended in the water.

For Dianne and myself this is a special place, as mentioned elsewhere, mainly because of the church on the shores on the lake. Here you'll find the beautiful Church of the Good Shepherd, where the altar window frames a perfect view of the Southern Alps beyond the lake. The church was built in 1935 for the pioneer families of the Mackenzie district and is still used as a place of worship. The church is a very popular place for tourists to take photographs.

We will always remember Sheila Watkins playing the organ in the church and the rest of the party forming an instant flash-mob, singing Welsh hymns and classic songs like Calon Lan. Nearby is a bronze statue of a sheepdog, sculpted in recognition of the reliance placed on the sheepdog in this mountainous region where sheep graze on the slopes.

Our time was going too quickly and, before we knew it, we were heading back to Christchurch and then flying to the middle of the north island and Rotorua. Rotorua is a totally different world, almost lunar around the various geyser sites. Even our hotel reeked of sulphur. It was quite off-putting, worse than a room where someone has been smoking, but very strangely you soon become accustomed to it. You think your clothes are going to carry the smell, but, somehow, they don't.

Rotorua is a place that grows on you. Apart from the bubbling mud pools and active geysers, it's surrounded by stunning lakes, mountains and bush. There are lots of natural spa baths here, too. Most tourists like us head for the Polynesian Spa, very therapeutic, and even swimming in the thermally heated swimming pool is an experience. Of course, being covered in the bubbling mud is a definite must! I do recall, though, that it tended to stain people's jewellery. There's hydrogen gas seeping everywhere from fissures in the ground, but we soon got accustomed to its familiar aroma.

We spent a good hour or two exploring the Te Puia and Whakarewarewa thermal wonderland, a recreated Maori village with its bubbling mud pools, geysers and silica terraces, guided by a local Maori. Incidentally, lots of place names in Maori beginning with 'Wh' are pronounced 'F', so 'Fakarewarewa".

Some other places worth a visit are the Agrodome, which puts on a fabulous sheepdog and sheep display. This was ideal for my farming group but was also great family entertainment for anyone! We heard lots of interesting facts and had a few laughs as the people conducting the show were well versed in the presentation and introduced us to many breeds of sheep. We then watched a live sheep-shearing demonstration and a fun sheep auction with bidding from the floor and with audience participation too (we didn't bring any sheep home though). Not just sheep, but cute and remarkably well-trained farm dogs brought the event to life, as they demonstrated their ability to respond to commands and keep those sheep and even a duck or two in line!

Some of the visitors fed the baby lambs or even hand-milked a cow. Strangely, our farmers decided to stay away from that opportunity, happy to observe and smile knowingly!

We also headed to the summit of Mount Ngongotaha on a ten-minute gondola ride, where we had some lunch enjoying the fab views and some of us took the Skyline Luge back down, which was great fun on a three-wheeled cart using a unique braking and steering system.

We enjoyed visiting Rainbow Springs Nature Park, where we were guided through the park by a knowledgeable local guide. We saw a kiwi enclosure and were able to see the birds up close, as well as pools full of rainbow trout. Then we moved on to enjoy our 'Hangi' dinner, learning how this traditional feast of meat, poultry and vegetables is prepared and cooked in a pit in the ground before having a brilliant evening eating it whilst watching an entertaining cultural performance. We were regaled by songs, dances and games, complete with the spine-tingling Haka Maori war dance, in which I have to say I was cajoled into participating. I looked so tiny beside these huge warrior dancers. Still, they were impressed by my tongue and the ferocity of my dance moves, apparently.

The next day we continued north, on to Matamata Farm for a milking demonstration and a splendid brunch, with a wonderful welcome as always before arriving at the famous Waitomo Glow worm Caves. The caves were formed over 30 million years ago and are explored through two levels. The upper level is dry and includes the entrance to the cave and formations known as the Catacombs, the Pipe Organ and the Banquet Chamber. On the lower level are stream passages and the Cathedral. We started by following our local guide on foot as he explained about the caves with stories about their history. Then we boarded a boat and were bewitched by the light display as the boat glided silently through the starry wonderland of the Glow worm Grotto … until some wag decided to start splashing everyone. You know who you are!

We ended our New Zealand sojourn in Auckland, the largest city, where we visited the fine museum and the memorial on the volcanic peak, One Tree Hill. Auckland museum displays artefacts from New Zealand's natural and military history. Its imposing neo-classical building sits in a prominent elevated position close to a large public park and is well worth a visit!

In later years we enjoyed Devonport, just across the harbour on North Shore, a ten-minute ferry ride across the bay and worth a trip to look back at the Auckland skyline. It's a little place in a Victorian time warp, with art galleries, small independent bookstores and unique boutique shops. Packed with Maori and early colonial history, Devonport offers some of the best views in Auckland too. At the top of the main street we followed a path that took us way above the town centre to Takarunga/Mount Victoria, the highest volcano on Auckland's North Shore at 87 metres. Takarunga means 'the hill standing above'. I am told it offers the best sunset views of Auckland City and the harbour, a steady 20-minute uphill walk past some interesting little homes on the way. We met up here with a former Salopian and nursing colleague of Dianne's, Meryl Jones, who was enjoying the life here and especially the lawn green bowling scene.

Auckland has lots of restaurants and a thriving café culture. The historic Vic Theatre is Auckland's oldest surviving purpose-built movie theatre and now operates as an independent theatre offering the latest movies, local music and live performance.

Over the years we have explored more of the North Island including its capital Wellington, where I was fortunate to see Brendon McCullum score the first 300-plus score for New Zealand against India at Reserve Basin in a Test match on Valentine's Day 2014. We've seen the giant Kauri trees heading north towards the Bay of Islands and visited the place where the Waitangi Treaty was signed by Captain Cook with the local indigenous population.

On our round-the-world trip we were now on to somewhere totally different. We were due to fly to the South Pacific islands of Fiji, where our home was to be the elegant Fiji Regent Hotel at Nadi.

As we were checking in for our flight at Auckland airport, I became aware that a couple of our farming group, Ray and Flora, were having a tricky time with the check-in agent of Air Fiji. I wandered over to ask what the problem was and the airline was refusing boarding on the basis that there was now less than 6 months' validity on their passports, which was unacceptable to the airline and the Fiji authorities. This was going to be a huge spanner in the works as we would need to go back into Auckland and find the British consulate to have new passports issued, but of course the real hassle was the couple would miss the flight and the next availability was three days later when we would be leaving Fiji for our onward journey. A lengthy and at times fractious debate ensued between the checking-in agent, subsequently the airline manager and finally the Fijian authorities by telephone and after much discussion the airline allowed the couple to travel but would take no responsibility if they were not accepted into the country when they landed.

We decided to take a punt on it and all boarded, we travelled the short flight of just over two hours to Fiji full of concern and as we landed and approached the customs and passport area we were naturally full of trepidation. As it happened, they barely raised an eyebrow and we were all allowed to enter Fiji. Phew!

We took the short bus ride from the airport to the Regent Hotel (now the Westin) on Denarau Island, Nadi, which, apart from its fabulous location has an 18-hole golf course. The rooms or 'bures' as they were called (meaning bungalows, a little misleading as many were two storey) are spread around the spacious grounds and decorated in Fijian style. We were greeted with a welcome drink and everywhere you went you were acknowledged with the Fijian welcome of 'Bula', which is the shortened version of 'Ni sa bula vinaka', meaning 'wishing you happiness and good health'.

We were shown to our spacious rooms, mine being on the second floor of the large vales (vah-lays meaning house) units. Little birds, some of them colourful, and one big black bird with a white scarf-like marking and breast similar to a thrush would come and wait on the balcony rail, hoping you would share some snack or crumbs with them from the breakfast you might take on the terrace.

We loved the swimming pools, one with a swim-up tiki bar! All were bordered by palm trees and as the sun went down, men dressed in native anklets, bracelets, headbands and waist-hugging apparel made from palm fronds ran along lighting lots of oil lamps with flaming torches to the sound of native drums filling the balmy evening Pacific Ocean air and serenading our gathering of lobster complexioned westerners who had not paid enough attention when protecting themselves from the powerful sun rays of the day. As dusk descended and the blue skies became red, silhouetting the trees, we sipped sundowner cocktails, almost wanting to pinch ourselves to see if this was real!

Next morning, we travelled inland just a short distance to the Garden of the Sleeping Giant, sold to Raymond Burr by my distant cousin Keith Watkins who moved to Australia to live in Sydney where my boys live now. Keith actually bought a small Fijian island before he finally made his move to Oz!

Founded by actor Raymond Burr (remember him in the popular television show Perry Mason, the lawyer with his attractive secretary Della Street?) in the year 1977, the Garden of the Sleeping Giant is home to about 2,000 different types of Asian orchid and Cattleya hybrids. It was initially started as a private collection. We enjoyed a relaxing morning walking along landscaped lawns, lily ponds, fountains, the magnificent orchids and a short rainforest walk.

You cannot come to the South Pacific without visiting a small Robinson Crusoe-type island. To get a real feel for the Pacific we took an excursion by boat one morning to the Mamanuca Islands and specifically the exotic Castaway Island, surrounded by white sand, palm-fringed beaches and splendid coral reefs bathed by turquoise waters in which you can snorkel to take in the colourful plethora of marine life. We passed Plantation,

Ron and Dianne on Castaway Island Fiji

Beachcomber and Naistase islands on the 90-minute trip, serenaded by guys and gals in traditional Fijian dress singing Fijian ballads and plying us with beers and soft drinks.

My lasting memory was my merry band of quite elderly farmers complete with red noses from exposure to the Pacific sun, some in white vests with knotted handkerchiefs on their heads trying to step off the boat into the surf and wade ashore as we were greeted by a small band of Fijian native musicians and singers proffering a welcome drink and leis being placed around our necks.

Our penultimate leg was to Anaheim in Los Angeles for a couple of nights taking in Disney. Quite frankly, whilst a great place to break the journey, it was a bit of an anti-climax after New Zealand and Fiji. Still, a great experience and we all got home in one piece following our long flight from Los Angeles to the UK.

*

As the business had developed really well in 1998 the company decided to buy its own property rather than renting, so we could have control over capacity and costs.

During those early days I took a group of what were then termed 'learning disabled' to Disneyland Paris. Our clients travelled in a large coach which had a wheelchair lift and I followed in a transit van full of walking frames/wheelchairs and commodes. Guess who got stopped at customs whilst the coach was allowed to go on its merry way! Ah, the joys of travel! It was a little difficult trying to explain I was not importing these rehabilitation accoutrements, and saying that all of them were for my own use would have been stretching the plausibility for the customs official.

I also took the same party to San Francisco. Two of the ladies in the party typified their lack of inhibition. Hearing some music in the street in Union Square, they stopped outside a big store there and started dancing. They dragged me into the fray and, before long, we had a crowd around us thinking it was entertainment. Taking it in turns they would spin me around until I was exhausted. Lovely memories, I should have had a hat on the ground, we could have made some spending money!

CHAPTER SIXTEEN

Divorce and marriage, two contrasting emotions

Sadly, my marriage to Wendy had ended in divorce in early 1996.

This was a tumultuous time of change. We separated in 1995, but life as a father went on. Initially I was carrying on with my home duties, calling by the former marital home, doing chores, cooking meals and putting the boys to bed or taking them to school. Understandably Wendy found this more and more difficult and in due course I did not enjoy as much time as I would have liked with the boys. Divorce is such a difficult time for all concerned. It is full of pain and heartache.

As time passed Dianne and I got together and in late 1997 decided to get married.

In November 1997 on our honeymoon, or should I say pre-wedding travel on the way to New Zealand where the ceremony took place, we stopped off at the beautiful Pacific island of Raratonga, which is situated just north of the Tropic of Capricorn. We were greeted at the airport by an older guy sporting a headband of flowers and a colourful island floral shirt, playing a ukulele and serenading all of the arrivals whilst other islanders bestowed garlands of flowers on us. We truly felt we were in the Pacific.

There is a good bitumen road that follows the coastline around the island just as if someone had taken a broad brush to create a perimeter to the island's delights. We enjoyed exploring this small piece of tropical paradise by car and sometimes on a motorcycle but it didn't take long to get around as the circumference is only about 20 miles.

We stayed at the Pacific Resort and apart from the beautiful location on Muri beach I remember a darn cockerel used to wake us up every morning rather early and then kept crowing all day long as if he could not turn off his own alarm.

We used to have our breakfast in a delightful thatched rondavel with a stunning view to the reef beyond where we could see the water breaking on the coral. We hired a kayak one day to paddle around a distant island and back. It was a lot further than it seemed! Happy days, though!

One evening we decided to stop and see a cultural show at one of the bigger hotels on the island. Those entertaining asked where everyone was from. Lots of hands went up for some countries like Australia, and there was a smattering from all over the world, but only four of us from the United Kingdom.

After the show we made our way to the front, where the other couple were sitting. Dianne went to speak to them while I went for a comfort break. As I went into the bathroom I saw a chap coming out who I thought I recognised, but they say we all have a double. I thought no more of it, then returned to Dianne to see this same chap and his wife together. It turned out that not only were they from the UK, but they were from Shrewsbury and I used to work with them as director of Travelon about eleven years earlier. How is that for coincidence?

We went to the cinema one evening in the capital of Avarua. The Empire Theatre is a small air-conditioned theatre and we loved it. As I recall we watched 'Con Air' which was a strange one to watch on an island paradise, before flying on to New Zealand and crossing the Tropic of Capricorn. There was an intermission where the film ran down as if it had broken and we all went outside and had a snack or ice cream as it was warm and humid, then went back in and the film started up again … priceless!

We were blessed to marry on New Zealand's South Island at the little village of Tekapo on Lake Tekapo, a fabulously turquoise blue under the distant peaks of the Southern Alps and Mount Cook National Park. Mount Cook is New Zealand's highest mountain at 3,724metres/12,218 feet and is known as Aoraki by the Maoris meaning 'Cloud Piercer'. Sir Edmund Hilary

Church of the Good Shepherd

Ron and Dianne wedding day at the Church of the Good Shepherd!

Bora Bora Cultural beach dance

Another famous guest at Bora Bora's Bloody Mary Restaurant!

first climbed this mountain in January 1948. The name Tekapo derives from the Maori words taka (sleeping mat) and Po (night).

We married in the charming Church of the Good Shepherd, which we'd remembered from our previous visit with the group of farmers and which had stolen our hearts. It's a very small church with a picture window behind the altar looking out over the lake and the rugged mountain slopes covered in golden tussock grass and dotted with Matagouri, or 'wild Irishman' as it is sometimes called, a thorny bush or small tree some of which are near the church and lake.

Dianne walked down the aisle to the lovely track by Alison Krauss called 'I Will' and this is inscribed in our rings. As we were saying our vows a group of Japanese tourists arrived and peered through the windows witnessing our marriage and needless to say photographing us together and with them after we exited the church.

We love New Zealand and following the low-key ceremony we had photographs taken in the delightful lakeside garden of the organist and then drove on the old unmetalled backroad used by four-wheel drives to Queenstown the fabulous resort set on Lake Wakitipu. We had just a small-town car, no four-wheel drive, and soon realised once we had passed through a stream and ascended a road onto the pass through the mountains that the gravel road was very dangerous, one false move in a braking manoeuvre on this narrow road and we would be over the precipitous edge and plunging into the rock-strewn valley, and woe betide us meeting a vehicle coming the other way. Fortunately, we made it in one piece and had a fabulous honeymoon exploring many new and familiar sights like Milford Sound.

On the way home we stopped over at Tahiti then by boat to Bora Bora, where some of South Pacific the musical was filmed. A fitting honeymoon finale in a tropical paradise!

CHAPTER SEVENTEEN

———

Selling the business

Ron Morgan Travel moved to St Mary's Street in 1998. We had a lovely big new office and we went upmarket yet again, perhaps losing some of our less lucrative business. It's always a difficult decision to make, but some of the travel we had been selling was not profitable, more a loss leader and a service. We had to find something that gave us more margin. We decided to concentrate on tailor-making holidays and tour groups, essentially creating our own tours which I escorted.

We were offering a different booking experience, a more spacious and luxurious ambience. We invited people to come in, relax and have a coffee in a dedicated lounge to watch a video. It was a special and exciting occasion where people were spending a lot of money with us.

We had a grand opening, which included having a Richard Branson lookalike to open the office and cut the ceremonial ribbon. He looked so much like the real man, many attending asking for autographs and believing it was him. Apparently even Richard Branson uses him at events he hosts including dinner parties. One of our local restauranteurs attended with a culinary offering for Richard's double and such was his likeness, she believed it was him. We kept trying to interject and explain but never managed to, so the lovely lady left believing she had met the real Richard Branson. We felt a little bad, but also vindicated in our choice of guest.

We started a specialist Florida department, developing this for the sale

Richard Branson -lookalike with Ron at the official opening of Ron Morgan Travel

of flights, car hire and attraction tickets, alongside villas. The year before we opened the new office, we had bought a six-bedroomed villa in Davenport, Florida near Disneyland, the biggest property on the development at the time. This boosted our ability to sell flights, car hire and attraction tickets as we rented out the villa to parties of twelve to fourteen people. We also enjoyed staying in it ourselves. Doing maintenance on it was strangely cathartic.

We rented out some of our office space to a little start-up sports company run by Steve Scott, Ken Dyas and Charlie Hart, father to our former England goalkeeper and local footballing legend Joe Hart. I cannot believe Joe would drift though the office up the stairs to see his dad, never thinking he was destined for football stardom. Like his dad he was a nice fella, very grounded, and a good cricketer, too.

For a short period of time, we had part of the office as a Shrewsbury Town FC desk where town fans could call in for various information. Arthur Rowley, the prolific goalscoring forward, spent an afternoon with us signing autographs for fans who called in. Wolverhampton-born Arthur was a boyhood hero of mine as I was fortunate to watch him score many a goal for Shrewsbury Town and indeed he went on to manage the club, too. He holds the record for the most goals in the history of English league

Ron Morgan and Arthur Rowley sharing a chat at Ron Morgan Travel

football, scoring 434 from 619 league games. Arthur still holds the club record for the most goals in a single season at both Leicester City and Shrewsbury Town, scoring 44 goals in 42 league matches at Leicester in the 1956/57 season and 38 goals in 43 games for Shrewsbury in the 1958/59 season. He is also Shrewsbury's record league goal scorer with 152 league goals.

At Ron Morgan Travel we continued our farming tours with the inimitable Art Barnes as guide. He had escorted us to Canada a year or two earlier and now took us to the western states of Colorado, South and North Dakota, Wyoming, Montana, Utah, Arizona, and into Nevada and California. What a trip!

After selling the business we retained the villa in Florida for a few years and on one trip to the villa in February 2003, we drove to the Gulf Coast to watch the Columbia space shuttle take off from Cape Canaveral. It was a truly awesome sight. Sadly, it never returned, disintegrating on re-entry to the earth's atmosphere.

As we spent many years visiting Florida it was perfectly placed for Caribbean cruises and it was so easy to get some last-minute deals and depart from Fort Lauderdale or Miami for a week or more of floating luxury. We enjoyed the life on board as much as the ports we visited, especially the entertainment when it involved familiar stars of stage and television. The lectures were my favourite, often given during the day, sometimes by someone well known but nearly always very interesting whether they were autobiographical in nature, historical or just educational. One I remember was a talk by the Chelsea footballer, John Hollins

As usual we met some characters. There tended to be a lot of Americans and Canadians on these ships and one day a larger-than-life African

Carnival Cruise ship

American lady burst out of the lift as we tried to enter and said to us, 'Excuse me, can you tell me which way is up?' Puzzled as we were by the question, we were able to point her in the right direction.

Dancers and the show productions, often using material from well-known musicals, were very enjoyable. As the ships became bigger, prices seemed to get cheaper and of course more people cruised. Gone was the first- and second-class divide on ships of yesteryear, and the 'one-class' ship developed a dominance on the open seas with its swimming pools and spa facilities, many restaurants and even a choice of entertainment theatres plus cinemas. The main talking point was the quantity of food and the 24-hour opportunity to consume it.

The midnight buffets with their wondrous ice carvings were legendary as were the waiters' processions to tunes like 'Feeling hot, hot, hot' as they paraded with their flaming Baked Alaskas as the sound of clicking cameras came upon us like a symphony of cicadas!

We also bought some land in the Rockies in Canada with a view to building a house on a golf course above Canmore, a less fashionable but up-and-coming neighbour to Banff. I was not feeling well at times with my ulcerative colitis, probably brought on by the stress of the business. We sold the land far too early, much to our chagrin, before building. Today it would be worth three times as much at least!

Keeping the property theme, Dianne and I bought a townhouse in Queenstown, New Zealand, a place we loved.

We had expanded our staff to eight. My dear old mum had been with us from the start and was a big help in making Ron Morgan Travel successful. Mum was our bookkeeper and had become a partner with Dianne and myself, because by then David and Sue, our first manager, had both moved on.

One sad occurrence happened right outside the office. I happened to glance through our window at a lady as she crossed the road from the hairdressers towards us, then, in the blink of an eye, she was knocked down by a vehicle. Instinctively I rushed out and could see she was unconscious with her head bleeding.

I put my jacket under her head to raise it off the cold hard surface of the road. Someone called an ambulance and I just talked to her while stroking her forehead, as I had been told the last sense that goes is the hearing. Alas, she passed away.

A few days later, her daughters came in and thanked me and wanted to know what I recalled. They wrote a lovely letter to me, and I was invited to the funeral and the wake at Drapers Hall which I duly attended. It was a very sad time and a profound time for reflection on how quickly your life can change or indeed be taken.

We were fortunate to have had little turnover in staff, which was unusual in the travel industry, Debbie Davies had been very loyal, staying with us from the start to when we sold. We had some lovely, caring staff, reflected by the loyalty and satisfaction of our clients. In 1996 Claire Butler (now Moore) had joined us straight from college and she went on to be our manager and indeed is now a partner in Peakes Travel, an achievement so well deserved. A really hard-working, caring and very knowledgeable travel and people person.

Running the business had been a wonderful experience, but the stresses of business and going through divorce and subsequent illness took its toll. Wendy and my youngest son Tom moved to be with Wendy's new partner in Surrey in 2001. This upset me very much that I would not be able to see Tom as often as before. Wendy needed to start a new life and I wished her well but it was a traumatic day when they left. I can still see Tom, his face and hands on the rear windscreen of the car as they drove away.

Queen Mary group

Our Florida Villa

Illness intervened and a medical consultant advised me to pull back from the business. We sold a very successful company in 2001 to Peakes Travel, who wanted another edge to their business and one that was more profitable. Unfortunately, while they were purchasing the business, they reduced the agreed offer on the date of closing, giving us no time to react, in a 'take it or leave it scenario' and as I was in a vulnerable position health-wise we just rolled with it. I would normally have just pulled the plug on the deal but was too physically and mentally exhausted and had to accept that some people do business differently.

My staff at Ron Morgan Travel were outstanding, chosen as much for their character as ability. I insisted they all be retained by Peake's and fortunately Peake's agreed. Only one remains there today, the very capable and affable Claire.

We retained the property on St Mary's street for a while before selling. Peakes moved the business to their office in the Darwin Centre and then relocated to Mardol, where, not surprisingly with Claire's leadership, it is flourishing.

CHAPTER EIGHTEEN

———

Teaching, adapting and 9/11

It was awful selling the business and giving up my travel career. I felt a real sense of loss, like a bereavement, but we knew it was the right course of action. Eventually, after a period of taking stock, I was feeling a little better and stronger and I looked at what else I could do.

For a short while I worked for a day or two a week at Owens Travelmaster in Welshpool, but in reality, I was just trying to hold on to travel as it had been my life, and I knew this was not the way forward.

I had been teaching at Shrewsbury College, the Tech as most of us oldies knew it, on London Road. I had been teaching Travel and Tourism part time since the late eighties, but legislation changed and required education qualifications to teach, even if you had the vocational experience, which I had relied upon for over a decade. I needed to improve my qualifications or get out.

I decided to carry on and took on this new challenge. I sat a PGCE course whilst teaching full time. It was quite an arduous undertaking at fifty years of age, but no doubt it improved my teaching and understanding of education delivery. I met some great fellow teachers too. It was a proud moment having my mum and dad and Di's folks in attendance at the Abbey to witness me receiving my PGCE whilst I and others paraded by dressed in cap and gown.

I had also started doing town guiding for Shrewsbury on a regular

*Ron with his Mum & Dad –
P.G.C.E. award ceremony*

basis, having gone through a training programme and examination for the local tourist office. I loved doing this and sharing my knowledge and indeed continually gained information from reading and picking up anecdotes from other people. Stan Sedman came through this course in the same cohort. He continues to show people around his adopted town, now doing it for donations to good causes. He is a splendid guide and a thoroughly nice guy.

Shrewsbury is an incredibly attractive medieval town, the birthplace of Charles Darwin and with many fine buildings. It's a great place to stay as a gateway to Wales and in particular the rugged peaks and beautiful valleys of Snowdonia. Of course, we have our own rather attractive castle sited on high ground where the River Severn forms a horseshoe around the town, perfectly placed to protect the town from invaders.

Nowadays we have some fabulous festivals, the International Cartoon Festival, Shrewsbury Folk Festival, Food Fairs, Literary Festival and of course the world-famous Shrewsbury Flower Show, which has been going for well over 100 years and in my opinion second only to Chelsea! Percy Thrower, the first major television gardener, helped put us more on the map in the sixties and seventies. In fact, at Luxitours, when I worked there, we used to sell Percy Thrower Floral Cruises, where our Managing Director was accompanied by Percy on a themed cruise with Percy giving lectures. He was always very pleasant as he drifted through the office at 75 Mardol, always saying 'hello' en-route to his meeting with Brian Bass.

As time passed, the Shrewsbury tourist office added ghost tours to our repertoire, which were great fun, especially when we were allowed into

Ron taking a Ghost tour dressed in hat and drizabone coat

old hostelries like the Prince Rupert, the Lion and Nags Head. We had some interesting atmospheric experiences on these tours.

Through the guiding, I met with a production team of former BBC television employees and recorded a 'Haunted Shrewsbury' DVD, including seances and mediums, which, whilst stretching plausibility, was really enjoyable and at times beyond comprehension.

During the late nineties I had qualified as a real estate licensee in Florida and advised and helped facilitate a fair number of holiday home purchases in the Orlando area

When I lived in North Carolina in the eighties I had taken a real estate course and passed and qualified as a real estate salesman (we call them estate agents) whilst in the States you have to pass examinations and continue with examinations all the time you practice in order to retain your licence and continue working in the industry as mandated by law. I am not sure this is the same in the UK real estate business.) I let my licence lapse when I returned to the UK.

However later in life after a brief dalliance with the Portuguese Algarve property market possibilities. I had learned many lessons and saw the burgeoning Florida market as an ideal springboard for me to enter the realtor business as another string to my bow. Sure enough, in 1997 I went over in the winter to take another series of courses and examinations and after a few visits and by passing the quite tricky examinations I qualified as a Florida real estate realtor!

It was a lot tougher than I thought it would be but the bottom line is I learned a lot about not just the machinations of the industry but property law, and all that was needed to make me a knowledgeable and indispensable professional to any would be buyer and indeed seller of real estate. Of course, because we still had the travel business at the time, we also benefitted from selling the flights, car hire, attraction tickets, insurance and other ancillaries that all would be renters would need in Florida.

For a good number of years, the business synergies worked very well and I learned a lot, met some lovely people and sold and rented some fabulous properties, we also sold some properties elsewhere like Spain but Florida was our main market. Every two years I had to take a course online then take another examination to prove I was up to date with any changes in law and other important developments in the real estate industry!

I must admit I failed to conclude some sales when I felt the purchasers were going to be stretched a bit too far financially. It's easy to sell such a wonderful dream, but I figured it could end up a nightmare for them. Perhaps I'm daft, but I need to sleep at night.

Teaching had become quite a drain as more and more paperwork was demanded. I spent most of my time preparing schedules, lesson plans justifying content and pointing out the inclusivity, then reflecting on the results of the lesson plans. I enjoyed teaching and engaging with the students but I fell out of love with the formalities that went with it. It was time to call it a day.

Even after finishing teaching, former fellow lecturers would ask me to come in and talk to the travel and tourism groups to share my experience in the travel industry and do a Q & A. My good friend Angelina invited me on one such occasion to the Shrewsbury Sixth Form College.

Before I started my talk, Angelina said, 'Today we have a very well-known travel personality who has worked in the travel industry for many years. He lives locally and used to have a travel company here.' I was basking in this kind and generous introduction when Angelina asked the question, 'Any ideas who this might be?' Well, I was not sure whether to be flattered or just very amused or both when a young lady at the back of the group put up her hand and offered her thoughts. 'Is it Thomas Cook, miss?'

I never thought I would ever be confused with the legendary Thomas Cook, who, I informed the students, had lived in the nineteenth century. He was credited with being the first package holiday operator when he bought rail tickets from different rail companies for the journey from Leicester to Loughborough for a temperance meeting in 1841. He came to national attention when he conveyed 165,000 people to the Great Exhibition at

Crystal Palace in 1851. Ultimately Thomas Cook offices could be found all over the world, thanks in great part to Thomas Cook's son, John Mason Cook. Thomas Cook ceased operation in 2019, a very sad occasion for the travel world and its many clients.

In 2001 we were booked to take a group on tour to north-east USA. We were due to arrive in September, then 9/11 happened. It was one of those times in life where you probably remember where you were.

I was in the staff room at Shrewsbury College when I heard the news. At that time, having sold Ron Morgan Travel, I was teaching full time. One of my IT colleagues in the room next to the staff room invited us in to watch on television what was unfolding. What we witnessed was beyond words, it was like a Hollywood disaster movie.

At the time one of the chaps there in the room with us did not realise his brother was on one of the planes that had been flown into the Twin Towers. Now to make it even more real we were flying into New York less than two weeks later. We were booked into a hotel which was devastated when the towers collapsed, so we were moved to a property nearer Central Park.

I have been to New York many times and always found the New Yorkers can be a little hard and brash at times but this time they were so receptive, we were even being thanked for visiting as they felt under attack and were astonished that we had still travelled to stay there.

One of the tours was up the Empire State Building, and from the top we could see the remains of the Twin Towers, still smouldering. We could not get close to what we now refer to as 'ground zero' but we did see many of the fire stations and the tributes to those who had responded to the crisis and paid with their lives.

At the end of the tour having enjoyed the New England delights from Paul Revere to the Pilgrims arriving on the Mayflower and lots of lobster and seafood later we put our group back on the plane to the UK where they were met and transferred home whilst Dianne and myself continued on to Florida.

Well, we almost didn't because, unbeknown to me, Di had put two pairs of scissors in my backpack which I took on as hand luggage and, guess what, security, which was on high alert, picked up our faux pas. We were mortified, expecting to be arrested and interrogated but fortunately all they did was confiscate the offending articles. Phew!

We decided to do some travelling so we flew to Chicago and explored this wonderful city before heading on a road trip south following route 66, made famous by the well-known lyrics from the Stones song Route 66 to 'Get my kicks on route 66'. We stopped off in Springfield Illinois to see the home of Abraham Lincoln before visiting famous diners renowned for their 'wieners' then heading west off the 66 towards Hannibal Missouri on the Mississippi River where Samuel Longthorne Clemens was raised better known to many by his pen name Mark Twain. Hannibal was the inspirational location for books like The Adventures of Tom Sawyer and Adventures of Huckleberry Finn. We then headed back to route 66 moving south to visit St Louis where we took the elevator which is very small and claustrophobic around the famous St Louis arch then travelling through Kansas and Oklahoma.

We headed to Dallas to explore the events of the JFK assassination taking in the book depository and the grassy knoll. We also explored the delightful South Fork Ranch, home to the fictional Ewing family in the television series Dallas. Austin capital and famous for its country music bars was a lovely city. On to San Antonio home of the world famous Battle of the Alamo, which took place on March 6, 1836. On a lighter note, the walk along the Paseo del Rio studded with restaurants and bars is a delight especially at night. Next stop was Houston where N.A.S.A. have a facility and immortalized in the Apollo 17 contact with them 'Houston we have a problem' and we were fortunate to watch a rodeo followed by a Willie Nelson concert.

From Florida we took a road trip to New Orleans, taking in the delectable St Charles Avenue home to many plantation houses and well explored by the tram which operates the length and well worth the experience in itself. Bourbon Street and all the jazz and blues clubs are a must in the French Quarter, I would recommend Pat O'Briens for the fabulous and very potent Hurricane cocktails, be careful one might be enough. Memphis was a must to take in Elvis and his home at Graceland before exploring the music scene on famous Beale Street and a visit to Sun Studios A bonus was going to and through the small childhood home and birthplace of Elvis at Tupelo. Nashville was a mecca for us to see the home of country music. Rymans auditorium the original venue for the 'Grand Old Opry' did not disappoint nor did the many bars such as 'Legends bar' with country music of up-and-coming artists which lured you in from the sidewalk. Dollywood like it or loathe it just had to be done before finishing up at Chattanooga home of Glenn Millers well known tune, 'Pardon me boys is that the Chattanooga choo choo?'

Includes: Cozy Dog Diner route 66; Hannibal Missouri; South Fork Ranch; Book depository Dallas where JFK was shot from; Ron at the Alamo; Ron in cowboy hat Houston; Graceland; Ron at birthplace of Elvis in Tupelo; Tennessee Sun Studios where Elvis recorded; and Chattanooga Choo Choo

CHAPTER NINETEEN

My true calling!

After a year or two's break from travel arranging, some of my former clients, now friends, requested I put together some overseas jaunts again. So I began creating some ad hoc tours on a regular basis.

With this continuing demand, it seemed logical for me to arrange some more tours and finish teaching. It was the best thing I could have done. I was feeling a lot better as I had less stress. I started doing one short-haul tour a year, sometimes two, and added a long-haul tour but kept the groups small and manageable with about fifteen to eighteen people.

It was high specification and very personal, while visiting lots of new destinations and some I had previously visited. It was brilliant travelling with such a like-minded and lovely group of people!

Not only did we travel to wonderful places but we had some really challenging experiences. Climbing Mount Kilimanjaro, the highest free-standing mountain in the world and the highest mountain in Africa at 19,340 feet above sea level, was one of them.

We were a small group of eleven, aged thirty-eight to sixty-five, most of us in our late fifties or early sixties, and one guy had undergone a triple by-pass a few years before. Our first sight of 'Kili' was jaw-dropping. As we took in its majesty rising from the savannah below the cloud cover cleared and the glacial area high up on the mountain came into view. We were really excited to get started.

We had done no major training for this, just a fair bit of walking over the Shropshire hills with the odd foray to climb in Snowdonia, North Wales. In fact, it was only decided over dinner one evening and probably too much alcohol on the part of some that we should go to Africa and then somehow, we decided to climb Kilimanjaro. Within a year we were there.

As plans for this tour had developed, I had expected the enthusiasm to wane and reality kick in, especially when I shared the facts that on average someone dies every three weeks on Mount Kilimanjaro, seventeen deaths a year and around 40 per cent of those who attempt the climb fail to reach the summit. On those odds at least four of us would not reach Uhuru Peak.

I had not felt very well the few days before the ascent, having picked up a bug. In addition, I had damaged my Achilles tendon before leaving England and been to a physio, who helped immensely, but I was wary as to how I would cope.

After the long flight via Amsterdam, we spent a night at the Kilimanjaro International Airport (KIA) Lodge. This hotel was a short, bumpy drive from the airport in two rather battered vehicles. As I pondered the mountain trek ahead and after reading the statistics of casualties and even deaths on the mountain, I hoped KIA did not stand for killed in action.

The next day as we drove along the open road, we passed people driving their livestock to market. These locals were dressed in all manner of colourful garments going about their daily business nonchalantly and slowly as the heat of the day increased. Initially the sunburned savannah land seemed as though it could barely sustain the herds of animals, but before long the vegetation became more verdant. As we neared Marengu, banana plantations appeared. In the distance we had spotted what we thought was Mount Kilimanjaro again but it was in fact Mount Meru, the second highest peak. Then gradually as the cotton balls of cloud dotting the blue sky cleared, we had our first close-up sighting of Kilimanjaro as she shyly showed her snow-capped peak before hiding it again behind the clouds, as if luring our intrepid bunch of trekkers towards our challenge.

After about ninety minutes' drive, which included a short stop in Moshi, the capital of the Kilimanjaro region with a population of around 200,000 where we exchanged our dollars for some Tanzanian shillings, we reached the Kilimanjaro Mountain Lodge, a fine hotel with a swimming pool and an attractive African bar and small restaurant.

Here we had a pre-climb meeting with our tour guide John, where some

Marengu Gate, starting point for the Kilimanjaro climb

tips and information were offered. Based on John's advice at this stage, we decided against using Diamox, a drug used to alleviate the effect of altitude sickness, yet most of us took a supply up the mountain with us. Any dietary requirements were logged to pass to the trek cooks. Extra water was going to be carried by the porters as there was none available on the mountain.

We awaited a local supplier who was bringing some mountain gear people could hire if they so required. Meanwhile we consorted with two very old, huge tortoises which were local residents on the lawns of the hotel grounds. I felt like Charles Darwin.

Eventually the kit supplier arrived with a motley assortment or rather poor and worn gear, sleeping bags, duffle bags which porters would carry or in which you could leave stuff to be collected on our return. There was walking poles and down jackets, but some were just like ski jackets, not really fit for altitude. I'm glad I took my own.

We now had the rest of the day free before heading to the mountain the next day. We had a team photograph taken on the hotel steps which looked for all the world like an old colonial hotel and in sepia perhaps we could have aged the picture to a bygone age.

We met with two of the mountain guides who do this mountain guiding to supplement their income for their family, and they proceeded to take us through the thick vegetation in the immediate locale of the local village called 'Kyalla' and down to falls, tramping through a banana plantation, where we were told about the local vegetation and discussed what to expect on the mountain.

We paid a small entrance fee to the area. It was quite a steep descent to the foot of the falls and the walk back up stood us in good stead to start the legs working for the trek tomorrow. There were lots of steps and tricky rocky areas to scramble over in places.

On arrival back at the hotel a local family pre-wedding celebration was taking place around the pool. It is tradition there to have a send-off for the bride before she marries the following day. A walking band played them into the hotel in ceremonial style.

For us we had more serious planning to take care of with the big ascent beginning tomorrow. We packed and repacked our kit bags, which the porters would be carrying up the mountain for us, as well as our day packs, which we would carry. We had to discard and leave in the hotel more than half of what we had brought with us due to space constraints. Big decisions were on the three pairs of footwear I had brought with me. I decided on the most robust boots and a pair of trainers for downtime at camps.

Finally, happy with our packing, we went for a swim, followed by a three-course dinner and an early night.

Rooms at the lodge were spacious and well-appointed with large bathrooms, a fridge and a flat-screen television, hairdryer and kettle, luxuries we were not going to enjoy for at least another week after that day. The room had an attractive diamond-window effect with opaque glass between the bathroom and bedroom.

The bed was without a headboard and had a wooden lamp with cream shade either side on small wooden bedside tables that matched the desk set below a mirror beside the luggage bench. The wooden parquet floor led through the brown and cream curtains with elephant motifs to a small balcony with wicker chairs either side of a glass-topped wicker table. It felt very African colonial in style.

I stared out of the window from the comfort of room 204 across to the fabulous yet daunting vista of Mount Kilimanjaro and wondered what the coming week would bring.

After a poor night's sleep and some stomach cramps, I had breakfast and we readied ourselves for a 10am departure. We had started to pick up some of the Swahili language, words like 'asante' meaning 'thank you', 'hakuna matata' made popular by the Disney Lion King movie, meaning 'no problem' and, of course, 'jambo' meaning 'hello' or 'welcome'.

John, our local liaison guide was proving to be more and more unreliable. At first we took it as the laid-back African way, but gradually, disappointed with the quality of the kit he'd arranged, we were starting to be concerned about his stewardship. Our concerns were heightened when he arrived 45 minutes late.

He had promised to bring certain small items for the trek but failed to deliver these, he made calls and procrastinated, but we had to get going so pushed on towards the entrance to the trek at Marengu gate at the foot of the mountain.

Once more pfaffing around delayed us and culminated in John asking us for another $606 for the guides, porters and cooks because the office had only sent him money for ten of us, not eleven. Not wishing any further delay I paid on a credit card. John said he would make sure the money was sent up to me in cash by the time we reached our first night destination, Mandara Huts, but he and the money failed to materialise.

Finally, our guides Alex, Djuma and their leader Bariki introduced themselves and gave us a pep talk. We took photographs and then at last set off on our six-day trek at 12.45pm. We were preceded by a group of porters who had our kit bags and water and the cooks who carried all the utensils and food. There were no animals, everything was carried by these men.

So a long human chain ascended through the jungle. The rocky and soil paths were well defined as we passed small waterfalls in the rainforest. I felt awful but tried not to let that show and affect the rest of the group. I dropped in at the end of the line, which was an excited snake of chatter but I soon dropped off the pace, falling back, and this was just the first day.

I really lacked energy and tried to infuse my body with water from my camelback water bladder but it tasted foul. I was so relieved to reach the lunch break point at 2.15pm.

The rest and fabulous lunch the cooks made on the trail reinvigorated me. An excellent spaghetti bolognese, followed by some banana and mango and a fruit drink, were just the ticket. I struggled to eat, but what I did consume proved very important. There were some long-drop toilets here as this site was a designated picnic stop.

I set off with renewed energy, and with the aid of a sip of water here and there and the occasional snack bar, I stayed with the group, who stopped occasionally to take in some of the wildlife, like the blue monkeys we saw swinging in the tree canopy, their raised foreheads exaggerating the monkeys' inquisitive gaze at us but they were too skittish to approach. Finally, the wonderful sight of Mandara Huts hove into view as the narrow trail opened into a jungle clearing at 4.45pm. We had reached camp one!

I was so happy to have made the first camp. The climb had seemed to get easier for us all, we had terrific guides led by Bariki and a brilliant team of cooks. Our party consisted of over thirty, with porters, cooks and guides as well as ourselves.

We were located in an upstairs dormitory on bunks in hut eleven and the camaraderie was excellent. We were all tired but the banter was good. A couple of us had started taking Diamox which is supposed to help with altitude sickness, but unfortunately it is also a diuretic. We used the toilet facilities which surprised us with two western toilets and a shower, but in reality, we were all on wet-wipe routine by now as far as washing was concerned.

Our group had some popcorn and biscuits and a cup of tea and then much to my surprise went on a walk to the nearby Maundi Crater. I could not understand the wisdom of this until I was taking in the vista of two countries gazing down over the Tanzania/Kenya border. We then took a short stroll into the crater and back to the huts.

Dinner was so welcome, a hot soup, followed by boiled potatoes with fish in a vegetable sauce. These guys could cook, it was nutritious and tasty. Solomon our cook was a star, all dressed in a white apron and chef's hat with the 'Seven summits' motif on them. Djuma had brought me a walking pole up from Marengu and he taught us some more words, 'araca, araca', 'hurry, hurry' and 'twendi', 'let's go'.

There was a mix of nationalities staying here, Australians, Americans and Canadians amongst them. We completed our ablutions, wrote diaries and turned in at 9pm hoping for some sleep.

As we settled down after a hilarious exchange of observational comedy from the day's activities, it was all going quiet when someone broke wind. No one said anything for a few seconds, then one of the party said in a sheepish voice, 'it wasn't me'. Well, that was it, we all broke into howls of laughter.

During the night Matt and I, the Diamox imbibers, made more than one visit to the loo, which was a trek in itself. We had to negotiate a ladder down to the ground floor then walk down a jungle path to the little toilet block, not knowing quite what we might meet on the moonlit strip of land, then return as quietly as possible trying not to disturb our fellow walkers. By the fourth nocturnal expedition we were well practiced, even if it was not easy keeping quiet wearing heavy mountain boots.

We heard the staff making breakfast below around 5.30am and all of us were up for 6am. None of us had that much sleep. By 8am, all packed up and refreshed and recharged after breakfast, we contemplated the day ahead. We planned to ascend another 3,346 feet to our next overnight at Horombo huts, 12,200 feet above sea level.

We had decided to take the Marengu route, often referred to as the 'Coca Cola' route. All other routes involved tented camping. It was nice having a solid structure to sleep in , even if it was basic, as you would expect. It was certainly good to reach camp each evening. The meals prepared by our cooks were excellent. Even on the treks the cooks would go on ahead and cook a great meal on basic stoves. They set out a table with napkins at picnic sites en route and made our ascent as comfortable as possible. Using the toilet, of course, was al fresco, and even at the camp areas after our first night they were long drop. One just before the last camp at Kibo was in the middle of nowhere and the drop was over the side of an escarpment, which gave the description of long drop an added meaning!

So, our line formed and at 8am we headed upwards, stopping every hour or so for a drink and a snack bar. By midday we were ready for lunch. As per the previous day's pattern the cooks and porters went on ahead, so when we arrived at a well-rehearsed picnic spot, everything was laid out. A terrific buffet of chicken, pancakes, boiled egg, bread and fruit, all freshly prepared and followed by tea and coffee. How good is that for a trekking meal? It beats trail mix and a snack bar and lifts the spirits, but also fills the bladder and bowel. With no public toilets available, it also makes you very aware of searching for a suitable spot to relieve your inner tensions, shall we say!

On the way from Mandara we witnessed the terrain change from thick jungle to small scrub-like bushes and lobelias, of all things, like some sort of huge red-hot poker here and there probing upwards into the thinning atmosphere as if searching for oxygen.

There was not much wildlife in evidence. The noisy, white-necked, almost vulture-like raven scavenged around the camp areas. We saw some swifts and some small brown birds. There was the occasional glimpse of the shiny green coat of a malachite with its scarlet tinge. Sometimes a long-tailed bird would swoop down onto the trail and the surrounding area. Much to the concern of one or two in the party some grass mice appeared, small brown furry creatures with four black stripes and a tail. Thank goodness we didn't see the mole rat or climbing mice.

We trekked steadily, albeit 'pole pole' pronounced 'poley poley', which meant slowly slowly, as per our guides bidding. After six-and-a-half hours we turned a corner and there were the Horombo huts, an oasis of relief. We crossed a small bridge over a gulley to reach them and in this depression saw a fabulous plant called a senecio, a cross between a palm tree and groundsel and only found here and Mount Kenya.

We signed in at reception and walked past the helicopter landing area and some huts to reach the huts allocated to us, two sets of bunks in each, shall we say cosily laid out, not offering any privacy.

We dropped our bags, visited the toilet blocks and reunited in the communal refectory to replenish our reserves with some popcorn, nuts and the ubiquitous cups of tea. We took in the thinning atmosphere, as clouds rolled in and out and we started to realise it was getting a lot cooler. It is often five days this way to the summit. We decided to stay an extra night at Horombo Huts and climb up to Mawenzi saddle as a practice and return to camp to help us acclimatise to the altitude.

One or two were feeling a little nauseous and had headaches by now. A few more had taken the drug called Diamox, which was supposed to help. Of course, you never know how you would have been if you had never taken it. It was getting pretty cool at night and you tended to sleep in your clothes as well as tucking yourself inside a sleeping bag.

We headed back to our huts to organise ourselves for a two-night stay before regrouping for dinner at 6pm. Dinner in the gloaming was the order of the day, as the power was out in camp. However, we still had a fabulous supper of pumpkin with garlic soup, followed by rice, green beans and carrots which accompanied a tasty beef stew. Watermelon and a cup of tea set us up to recover and rest. The generator still had not kicked into life, but before darkness fell, the clouds cleared and blue sky appeared allowing us to clearly see the Mawenzi saddle and Kibo peak above us. Our guides Bariki and Alex answered our questions. How far? How long? What to wear in the days ahead?

Tired, we left the candle-bedecked table of bonhomie and retired to our huts to read and write our diaries, before readying our kit for the following day. It was just 7pm. It was at this stage that we really needed the warm sleeping bags. Only then did we find out the ones we had hired were useless, not only were they second-hand but they were certainly not clean and to say my sleeping bag was child-sized was an understatement; it barely came to my waist. Di and I swapped and made the best of it. A lesson learned was to test all the kit before departure.

During the night I had to make a visit to the toilet block and was blessed with a superb moonlit silhouette of the mountain top against the blue velvet sky dotted with thousands of stars. It was magical, the stillness and solitude and the cold clear air added to the special moment.

As we were only going to do a day return jaunt the following day, we could just take a light day pack. We watched as the sun rose over Horombo Huts and looked skywards towards Kibo as its glistening snow-capped peak came into view. Below us was a feather bed of clouds blocking out any view to the mountain base.

We picked our way over the rocky terrain towards the canteen, to be fortified with a breakfast of maize porridge and honey, bread, jam, pancakes and eggs. By the time breakfast was over and we assembled to ascend, the mists had rolled in again and the constantly changing microclimate tested our resolve.

Feeling like true adventurers we fell into line and entered the mists ahead, walking relentlessly up the well-defined path, heads down, watching our boots, one step after another. It seemed very slow progress to a point that initially had looked quite near our camp earlier that morning, but we understood this was part of the plan and training to cope with the altitude. We finally reached Zebra Rocks with their stripes of different colours after about twenty minutes of steady trekking, a midway point to a ridge at 4,100 metres, 13,325 feet, which took about another twenty minutes. Before taking another rock- and shale-covered trail back to Horombo Huts. We had taken a snack with us and with stops punctuated with a little fauna and rock inspecting. The whole foray took about three hours at about 1,000 feet higher than we were sleeping. It was not an arduous day and formed part of the rest and recuperation time. It's just a pity it was so misty, but that's mountain weather.

It was time for lunch by now and the crew brought us bowls of warm water to wash outside, Which was bracing given the temperatures at altitude. Chicken, chips and coleslaw was perfect followed by orange, watermelon and tea. Our afternoon consisted of mooching, resting and repacking our gear ready for the next day's ascent.

By 4 pm we were eating again, such was the desire of the guides to make sure we had our bodies well fuelled. It was also a great time to all get together and shoot the breeze and excite each other with tales and expectations of the climb. We had our popcorn, nuts and tea and I had taken some questions for a quiz. We duly split the party into teams and had a sort of trek team challenge, which proved fun and before we knew it, we were having dinner.

By 7.30pm we went back to our huts full of trepidation for the day ahead, knowing we were reaching the sharp end of the challenge literally and metaphorically speaking. We checked our packs yet again trying to gain every advantage we could from working out access to the things we would need most on the test ahead. We readied our water bottles.

I had a fitful night's sleep and had two toilet breaks, more through nerves than anything else, although the cold wind around your nether regions often left you feeling like you needed to go again before you returned to the hut. You soon get used to sharing dormitories with your teams, oblivious after a while to all the sounds and smells of bodily functions, be they snoring or from a more odorous cavity. Unlike back home you do not have to think about what to wear as you often do not undress and sleep in all of it to keep warm. We would wash with wet wipes in our sleeping bag, just to smell as sweet as possible.

A cold misty atmospheric early morning greeted us. Warm bowls of water were brought to us by the porters and we bathed, before using the primitive privy, or the western-style toilet albeit with no lock on the door. The hole in the ground usually had its own considerations as you had to avoid standing where people had misjudged their bodily evacuations

Onwards and upwards, we headed to the last planned uphill night on the mountain at Kibo Huts. Departing at 8.30am, for some reason I felt fatigued, perhaps the altitude was affecting me. We reached an area called 'Last Water'. We now knew there was virtually no water other than what had been carried up there in bottles. We became well practiced in eking out our water, drinking steadily which was important in staving off the altitude sickness but also with conservation in mind.

Terrain had changed again to alpine desert and knew we were heading towards an icecap. We stopped by a group of big boulders by the Mawenzi saddle and had some drink and snacks to keep up our strength even though we felt nauseous. We watched as the grass mice scurried around our feet.

I was fine until we stopped to eat at some picnic benches just within sight of Kibo Huts, when something came over me. It had been a lovely meal with terrific views down the valley. On the last little ascent up the trail, I started to feel unwell.

Perhaps seeing the stretcher pass us evacuating a lady with a very swollen head and face, presumably with a cerebral oedema, and remembering others we had seen at Horombo who had had to turn back was sub consciously getting to me.

The stretcher had a bicycle wheel centrally placed and was being rapidly wheeled down the mountain by two very athletic fellows to assist in the recovery from the altitude effect and being taken down to a level where a patient could be extracted off the mountain to a hospital at Horombo Huts just over 12,000 feet where there was a helicopter landing area.

This one event reminded us all of the daunting task ahead. When I reached Kibo Huts at 15,520 feet (4,730 metres) above sea level I felt awful and just lay on the bunk. Others were feeling a little worse for wear too. For the first time I considered whether it was wise to go on. When I did stand, I was swaying a little, I was struggling to think straight and put things in perspective, but I forced myself to eat something and drink some water and tea even though I did not feel like it. It's very difficult when you feel so sick.

After a few hours' sleep, the lights snapped on and we rose rather groggily around 11pm. Towards midnight we had checked all our gear and rucksacks yet again although it was hard to concentrate given the altitude effect. Our head torches had fresh batteries. We ate and drank once more.

In a zombie-like state I had found the energy to put four layers of clothing on my lower half and five layers on my top half. We tried to help each other and I was glad of Dianne's and others' assistance. Fully clad I sank back onto a bench trying to rationalise the wisdom of what I was about to embark upon…

Then without thinking we filed out into the bitterly cold and strong wind that was snapping at anything that was loose. We could hardly walk, I felt like the proverbial 'Michelin man' because of the number of layers of clothing, let alone the lack of oxygen, which seemed to have forsaken our lungs.

We lined up and in a long snake of eleven hopefuls and three guides, Bariki, Alex and Djuma, we started the final ascent to the summit. I was on autopilot, I'm sure if people had looked at my eyes that would have seen a glaze. Dianne was thinking I should not do it, I'm sure. But heck, do you think I was going to allow the other ten to summit without their leader? No way!

Every minute seemed like an eternity. Dianne had asked the biggest man mountain of a guide, Djuma, to look after me and keep an eye on me. He nodded and said, 'I will look after papa!'

The stars twinkled above in the dark canopy and a full moon started to brighten our path as we reached the start of the immediate ascent. The wind seemed to drop, our head torches appeared ahead like a voluminous caterpillar, like a necklace of seeing eyes. Soon the torches became superfluous as our eyes adjusted and mother nature's light from the moon lit up the lunar landscape ahead.

It felt like I was in a long line of prisoners being sent off for hard labour. I got into the zone, head down and plodding upwards one foot after another, slowly but surely.. My bowel was in inertia, my head the same. The rarefied atmosphere took its toll on all of us, the altitude is a silent killer.

It does not help the mind when as we journeyed towards the ridge, we witnessed people being rushed down the mountain on stretchers to a lower altitude. Occasionally our line would stop and the guides would insist we had a drink. We looked up and watched as a party who had started ahead of us weaved back and forth above us, just the line of their head torches visible against the night sky.

However, the huge full moon seemed to gain in luminosity and became so bright, once our eyes had adjusted to the ambient light, we didn't need the head torches at all.

We passed another small group who had stopped. It was evident a woman in their party was incredibly disorientated, but we never found out if she made it or not. We reached Williams point at 5,000 metres and in spite of having trekked for two hours we had only ascended 230 metres. I had lost track of time until then, the constant zig-zagging on the difficult terrain, stepping forward and occasionally sliding back mind-numbingly discouraging.

After some time, we rested at Hans Meyer Cave under a large rock overhang. The guides brewed up on a little stove they had carried with them along with the water and made us all a welcome cup of tea and encouraged us to have a biscuit or some sort of snack. Hans Meyer was credited as being the first European to reach the summit of Kilimanjaro in 1889.

We set off again after being told we were only half way to the ridge called Gillman's Point. It felt like we should be near the top by now, but heads down again we gradually ascended the narrow hard ash and stone path, 'pole pole', the mantra ringing in my head as I watched my boots being placed one footstep after another. It was becoming a real mental game more than physical.

The higher we went the colder we became. It was so cold the plastic lines from our water bottles were freezing up. If we wanted to get something from a pocket of our rucksacks we struggled as we didn't want to take off our gloves. Most of us had an inner thin layer glove, a thick glove over that and finally a large mountain mitten over those. It was freezing, we were so fatigued, it was becoming a gargantuan effort to do the simplest things. Our guides often helped us warm up the line from our water bottles to enable a little water to flow. As well as helping us, as they constantly encouraged us to drink and snack as much as possible to keep our energy levels up.

My nose would not stop running as we went higher. I was even dribbling like a geriatric, initially you try and wipe the excretion away with your mittens with limited success, then you just give up, exhausted by the effort. Every now and again the lead guide would shout YO! And we would have to respond, monitored by the other two guides strategically placed in our procession of pain. I barely had the energy to respond.

Eventually, we stopped to rest standing on the trail. Still in line, we gazed upwards and like an epiphany we saw the night sky appearing over a foreboding wall of rock, which we believed and hoped was the summit.

With renewed vigour and filled with hope we plodded on to this summit. We summoned every bit of strength we had, and as the terrain became more unpredictable, we stumbled often on the rock-strewn path. To make the steepness easier to deal with, we zig-zagged more and more along the trail, leaning into the mountain on the track, just a boot-width wide now. On one of the twisting turns, however, one of the party was so mesmerised by the constant grind and debilitating altitude, they just kept walking off the track, oblivious to the rest of us who had turned the other way. If it weren't for the guides she could easily have walked to her death.

Finally, we stumbled over the last few rocks onto the ridge and entered a nirvana of success. We were at the top, there was euphoria and hugs all round from this small band of humanity focused on a common goal. Triumphant photographs were taken with fumbling, cold hands released from the cover of gloves but quickly reinserted as we stood in front of the Gillman's Point

Group picture at the Summit of Mount Kilimanjaro

sign, which proclaimed 5,681 metres (18,638 feet), over 1,000 feet higher than Everest base camp.

As I had become ever more exhausted mucus had hung in long strands from my nose and I had dribbled onto my neck scarf buff as my mouth hung open gasping for breath. One photo showing me with an icicle ¾ inch long hanging from my lip where I had dribbled without being aware has become legendary.

We had reached this peak as the sun was rising. We had managed to summit before sunrise … or so we thought.

Djuma started walking off along the ridge, saying, 'Please follow me, we must keep moving, we will get too cold.' We all looked at him and someone said, 'But we're at the top.'

'No, no,' he said. 'This is Gillmans Point, we have another ninety minutes to go.' With that he turned and walked away leaving his message on the wind.

One of our party just stamped her feet and refused to budge like a precocious three-year old saying 'I'm not going'. I wanted to do the same but was too exhausted to summon any sort of response. One by one, we all fell in line and headed across the ridgeline towards the real Mount Kilimanjaro summit of Uhuru.

I was moving so slowly, Di and Chris Mc stuck with me. I was determined to make it, but it had to be on my own terms. I counted eleven slow paces forward then rested then repeated. I had little left in the tank.

My fellow trekkers were brilliant, Chris started shoving Kendal mint cake into my mouth and with gloves on it was difficult to get to my mouth. Gradually the sugar rush helped and I became reinvigorated and stronger, warmed by the sun's rays and the thought of success. Edmund Hilary had famously taken Kendal Mint cake to the summit on his conquest of Everest Ron Morgan was now doing the same on Mount Kilimanjaro!

At last another green sign came into sight. My spirits rose only to be dashed, as this was Stella Point, only 705 feet from Gillman's Point. It had felt like an eternity getting here.

Some trekkers who were descending passed us and offered words of encouragement. I thought of my dear old dad who had passed away a couple of years before and asked him to give me the resolve and energy to carry on.

Finally, like a mirage, we saw the end of the rainbow, a sign signalling our journey up towards the stratosphere was almost over. A few steps from the top we witnessed the morning sun shining resplendently upon the glacier down to our left. Djuma came back and wrestled my pack off my back and we stumbled to victory. I allowed myself a few tears as I assimilated the wonderment of the vista and marvelled at the achievement of standing on the rooftop of Africa with my wife Dianne and good friends. We had conquered the highest free-standing mountain in the world! As the photograph on the top of Mount Kilimanjaro shows, we were exhausted, elated, proud and emotional. All eleven of us had made it!

On the way down we felt like victors and revelled in our achievement. We were conquerors!

About thirty minutes from the bottom, I had this sudden urge to evacuate my bowels, but despite heading into the bush in a burst of panic nothing happened. As time passed my bladder and bowels were telling me the matter was more and more urgent, I felt like I was going to burst, yet the body had reached inertia.

In distress, I started to run with my pack on, leaving the guides behind. A couple of young African lads appeared on the trail and tried to sell me something. I frantically said, 'I can't stop, I'm in a hurry.' These lads tried to run alongside me and were very persistent, but even they were struggling to stay with me, such was my desperation.

Eventually I threw a couple of dollars towards them and they stopped to pick them up. I hurtled downhill, finally reaching the park headquarters and a western toilet, the relief was immense, but it was short lived as I still could not go. All the time our guides had told us to stay hydrated going up the mountain, but in the euphoria, I had forgotten this advice which became a mantra for the group going up but not coming down and now I was paying the price.

Taking copious photographs back at the Kilimanjaro Park gate and after being serenaded by the whole crew we tipped them well for their fabulous guiding and for the porters and terrific cooks in Freddie and Suleiman and said our goodbyes, even sharing a glass of bubbly.

We travelled on the coach an hour back to our base hotel near the airport. I still could not urinate and that hour felt like forever as I felt like a balloon was expanding inside me.

On arrival at the hotel near Kilimanjaro airport, I drank bottle after bottle of water, hot tea, then started on the beer, I thought I was going to explode. Then finally the dam was breeched and the bladder was munificent. I knew how a camel felt now! So the moral of the story is drink plenty at altitude going up and coming down!!

We were joined by another nine friends who flew in from the UK. We went on to do a safari of the Tarangire, Manyara, Ngorongoro Crater and Serengeti before taking a small plane out of the bush to Zanizibar and a delightful time of relaxation on the beach.

I was exhausted after the Kili trek, my face and body clearly portraying this. I even lost a massive filling out of one of my molars during the safari. It was probably all that bouncing around in the Toyota Landcruiser, but what a wonderful experience it was with three excellent guides, Mo, Dennis and Nas. Of course, we saw lots of wonderful animals and not always from the land. We soared over the Serengeti in a balloon, which was magical and highly recommended. As we floated quietly and serenely above the many animals below, witnessing them going about their daily business, it was as if we were spirits watching over the earth. We flew silently over herds of wildebeest, prides of lions and so much more, the animals occasionally startled by the whoosh of the gas from the balloon.

Serengeti entrance

On landing we were directed to a colonial-style camp which had been set up in the bush by the balloon company where they served us breakfast and even supplied a toilet tent.

In fact, I could have done with that tent on one of the game drives. We had stopped really close to a pride of lions, who were lazing in the midday sun. I was really desperate to go to the toilet again and checked with the guide how far it was to the nearest facilities and he said about fifteen minutes. Relieved, I thought to myself I can hold on.

A few minutes passed, then our guide said, 'You're all enjoying this so much, shall we have our packed lunch here watching the lions?'

Well, this distressed me on two counts. One, I was not sure if us fattening ourselves up in front of the lions was a good plan. But two, the bladder predicament meant I was contemplating getting out of the vehicle to relieve myself, such was my desperation. I suffered in silence as everyone excitedly devoured their sandwiches whilst viewing the lazing lions, taking many photographs.

Fortunately, the old Kilimanjaro spirit kicked in and the seal stayed that

Balloon landing on Serengeti

way, until relief from Mafeking arrived in the shape of a rest stop and tourist facility almost an hour later. I did fear inertia once again but instead did a version of 'fireman Sam' arriving at a blaze.

One of the things I love about travelling is hearing words that we often use at home and finding out their true meaning and origin. Safari is a case in point, meaning journey.

Our safari was fantastic, providing so many memories, such as the prolific birdlife dominated by the beautiful lilac-crested roller with its long tail plumage that we saw these in abundance. I also loved the prim and proper secretary bird, a large stork-like bird with very long tail feathers. But it was the animals who were the stars, whether it was a pride of lions chasing a warthog, a ring-tailed bush baby, giraffes, the leopard lying languidly in the tree with its kill beside it, the large herds of elephant, there was so much. In the night we would hear animals below our hut which was perched on stilts. One morning Dianne was in the rather quaint lodge-like accommodation built of local materials when she realised a giraffe was at eye level looking at her through the break in the thatch that doubled as a window.

*Breakfast alfresco on the
Serengeti plains*

We loved all the parks but it was particularly special seeing a black rhino in the Ngorongoro crater. The amazing African extravaganza continued when we were driven to a dirt airstrip in the Serengeti and flown out by small plane, changing in Arusha before flying to the exotic island of Zanzibar where we stayed at the delightful Pongwe Beach after a transfer of 45 minutes.

On the flight dusk was approaching and we were sitting right behind the pilot and co-pilot. After a short while it was quite clear to us the plane was circling. We thought we were in a holding pattern until we realised the pilots were having trouble finding the airport and landing strip. A debate was taking place trying to decide if it was a road with streetlights they could see or a runway. Strangely their radio communication with the control tower did not seem to be helping. We kept an eye on things in silence until we felt we should become involved and offer our suggestions. Finally, agreement was reached as to which area was the runway. Ten minutes later we landed safely on Zanzibar. A rather bizarre and somewhat perturbing yet humorous experience, as it all ended well. Most of the passengers on our small plane were oblivious of the confusion.

Our Zanzibar hotel was situated on a picture-perfect beach with white powdery sands, turquoise waters and shady palm trees – some of which had hammocks strung between them.

Whilst there were some suites our beachfront and garden villas were rustic and comfortable. The rooms had a coconut palm thatched roof, the leaves sourced from their own beach, lime-wash walls and hand-crafted wooden doors. Within the rooms themselves there was hand-crafted furniture and a large en-suite bathroom with walk-in shower room. All had mosquito nets and ceiling fans, which I always prefer to chest-aggravating air conditioning units.

This was only a small resort property with about twenty rooms but we took full advantage of the relaxing location, reading, snorkelling, kayaking and some even went fishing. Each day we would gravitate towards the infinity pool and spend time in the relaxed restaurant bar. It was a perfect finish to our African odyssey.

On the last night it was out fifteenth wedding anniversary so we had arranged a table on the beach where we were given some special attention by the staff. All very romantic, made even better when the rest of the group who had been conspiring and rehearsing came along and serenaded us with that fabulous track 'Islands in the sun'. A perfect end to a wonderful experience with fabulous people!

After a transfer back to Stonetown and Zanzibar airport we checked in to our small Cessna Caravan aircraft for the flight to Dar es Salaam. There were only nine seats on each plane and once again we found ourselves sitting right behind the pilot and could see his every move and the instrument panels.

When we were flying into land the pilot was tipping the wings of his plane and was craning his neck. We wondered what he was doing then realised he was looking for the runway, so, with a feeling of déjà vu, we joined in and pointed it out to him and sure enough we landed safely.

Whoever would have thought I would be in the wilds of Africa ascending the highest mountain on the continent and then spotting animals in the iconic Ngorongoro Crater and on the Serengeti? The nearest I got to Africa in my childhood was when I asked my folks where they were going and they would reply 'Timbuktu'. I had no idea in those days that Timbuktu was a town in Mali on the river Niger.

MOUNT KILIMANJARO

Like a sentinel over the land
The snowy peak looks down
Beckoning trekkers to their test
To reach its rugged crown

Phallic lobelias, monkey trees
Tall grasses and giant heather
'Pole pole', 'drink some more'

Changes in the weather
Lulling us, relaxing us,
The tree-covered base slips by
A mountain terrain bare of bush
Now meets our weary eye

Four days on …the top in sight
Our lungs bursting with pain
The final ascent at altitude
Is driving us insane

Finally the sunrise comes
As does the summit seen by few
A salty tear comes in my eye
As I scan the awesome view

Exhausted, elated
Emotional and weak
We sit below the famous sign
That adorns Uhuru peak

Photographs, congratulations
Amidst back slapping and good cheer
Our hardy party of eleven
Decide we won't be back next year!

CHAPTER TWENTY

Freedom to explore: Sri Lanka, Istanbul, Iceland, Budapest, Prague, Borneo, Malta and Morocco

It was wonderful to be arranging tours and travelling with friends again. We continued our diverse adventures, usually with a city tour or two and one long-haul trip each year.

A destination I had always wanted to visit was Sri Lanka, which means 'splendid land'. The island is 267 miles long by 140 miles wide, filled with lush jungle, an undulating landscape, waterfalls, vibrantly green tea plantations and palm-fringed beaches lapped by the Indian Ocean.

The people are very friendly and happy to see you. Personally, I thought back to the news programmes which chronicled the huge tsunami on Boxing Day in 2004 in the Indian Ocean affecting a number of countries, one of which was Sri Lanka. The south and east of the country were the worst affected, thousands were killed and over a million made homeless. I remember a train called the 'Queen of the Sea' packed with 1,700 people being struck by the tsunami and Yala Wildlife Reserve where a group of tourists sightseeing on the beach were enveloped by this powerful natural phenomenon. Of course it's not just the initial towering wave but diseases such as cholera and dysentery and the instant disappearance of field crops, jobs, belongings and shelter that caused continuing hardship and contribute to deaths.

All this was on my mind as we flew into the capital Colombo on the island which I grew up calling Ceylon and famous for its tea. Tourism was badly affected by the tsunami, too, so it was interesting for me to see how well they had recovered. We were staying initially at the majestic Hotel Galle Face, built in 1864 and set right on the Indian Ocean, one of those Grand Dame hotels that characterise the days of imposing hotels built worldwide in the burgeoning Victorian empire that coloured so much of my school atlas pink in the 1950s. The hotel didn't start as a hotel, though. It was a Dutch villa that colonial big wigs used to meet at then some of these British entrepreneurs decided to turn it into a business and it became known as the best hotel east of Suez.

We loved our stay on this tropical island and especially so as we headed

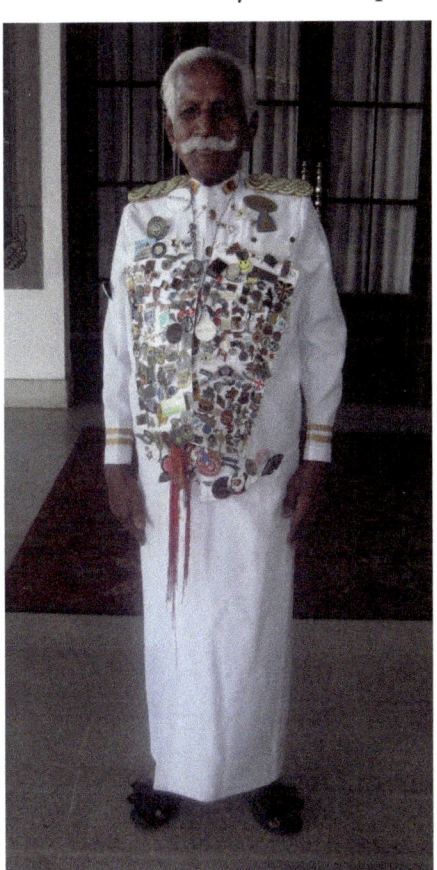

into the hinterland and visited delightful places like the Elephant Orphanage at Pinnawala, but the 55-mile journey there felt a lot longer on the winding narrow roads from Colombo. Pinnawala was started in 1975 to house abandoned and wounded elephants. There were about sixty-five elephants the day we visited including baby elephants brought from various parts of the island. We enjoyed watching the elephants come down to the river to bathe with their mahouts while we sat on a terrace and had a drink.

The impressive doorman at the Galle Face Hotel

John and Ron skirting the issue of attending a temple!

Pinnewala Elephant Sanctuary

We visited the Cave Temples at Dambulla. Dambulla is a vast, isolated rock mass 500 feet high and a mile around the base. Here is found the famous Rock Temple dating to the first century BC. The caves sheltered King Walagamba during his fourteen-year exile from Anuradhapura. When he regained the throne, he built the most magnificent of Rock Temples to be found and this is now a UNESCO World Heritage site.

For me, a real highlight was visiting Sigiriya Rock Fortress, which is a World Heritage Site, and the fifth-century 'Fortress in the Sky', which is perhaps the most fantastic single wonder of the island. It is also known as Lion Rock because of the huge lion that used to stand at the entrance to the palace on the summit of the 600 foot high rock. You can still see the foundations of the royal palace, water tanks to supply water and other buildings, including the guardhouses on the perimeter. On one of the stairways, the only known ancient work of Sinhala secular painting has survived in the form of frescoes of twenty-one life-sized damsels still shining in their original colours. It was a fair walk in the heat and humidity on uneven steps and narrow iron stairs which had been fastened to the rocks as we ascended to the top, but what a view over the lowlands below.

Polonnaruwa, another World Heritage Site, was the Capital of Sri Lanka from the eleventh to the thirteenth centuries and contains some splendid and spectacular statues. Lankatilake, Tivanka and Thuparama are the most beautiful and largest image houses and Tivanka has the best examples of frescoes of the Polonnaruwa period. Rankoth Vehera and Kirivehera are well-preserved large stupas, and Gal Vihare, a rock shrine, has four Buddha statues.

Eventually we reached Kandy, the hill capital and yet another World Heritage Site. It was the last stronghold of the Sinhalese kings during the Portuguese, Dutch and British rule and finally ceded to the British in 1815. To the Buddhists of Sri Lanka and the World, Kandy is one of the most sacred sites as it is the home of the 'Dalada Maligawa', the Temple of the Sacred Tooth Relic of Lord Buddha. Close by are the remains of the Royal Palace ('Maha Wasala'), the 'Palle Wasala' where the Queens stayed-now used for the National Museum, the 'Meda Wasala', where other close relatives lived. The Audience Hall, Natha Devala and Vishnu Devala are situated close by.

At Kandy a few of us just could not resist finding the Test match ground where Sri Lanka play their cricket. It was a nice setting and we strode out to the pitch to get a view back to the grandstands surrounding the oval. It all felt rather British as if we'd been transported back home.

Tea picking weigh-in!

Group in Tea country around Nuwara Eliya

Next day we headed on to Nuwara Eliya, often referred to as the 'Little England' of Sri Lanka, set against a beautiful backdrop of mountains, valleys, waterfalls and tea plantations. It is supposed to be one of the coolest places on the island, and it is really just like an English spring day, although the temperature does drop at night, making you reach for the duvet. All around Nuwara Eliya you see evidence of the British influence, houses like country cottages, bungalows or Queen-Anne-style mansions. Victoria Park, in the middle of the town, is a lovely place for a stroll or a picnic and is also good for any budding ornithologists as you can see some rare birds here.

We stayed at the Grand Hotel set on the golf course and the former residence of the Governor of Sri Lanka, Sir Edward Barnes. Built in 1891, another British Tudor-style property, it fits in well with the colonial bungalows with their managed hedgerows and attractive landscaped gardens. Feeling the cooler temperature as it is over six hours' drive from the capital, you can see why Nuwara Eliya was very desirable as an escape from the humidity and heat of the lower coastal areas and why many English and Scottish pioneers came to develop the tea industry in Sri Lanka.

We headed south-east across the Horton Plains towards a wildlife sanctuary called Yala. En route we stopped off at Ambewela Farm, where they make dairy products such as cheese. Witnessing dairy farming Sri Lankan style is mainly carried out with very small herds compared to the UK, traditionally no more than three or four head of cattle, which are hand milked.

On the journey via Ella and Bandarawela, we often stopped to enjoy Sri Lankan fruits from wayside stalls. Ella is a small village in the highlands surrounded by waterfalls and tea estates and fabulous views when the mist clears and the morning sun rises to reveal the fabulous mountains and the valley below. This is good hiking country for visiting highlights like the Nine Arch Bridge, the Rawana waterfall and the picturesque Ella Gap, which you can visit using the village of Ella as a base. Also well worth consideration is taking the railway journey from Kandy to Ella. If you do take this trip, alight at Haputale to visit one of the best viewpoints in the area called Lipton's seat, which is where Sir Thomas Lipton, of Lipton's tea fame, used to sit to check out his land and tea plantations. If you don't fancy the walk from the station you can hire a tuk tuk for a small fee

We reached Yala in the afternoon and after checking in headed out on a safari. Sri Lanka's most well-known and second-largest national park

The iconic Mount Lavinia Hotel where we had afternoon tea

is popular for elephant, leopard, bear, crocodile and wild boar, among others. Its open, undulating terrain made it famous for elephants for many years, but recently the park has also received much fame through publicity by National Geographic and the Discovery TV channels, which focused on a leopard research, conservation and identification project. Apparently Yala National Park has the world's highest concentration of leopard per square kilometre.

In reality we saw very little and were disappointed, so much so that the one thing that sticks in my mind is arriving at a beach where an elephant had been rubbing up against the bark of a tree but had departed. We were shown a memorial here commemorating the tsunami that struck on Boxing Day 2004. The memorial stone says: 'In memory of 47 lives taken by the tsunami at 9.20am 26 December 2004'. Apparently fifteen were Japanese and German visitors.

We continued our journey after an overnight stop to Galle, visiting the Dutch Fort. Founded in the sixteenth century by the Portuguese, Galle reached the height of its development in the eighteenth century, before the arrival of the British. We were told it is the best example of a fortified city built by Europeans in South and South-East Asia. We had followed the coast all along the south with its pretty beaches and witnessed the traditional stilt fishermen at work. Stilt fishing is a method of fishing unique to Sri Lanka. The fishermen sit on a cross bar called a *petta* tied to a vertical pole and driven into the sand a few meters offshore. From this high position, the fishermen cast their lines and wait to pounce. In Galle itself we stood on the fortress walls looking over the cricket ground where Sri Lanka happened to be playing England, so we had a grandstand view of the game and settled down to watch a few overs.

We were ready to relax after all the travelling so after a city tour of the less-than-inspiring Colombo we headed to the beach for a few days and thoroughly enjoyed kicking back, reading a book, enjoying the facilities at the Negombo Beach Hotel and strolling along the lovely sandy beach.

One day we popped over to Colombo for afternoon tea at the colonial gem, the Mount Lavinia Hotel, formerly home to the Governor General, set on a promontory with its private beach below. It was a real decadent delight to be sipping Earl Grey tea and eating dainty cucumber sandwiches and scones whilst sitting on the lofty terrace which overlooked the entire bay. The hotel was full of sumptuous public rooms and spacious suites in dark wood which gave echoes of the empire at its height. What famous souls have trodden these floorboards and quaffed a sundowner on the terrace, we wondered.

We found Sri Lanka a delight and loved the people, who were very friendly, even though at the time we were visiting there was still conflict between the Tamil Tiger rebels and the government forces.

*

Over the years I have been fortunate to travel to many countries and diverse contrasting cities like Iceland's Reykjavik and Morocco's Marrakech. Iceland was a revelation and I wondered why I had waited so long to visit. Perhaps because we envisaged a frozen wasteland, but it really is a fascinating and wonderful land of geysers, waterfalls and volcanoes.

Of course, you have to consider the time of year because it is also the land of the midnight sun. When we went our rooms had very thick curtains because it stayed light well into the early hours of the morning and it seems very strange, as if we shouldn't be sleeping. Volcanoes, mountains, glaciers, geysers, lava fields and hot springs – natural wonders certainly aren't in short supply in Iceland. The spectacular island, which hovers just below the Arctic Circle, is one of the most geologically active places in the world.

We travelled in May and the weather was very kind to us, staying dry with a smattering of blue skies and sunshine around 20°C, although you needed to beware of the occasional stiff breeze. We loved Skogafoss

Iceland Reykjavik Sculpture with group

and Seljalandsfoss waterfalls, Eyjafjallajokull volcano. This is the volcano which underwent a massive 2010 eruption causing air traffic around the world to be cancelled, with the wind carrying ash clouds thousands of miles.

Reykjavik's corrugated iron houses are painted in every colour of the rainbow. Its numerous independent restaurants serve great fish dishes, as you'd expect. Harpa, the multi-million-pound opera house dominates the city's harbour thanks to its huge size and unique geometric glass design. It was whilst in here that we spotted James Taylor was in concert but unfortunately is was sold out. Disappointed, we decided to visit the Blue Lagoon the next day, which is just outside Reykjavik. At reception you're given a clever wristband, which acts as an electronic locker key and in-water credit card. When you emerge into the icy air, the steaming 37°C water practically begs you to enter. The lagoon is home to a spa area and a swim-up bar, plus there's a separate massage waterfall and sauna. We were loving this, the weather was good, and then the icing on the cake was we spotted James Taylor with a female companion. How lucky was this, we could all say we had shared a morning with James Taylor at the Blue Lagoon. Of course, I really wanted to say hello and the introductory line was going to be, 'Hi, James, I love your music and I see "You've got a friend!",' but I just couldn't bring myself to disturb his obvious enjoyment at chilling out. Well, perhaps the wrong turn of phrase for a steaming natural spa bath!

We loved Snaefellnes peninsula. Its feeling of wilderness, Snæfellsjökull, lava fields, caves, waterfalls, volcanic features and fishing villages make up Iceland's smallest national park, the most wonderful place to visit. The 90-km-long peninsula is a legendary area of magic and beauty. Búðir is a small hamlet in Búðarhraun lava fields, important in terms of the history of trade and industry in Iceland. The most notable building in this area is the old and stunning Búðarkirkja, the Búðir Church. It was built in 1848 and reconstructed in 1987. The stunning isolation of the black church shows off the desolate yet beautiful Icelandic natural landscape as there is nothing around it but mountains.

A jewel in Iceland's crown is certainly Gullfoss This large waterfall is on the river Hvita, the source of which is the glacial lake Langjokull. Gullfoss means 'golden falls' because of the glacial sediment in the water. This gives the water a golden hue when the sunlight catches it. Gullfoss is one of the world's most beautiful waterfalls. The giant gorges and the powerful crashing water can take your breath away. The viewing platform at the top of the falls leave you mesmerised as you witness the plunging glacial water pound into the deep and dramatic canyon below.

*

In May 2014 we took a superb tour to Istanbul (formerly Constantinople) where east meets west, famous for the Topkapi Palace and Hagia Sophia and Blue Mosque on the Bosphorus.

We arrived in Istanbul at Ataturk Airport, named after Kemal Ataturk, a soldier, statesman and reformer who was the founder and first president (1923–38) of the Republic of Turkey. I learned a lot about him when researching my great uncle's military service in Gallipoli. Ataturk modernised the country's legal and educational systems and encouraged the

Istanbul - Rooftop group picture, Hagia Sophia in the background

adoption of a European way of life, with a Latin alphabet, which resulted in many citizens adopting European-style names.

We stayed at the fabulous small boutique Hotel Biz Cevahir located in the old town at Sultan Ahmet, a great base to explore Istanbul old city. We were able to walk onto The Ancient Hippodrome where you see the Obelisk of Theodosius, the bronze serpentine column and the German Fountain of Wilhelm II. Then the real gem, Sultan Ahmet Mosque known to many as 'the Blue Mosque', which was built between 1609 and 1616. This architectural marvel is unique with six minarets and its magnificent interior decoration of blue Iznik tiles. When you stand in it and look up you are filled with wonder. It really does leave you rather humbled and in awe of the builders who constructed this fabulous Mosque. We visited Hagia Sophia Museum, built by Constantine the Great in the fourth century and reconstructed by Justinian in the sixth century. Of course, we just had to stroll through the Grand Bazaar, one of the oldest and biggest bazaars in the world with more than 4,000 shops.

I was quite taken by Istanbul's Underground Basilica Cistern or 'Sunken Palace'. It was very atmospheric, almost eerie. This massive underground water cavern of beautiful arches and Roman support columns and their watery reflections used to be the Byzantine city's main water storage under siege or in tough times. We also took in the wonderful Topkapi Palace, the great palace of Ottoman Sultans from the fifteenth to nineteenth centuries.

You pack in as much as you can driving along the Byzantine City Walls that surround the Old City. We reached the 'Golden Horn', the river that separates the city's European side into two parts, a natural estuary that connects the Bosphorus Strait with the Sea of Marmara. The bustling Golden Horn provides many ferry connections to the northern end known as 'Old Istanbul'. We checked out the famous Orient Express waiting room at the terminus for the famous train and imagined Agatha Christie developing her character Hercule Poirot.

We jumped on a boat and cruised along the Bosphorus, sitting back and taking in historic buildings such as Rumeli Fortress, Dolmabahçe Palace and many other houses and palaces on the European and Asian sides of Istanbul. We took a drive to Camlica Hill, which is the highest point in Istanbul and where you have the best view of the city.

Florence Nightingale has an association with this city having nursed here around the time of the Crimean War. There is a museum somewhat secreted away and you have to make an appointment in advance as it is set in a working army barracks. We took a ferry from Sultan Ahmet Serkeci over to Hamed which was very cheap, then had a bit of a walk to the barracks. The museum is well worth a visit, with information, photos and statues about the Turkish Military and its major wars. It includes an actual machine gun and tank. The building where the museum is located is the hospital where Florence Nightingale was based during the Crimean War. Inside the museum, you will see the treatment room she used (which includes an actual chest tube as well as her original desk, chair, rug and mirror). A room upstairs was her living room with great views of the old part of Istanbul.

We made our way to the Galata Tower, a medieval stone tower in the Galata/Karaköy quarter, just to the north of the Golden Horn's junction with the Bosphorus. It is 219 feet high, a cone-capped cylinder that dominates the skyline and offers a panoramic vista of Istanbul's historic peninsula and its environs. However, even in the queue waiting to enter the tower we could sense some hostility and resentment against us, as quite clearly we were western tourists. Such a shame as it is a fabulous city but we were happy to return to 'our' side of the river in Sultan Ahmet.

Istanbul is still a melting pot, and the politics of the region keep bubbling over. There is no disguising the aggression and unwelcome attitude of some locals to Western visitors, more so in the new town, when we visited. We visited Taksim Square, the site of many a demonstration and also bombings and other atrocities. We had a walking tour when one chap took exception to one of our group taking photographs of a general scene down one of the narrow streets. He proceeded to follow us and became ever more heated and aggressive and you could see other locals were starting to get fired up by him. We tried to placate him but it was only after we came across a procession with police in attendance that he backed off. It was clearly an anti-Westerner reaction.

I just couldn't come to this neck of the woods without driving to the Gallipoli Peninsula, famous for the First World War battles fought here, memorials, trenches and battlefields. This was once a real hotspot for the hippies of the sixties who were journeying south to find themselves.

Gallipoli National Park is where the disastrous conflict took place. We visited Kabatepe War Museum, Beach Cemetery (John Simpson's Grave), ANZAC Cove, Ari Burnu (the first ANZAC landing place), Lone Pine Cemetery (the Australian memorial), Johnston's Jolly Cemetery, walked in the ANZAC trenches, looked down Shrapnel Valley and saw the Turkish Memorial, the Nek and Walker's Ridge and Chunuk Bair (the main New Zealand memorial).

I had promised my dad before he passed away that I would make the journey to Turkey and pay my respects to my great uncle on behalf of him and my nan Daisy Morgan, nee Boucher. At the main Helles allied memorial it was a rather personal pilgrimage as I sought out Great Uncle Warren Ivor Boucher, killed in action in Gallipoli. He apparently was sent to a well for water and was shot by a sniper. As it turned out the water had been poisoned by the Turks anyhow. His army number was #2026 1st company 1st battalion Herefordshire Regiment. His residence was shown as Huntingdon, Herefordshire at the time of enlistment. He died on 9 August 1915 and is commemorated on the HELLES memorial panel 198. It was very emotional holding my finger on his name and recalling how his sister (my grandmother) told me about the day he found out he was being posted and, although it was supposed to be a big secret, he confided it was likely to be the Dardenelles. He swore my nan to secrecy not to tell anyone, especially not their parents. Nan was very emotional as she remembered that farewell never to see him again. He was only just eighteen when he died.

We hired an English-speaking tour guide to explain all the history of the

Gallipoli Peninsula in World War I. As it's a long day we stayed overnight at the Hotel Anzac.

The next day we visited Troy, the ancient city famous for its Trojan wooden horse. The replica didn't really do it justice (although you can climb inside for a photo) but it's well worth following in the footsteps of those ancient warriors. The story of Troy is part mythology and part archaeology. The city walls date from the early Bronze Age up to the Romans, in all, nine different settlements. An excavation information centre holds exhibits from the work under way. We saw the fortification walls of Troy VI with tower and gateway entrance, the Temple of Athena, the focal point of a great annual festival in honour of the goddess Athena, the entrance ramp to Troy II, the sanctuary, an important religious centre of its time and the Odeon, where musical performances took place.

*

Budapest is an intriguing city on the Danube River, not the Blue Danube waters of Johann Strauss waltz fame but a rather muddy brown. Nevertheless, this is a fascinating meeting point of three communities of the nineteenth-century, Ubuda, Buda and Pest, which merged to become modern-day Budapest, divided by the Danube river and connected by various bridges like the Chain Bridge and Margaret Bridge. The Castle District can be explored on foot as we did, taking in the Matthias Church and Fishermen's Bastion, which affords terrific views across the Danube, but the best view, I think, is Gellért Hill from the Citadel. From here we headed back over the Elisabeth Bridge to Pest's Central Market Hall and the largest synagogue in Europe, plus the City Park with Europe's largest thermal spa. A neo-Baroque palace, Széchenyi was built in 1913, and is open all year. With fifteen indoor pools and three large outdoor pools it is something of a maze. Inside are saunas, steam rooms, aqua fitness equipment, whirlpools and jets, and outside pools heated to 33°C (91°F) and 38°C (100°F), or you can swim laps in the main pool. The thermal water is high in calcium, magnesium and hydrogen carbonate and is said to ease joint pain and arthritis and improve blood circulation. We were fascinated by the locals playing chess on floating boards, pondering their next move at the edge of the pool.

I was taken by Heroes' Square with its Hungarian statues of royalty. The Opera House and St Stephen's Basilica whetted the appetite before we arrived at the impressive neo-gothic Houses of Parliament. A splendid city and well worth a visit.

I enjoyed Krakow too with its castle and Cloth Hall and the intriguing Jewish quarter, once the ghetto in World War II and where you can still go to Schindler's factory, made famous by the film *Schindler's List*. Warsaw was an amazing rebuild of the original city. You would never tell it had been flattened during the war. We also visited Auschwitz the largest concentration camp the Germans constructed in WWII. It was haunting, the piles of shoes still stick with me now, the silence and lack of birdsong leaves a lasting impression of the sadness and horror created here.

I prefer Prague, though. It was refreshing that Prague airport was relatively small and efficient, and on the short journey into Prague city on our first visit I was surprised at the rolling hills and relative lack of built-up areas that usually sprawl out of city centres. It seemed we were dropping into a valley when entering Prague, and we were soon experiencing views of some interesting architecture, before our first view of the River Vltava. As we gazed to the right, we spotted the much-vaunted Charles Bridge with its easily recognisable statues. We edged down the busy thoroughfares and numerous narrow streets into the old town of Stare Mesto, and were taken to a check-in office just around the corner from Charles Bridge.

We stayed in an apartment on the small street of Husova which is very centrally situated just around the corner from Old Town Square and just a few minutes' walk from Charles Bridge.

With our driver, using the key cards we'd been given, we entered the property through the first door and then up some stairs to another security gate. Then we climbed the rest of the seventy-five steps to the penthouse apartment. No lift, I'm afraid, but the apartment was welcoming, large, with two bedrooms and two bathrooms, kitchen and lounge/dining room. So great lodgings, I just needed oxygen to get there.

Prague old town just oozes history. You must visit the Castle over the characterful Charles Bridge. The old town square where I stayed the second time has fabulous architecture like the Church of our Lady, a superb gothic fourteenth-century church with towers 80 metres high. The baroque St Nicholas Church is another church located in the square. People flock here for the medieval astronomical clock which is set on the wall of the Old Town Hall. Created in 1410, it is one of the oldest working astronomical clocks in the world.

Strangely, perhaps, I was more fascinated by a wartime story, the culmination of which took place at St Methodius and Cyril Church, situated on Resslov Street just up from the river and the site of a showdown of seven allied paratroopers following the assassination of Reichsprotektor Reinhard Heydrich. Their hiding place in the crypt of the church had been betrayed to the Germans. Sixty-six years on someone still regularly brings fresh flowers. The crypt is open to visitors for a small entrance fee. Don't miss it and watch films *Operation Daybreak* and *Anthropoid* with Cillian Murphy before you go. It really is a poignant place to visit, the reprisals by the Germans were horrendous, with whole villages like Lidice being raised to the ground and villagers being murdered or transported to concentration camps such as Mauthausen. Plaques proliferate all over the city in streets and courtyards acknowledging the deaths of residents shot or murdered in the purge to find the assassination squad.

There are lots of places to eat, some with character and history like a restaurant in a square in Mala Strana called Cervena Sedma, named after Sonja Cervena, an opera singer, whose picture hangs on the wall. We had great cheese burgers, goulash, strudel and honeycake here. Afterwards we walked up to the castle and back down via Waldstein Palace, which was free to look around.

It's worth taking a boat ride, quite cheap for a 50-minute round-trip. The boats have a bar and toilet on board you see the city scape from a different perspective and relax as you pass the parliament buildings, embassies and other splendid buildings.

We enjoyed small, relaxing eateries like U Dvou Kocek (Two Cats) on Uhelny Tch (street) with accordionists playing and good food and another good restaurant The Black Cats – by the Mozart Haus. Close to where we stayed was a fabulous cheese and wine bar, Monarch Vinny Sklep on Na Pristine, right by 'The Hanging Man', a sculpture depicting Sigmund Freud.

Of course, we had to visit Wenceslas Square, the name familiar to us because of the Christmas carol. It was on the balcony of number 36 Wenceslas Square, the Melantrich Building, now Marks & Spencers, where in November 1989 Havel and Dubcek appeared to announce an end to communist rule.

We crammed a lot into our four days but I really love this place, it's very atmospheric and one Einstein and other luminaries used to visit. The doors of the Café Louvre first opened in 1902. You could feel the history here and imagine the meetings of business men and friends, liaisons, and novels that may have started at its café tables. Life has moved on yet here time has stood still. The old café traditions have been preserved. You still soak up the unique historical atmosphere, and almost expect Franz Kafka or Albert Einstein to walk in.

*

One year we had visited the Far East and enjoyed a fabulous time in Vietnam, Cambodia and Malaysia and had the opportunity to fly cheaply on Air Asia to Borneo. This is a fascinating part of what has now become Sabah and Sarawak, boasting Mount Kinabalu's lofty peak of 4,095m (13,435ft). It was very interesting exploring the 'land below the wind', nicknamed because it is situated below the typhoon belt, also the title of a book by Agnes Newton Keith.

We stayed at a lovely guest house in grounds extending to the sea with a swimming pool too, all close to the capital, Kota Kinabalu.

Our room was comfortable but basic and the hotel staff very friendly. One evening the owner's dogs came calling in the lounge area. I was patting one when the other launched himself on me from a couple of metres away. I put my arm up and he bit my hand badly, sinking his teeth in.

Children in Borneo displaying their catch of the day!

I immediately washed the wound and bathed it with antiseptic and bound it, but that night I became a little feverish and awoke very concerned.

My mind began to do cartwheels, concerned about rabies. Morning came and I spoke to the hotel owner, who was a retired chief inspector of police. He was very good, reassured me his dogs had been vaccinated and were domestic dogs, and took me to the local clinic where I had tetanus jabs, although I'd already had these before leaving the UK.

We flew later that day down to the southern coast and the resort of Sandakan as we had booked to go to the Sepilok orangutan sanctuary and on a river trip up the Kinabatangan river and the Danum Valley.

I still felt strange having read on the internet about rabies symptoms, so I visited a local doctor and figured perhaps I should have an anti-rabies shot. I related the story to the doctor. He simply said. 'You probably have not got rabies and, anyway, you needed to have a jab within twenty-four hours to help and there is no vaccine on the island. You would have to fly to Kuala Lumpur to get one. So, the bottom line is if you have rabies you are going to die.'

I came out not knowing quite how to feel. I flew back to Kota Kinabalu and visited the local hospital, where I had some blood tests and then spent a few more days on the beach waiting to die. Of course, in reality, I did not have rabies, but the infection was enough to play mind games with my fears! I did feel like I was 'going to the dogs' though!

*

One year I took a tour to Malta. We stayed in the district of St Julian's and were having a lovely time. My mother and auntie were in the group, so whenever I was free, I hired a car and took them on a tour in addition to the scheduled group excursions.

A few days before we were due to come home, one of the group was taken rather ill. She had passed out and was rather delirious. We thought it may have been sunstroke or perhaps food poisoning. We called the doctor, who examined her, gave her some medication and prescribed bed rest.

Twenty-four hours later she was no better, in fact in her delirium she thought I was her brother who had died years previously. Mind you, by now I was looking rather wan with concern for her, so could forgive her for her misconception.

Again, I called the doctor and suggested we might need to hospitalise her. He declined the idea for now and gave her some different medication. We hoped it would work miracles as we were due to go home the next day.

Next morning came and fortunately the fever had broken, but she was still very disorientated. Once more the doctor came and we asked him to sign a form to say she was fit enough to board, but again he declined, saying he still did not know what was wrong with her and he just up and left.

I suggested to her husband that I saw the rest of the group off at the airport and then stay back in Malta with them until his wife was well enough to travel. I knew once I had put my group on the plane the coach was meeting them in Birmingham to take them home so the rest of the group would be in good hands.

But the husband was very upset and uncomfortable at the thought of staying in Malta. 'I just want to take her home,' he said. I reacquainted him with the doctor's thoughts and his reluctance to issue a fit for travel notice. However, the husband was adamant.

So, rightly or wrongly, I agreed to be part of a subterfuge, whereby we told the airline the lady needed wheelchair assistance as she was frail and could not walk far.

Our group checked in. Our sick lady was drowsy and invariably asleep because of the medication the doctor had prescribed. We oversaw that she was taken by wheelchair to the aircraft and they produced a narrow wheelchair to fit the aircraft aisle on to which she was transferred and was carried up the steps, then wheeled to a seat. The flight seemed very long but she slept the whole way. On arrival in Birmingham, we asked for medical assistance, and she was taken to hospital then eventually transferred back to Newtown hospital, where a few days later she was discharged right as rain and with no answer as to what her malady had been. Phew!

*

Morocco is barely four hours' flight away but you are transported to a totally different culture. Marrakech is such a magical city you feel like you have gone back in time, enveloped by the old city walls.

Wonderful riads like the Riad Assakina or Riad Clos Des Arts are oases of calm away from the hubbub of the narrow alleys full of a cacophony of traders and hustle and bustle.

Evenings can be mystical, whether they are spent in kasbah-influenced restaurants serving classic Moroccan cuisine or just chilling in rooftop bars like the Kosy Bar restaurant in the Mellah, the old Jewish district where you look across the rooftops at the nesting storks.

Marrakech is all about exploring the maze of narrow alleys full of souks where a bargain or two can be had after some hearty bartering, of course, walking around the Medina (old city), taking in landmarks like the Koutoubia mosque. The Jemaa el-Fnaa square is quite magical, with its storytellers, dancers, water sellers, street theatre and flute-wielding snake charmers.

The square is edged along one side by the large Marrakesh souk, a traditional North African market catering both for the daily needs of the locals and for the tourist trade. On other sides are hotels and gardens and cafe terraces where you can watch the world go by and of course people-watch to your heart's content. Narrow streets lead off into the alleys of the medina quarter.

Café Argana and Café Zeitoun are just two of the busy square's cafés we often frequented for a coffee and pastry as the world turned, yet stood still. Another favourite was Les Terrasses Alhambra, where we would sip mint tea and eat tarte au citron whilst watching all and sundry go about their business.

Jemaa el-Fnaa keeps people for hours as they make a number of stops in the souks for mint tea as the traders show off their colourful and varied wares. In the evening there is a huge choice of restaurants like Pepe Nero, Le Tanjia, or my favourite, getting a taxi to Le Foundouk restaurant, an experience in itself, let alone the great meal. The restaurant sends a host with a lantern to lead you from the car to the restaurant. When you arrive, you walk up three storeys to the roof.

There is lots to do for a four- or five-day stay, including famous gardens to visit and sightseeing outside the medina walls, perhaps to Menara Gardens, where the Saadian pavilion is mirrored in the still water of a carp-filled pool, with the Atlas Mountains and olive groves in the background. Sultans used the pavilion for their romantic interludes. We recommend you taking in the beautiful Majorelle botanical gardens, privately owned by the estate of Yves St Laurent. Or visit the impressive five-star La Mamounia hotel, where our former prime minister and wartime hero Winston Churchill was a regular visitor. You could have a cocktail facing its majestic gardens.

I highly recommend having a hammam, not in the local public one Dar el-Bacha, which is an experience in itself but very basic. Instead, spoil yourself and go to the Les Bains de Marrekech. We had been promised a 45-minute treatment of a hammam black soap scrub and ghassoul body mask followed by a one-hour body massage and finally a 30-minute recuperation in the relaxation room, all interspersed with mint tea and Moroccan nibbles which sounded a real treat.

All started so well, bearing in mind most of our party were in their seventies and from a farming background. Men and ladies enjoyed a refreshing complimentary drink and were allowed a short time using the attractive blue mosaic tiled swimming pool which was dappled by the open courtyard's sunlight. Then armed with a towel, hairbrush still in swimsuits we were all ushered into a relaxing lounge the serene ambience provided by soft background music and pleasant aromas emanating from decanters of perfumed rose petals and other floral concoctions.

After being provided with luxurious fluffy dressing gowns and some slippers, men and ladies were divided into pairs and we were shown into separate small tiled bathing areas all of which were furnished with wooden slatted benches, where we disrobed. Doors then burst open and two ladies in white tunics entered rapidly, they reminded me of the psychiatric attendants in the film 'One flew over the cuckoos nest' with Jack Nicholson.

Before we could even say anything, we were dowsed with a bucket of ice-cold water which totally took our breath away, we could hear the shrill screams of ladies in our party from adjoining rooms all no doubt suffering the same fate.

With no time to react, we were made to lie down on a bench and the ladies proceeded with a stiff brush in the form of a mitt and they proceeded to exfoliate our body all over, the brushing was excruciating and a cacophony of screams followed from the surrounding rooms. After a short period of time this torture was over and then another bucket of cold water cascaded onto our reddened bodies, before the ladies just disappeared, into the ether as jets of steam entered the small enclosed space and a real sense of well-being enveloped our beings.

A little while later the door opened and we feared another deluge of ice-cold water but to our relief we were told to shower and we were taken to an area which was blessed with an incredibly relaxing high-ceilinged space a beautiful salon with lots of marble pillars, silk curtains billowed around us in the gentle breeze from unseen apertures. We lay on a couch and were invited one by one to come for a massage which was incredibly relaxing before being allowed to lie quietly on our couches and being served with mint tea and sweet and savoury Moroccan treats.

Our faces were glowing our bodies tingled and my hair stood on end but we all concurred that it was one of the most exhilarating and memorable experiences we had ever endured.

One evening we were entertained by dancers and musicians while eating a fabulous three-course meal followed by mint tea and coffee. The belly-dancers were a revelation, not quite emulated by a small cast from our group. Although John Oakley and the Fez Four were a relevation the night afterwards.

The tanneries in Marrakech were amazing to explore but not a very health-and-safety conscious outing, and definitely for the more adventurous. In fact, when the local guide meets you, he says you have to wear a gas mask, which is most perturbing as you start to enter what appears to be a ramshackle and run-down part of the old city, then you realise with amusement the gas mask he is giving you consists of a bouquet of mint to help distract your senses from the overpowering odour of the tanneries and you have to smile!

For a change of pace and to explore a bit we visited the picturesque port of Essaouira, one of the oldest fishing ports in Morocco. Old Mogador was founded by the Romans and inhabited by the Portuguese. This typical Portuguese harbour, with fine examples of both Moorish and Portuguese architecture, was the setting for scenes in Orson Wells' film *Othello*. We explored the Jewish quarter, the harbour, city ramparts, medina, Spanish cannons and the local artists' quarter. It is worth the journey but you need an early start as it is about a three-hour drive each way.

I always have to explore mountains wherever I am in the world, so, true to form, we headed towards the formidable Atlas Mountains. En route you see a vast array of vegetation, multi-coloured cliffs, mountain streams, Berber villages and local mountain people. This seemed more like the real Morocco.

I decided to stay up in the mountains at a lovely village above Imlil called Tamatert, which is inaccessible by car. Our transport dropped us 150 metres away, where we waited for our next form of transport … then the faint sound of hooves preceded the appearance of a line of donkeys bedecked

Belly Dancer John Oakley and the Fez Four for one night only!

with brightly coloured panniers into which we loaded our luggage and then followed our donkeys up a rugged rock-strewn path to the riad. I hope this isolation never changes.

Douar Samra was our base for this mountain village experience. It is a charming, traditional, flat-roofed, thick-walled stone refuge built by the men of the village. The lovely staff, Mina the wonderful cook, Abdou, who cultivates and nurtures the garden providing all the fresh vegetables and fruit for the tagines, and Rachida, Mohammed and Omar made us so welcome. When I wasn't feeling so well one day, they made me a restorative brew of mint tea and herbs and within an hour I was feeling so much better.

The rooms were unique and quirky. Yan, Sin, Kran and Kouss are in the main house without electricity but lit by candles for a very romantic and special ambience. All have balconies with either a garden or mountain view. Mouss, Sdez, Sbat and the Tree House are situated in the garden, all with electricity and private bathrooms. A menagerie of delightful animals populate the Douar: Jules, the donkey, and a cute family of Shitsu dogs along with some ducks and other fowl.

It was such a relaxing stay and made very special by the call to prayer which was one of the most mesmerising and relaxing I have had the pleasure of listening to. It's a far cry from Branson's luxury Kasbah Tamadot, but for me infinitely preferable, family run after being developed by its owner Jacqueline Brandt, a lovely Swiss lady. Some of the hardy adventurous types in the group of friends I was with planned to climb Jebel Toubkal, the highest mountain in West Africa at 4,167 metres, approximately 13,675 feet. Our friends said it was rather challenging because of the very stony and rock-strewn narrow paths and not at all enjoyable. For the less adventurous there always seemed to be a market to visit, like the one at Amiziz. So many great memories … looking for goats in trees … using black soap infused with olive oil … sipping a cold Flag or Casablanca beer after a long, dusty day in the heat of the North African sun.

I just love the food, too: salads, briouattes (filo pastry parcels filled with a selection of vegetables; cheese; prawns; or lamb), lamb tagine with prunes and almonds, chicken tagine with olives and preserved lemons couscous. Then there's succulent and refreshing orange salad with cinnamon, dates and almonds for dessert as well as copious pastries, always followed by mint tea or coffee.

MARRAKECH

Have you been to Marrakech?
Where you can buy most anything
Where belly dancers ply their trade
And musicians rarely sing

The riads are so wonderful
Oases from the heat
Where mint tea refreshes tired souls
and you rest your weary feet

Traffic in the narrow streets
Is something we don't like
But our footwork has improved no end
Avoiding cart and bike

Ambling through the colourful souks
Traders so keen to sell
'Big shop, small price', we hear them cry
A good deal … well who can tell?

Chilling at the Kosy bar
Beneath the Marrakechi sky
Where sundowners cost five pounds a pop
And huge winged storks fly by

All in all we've had a ball
We've shopped until we drop
Now we're heading for the hills
To try to reach the top

Douar Samra, Tamatert
Berber village perched on high
Rachida's culinary delights
Black kites up in the sky

A friendly place to unwind
Sunsets are so amazing
Mountains tower above the land
where goats and sheep are grazing

Our intrepid trekkers leave our fold
Gone for a day or two
To summit Toubkal is their aim
We wish them fond adieu

Late next evening they return
An exhausted triumphant team
Relieved we welcome them all back
Though tired, their faces beam

Our stay complete
We pack and head for home

Replete with happy memories
Until the next time we roam

So, when we are curled up this winter
When it's cold and wet outside
We will think back to Marrakech
And all feel warm inside

Dianne and team summiting Mount Toubkal

CHAPTER TWENTY-ONE

India and Nepal

Nepal and India are definitely countries that draw me back. From the first time I visited in the eighties, I couldn't wait to get back. I just love the diversity in India as you travel around, the colours, smells and sounds that assault the senses and really excite me. There are sights that make me sad, but just as many that enlighten, excite and enthuse me. I love the people and their friendliness. The colourful ladies' saris, even when I see them working in a field or in some industrial situation, they always look so well dressed. I love the throbbing hearts of cities as well as the feudal efforts still employed in many of the agricultural areas.

I love visiting places of historical interest especially with British colonial links. Along the way you assimilate many different bits of information which illuminate and educate you. Some stories, if not historically accurate, are amusing or acceptable as they are credible.

Following my first visit to India, I returned about five years later, escorting my first private tour to India. This trip, for Shropshire and Mid Wales farmers, started with an amusing check-in procedure for the group at the UK airport. Staff asked the usual array of security questions of the passengers: Is this your luggage? Have you packed it? Have you left it unattended?

One of the older ladies in our party who had been chatting with her friends came to the desk and was asked the question: 'Do you have anything electrical in your baggage such as a hairbrush?'

Ron and Dianne at the Taj Mahal

Our elderly fellow passenger pondered once more, then shouted back down the queue to her husband, who was chatting to a fellow group member.

'Gwyn, is my hairbrush electrical?' To which he replied in a long slow Welsh accent, "Nooooo".

The check in agent just shook her head as she looked at the line of coming passengers and thought to herself this is going to be a long day!

We landed at Delhi where we were staying at the fabulous Imperial Hotel in Connaught Place. After a day of relaxation, we set off on a full day tour of the hectic metropolis of Delhi with its population of 22 million people, taking in all the many sights.

We first headed for the Chandni Chowk Bazaar, one of the oldest and busiest markets in Old Delhi dating back to the seventeenth century. We also took in one of the mosques close to the bazaar, escaping for a moment the mass of humanity which seemed to move as one like an amoeba among the market stalls.

We were taken so far by coach from our hotel before transferring to rickshaws, which put us right into the pulsating atmosphere of everyday Delhi. It was a real hubbub of activity and wonderful to experience, with the colours of the garments people wore and sold and the smells of exotic spices of all different hues.

We headed to the seventeenth-century Red Fort and former main residence of the emperors of the Mughal Dynasty. It was also barracks for the British, then Indian, armies. It was built 350 years ago of red sandstone beside the River Yamuna.

We finished the day at Raj Ghat, which roughly translated means 'Kings Bank', situated on the Yamuna River. We walked along a stone footpath set amongst green, well-tended lawns, which led us into a walled enclosure. We removed our footwear and followed a constant flow of people all heading towards the black marble memorial platform which is left open to the sky while an eternal flame set at one end burns, marking the place where the peace leader Mahatma Gandhi was cremated after his assassination in 1948. It's a peaceful and spiritual space.

By now it was late afternoon, heading towards dusk and suddenly a loud report like a gun went off, surprising some of the party, who wondered what it was. I just calmly said. 'It was probably one of the ceremonial guns at the Red Fort doing a salute to something local, a bit like a midday gun used in many places.' In reality I had just guessed, to avoid any concerns my clients might have had.

When we were safely ensconced back at the hotel. I put BBC World News on the satellite TV and the immediate report was of a series of bombings in Delhi, which had killed 62 people and injured at least 210 others in three explosions.

These bombings were two days before the annual Diwali festival, which is celebrated by many. The bombs were triggered in two markets in central and south Delhi, one close to the Chandni Chowk Market where we had been earlier that day and the other in a bus in the Govindpur area in the south of the city.

Indian authorities believed the Kashmiri/Islamic terrorist group Lashkar E Taiba were responsible. It was clear to me, as we had passed all three sites during the day, that we had been very lucky, and these bombs were meant to cause indiscriminate harm and damage. We could easily have been victims. There but for the grace of God!

We carried on, visiting wonderful places like the amazing Varanasi known sometimes as Benares on the holy river Ganges where just before sunset we travelled through a rickshaw thronged throughfare bustling with people and sounds and smells assaulted our senses as everyone headed for the Ghat on the River Ganges where full of excitement and expectation we were blessed to witness the famous and evocative evening prayer ceremony known as 'aarti' all this for us at the time of the Diwali festival. it was celebrated whilst we floated on the Ganges in rowing boats that our guide had hired. We were enthralled by the almost hour long activities taking place attended by the thronged masses on the banks of the Ganges Ghats whilst we floated serenely on the holy river. Varanasi is a city of temples and a great religious centre for the Hindu and Jain religions and has much to see but this event is unmissable.

The smell of incense sticks burning, the prayer chants and the intoxicating sounds of the daily mass gathering was mesmerising as the prayer bells chinked and jingled. The culmination of this beguiling ceremony led to lamps being lit and floated on the water and flowers being handed out as an offering to be cast onto the much-revered mother Ganges. We took marigolds from a plate of flowers and placed them alongside tea lights in coconut shells which we set carefully into the water accompanied by your private prayer or wish that you hope to come true signalling a captivating finale to the evening as priests dressed in vermillion cloth sarongs poured water into Mother Ganges whilst chanting and praying what a wonderful experience etched forever in our memories!

As we made our way back to land, we noticed lots of detritus floating in the river including one huge bloated dead cow. We saw funeral pyres and wailing mourners, then the remains swept into the river. We also saw sadhus, men of mystery, holy persons who give over their lives to Jainism and Hinduism and look rather unkempt, to say the least, but are revered.

We visited the amazing temple in Khajuraho in Rajasthan, the Kandariya Mahadev. It has over 800 sculptures, including warriors, gods and goddesses, animals, dancers and couples in very erotic positions. The erotic panels are supposedly to celebrate the marriage of Lord Shiva. You could buy love manuals called Kama Sutra in the local gift shop, showing you some amazing contortions.

One of them reflects Khajuraho's most distinctive panel depicting a couple locked in intercourse, with a maiden assisting on each side. It appears to defy gravity, with the male figure suspended upside down on his head. Needless to say, there was a lot of tittering amongst our group of tourists.

India is the source of many words we have adopted or adapted. Tiffin is one example, I remember a chocolate bar called this in the fifties and sixties. It is apparently a British slang term for second breakfast or lunch but sometimes just applied to a light meal. But a more likely and well-used meaning is the Indian-English when Indian custom was adopted by the East India Company and other British settlers in India, superseding the British tradition of afternoon tea and leading to a new word for the afternoon meal. To confuse the issue there is an old English word 'Tiffing' which means taking a drink or sip.

*Varanasi group in a
boat on the river*

*Looking back at the
Ghats*

Funeral pyres

Yak and Yeti Hotel in Kathmandu

Tickety-boo is another import into the English language we could have brought from India, as the Hindu phrase *thīk hai, bābū*, means 'it's all right, sir'). The phrase could have been picked up by British personnel in India when administrating before independence and adapted and used in the amended form, spreading in fullness of time to the UK and elsewhere in the empire.

We a travelled up to Nepal and had a wonderful time there and were very fortunate to see many wonderful monuments and architecture, many since destroyed by earthquakes, and lots of terrific scenery when we headed out towards the Himalayas on our day trips.

When we departed Nepal, we flew back via Delhi, where we had to change planes to go on to London. Whereas normally we checked our luggage straight through, this time we were going to have to collect our luggage and check it in again in Delhi, as there was a security scare.

Whilst queueing for the flight check-in from Kathmandu airport, we were at the front of the queue having arrived early to avoid any rush. To our left a group of Nepalese arrived and started queueing in a separate line

Preparing for my flight over Everest with Buddha Air

then pushed in ahead of us. I immediately took umbrage with them and told them in English of course. 'This was not on and to wait their turn.' Whilst having this debate two burly guys approached me, followed by a rather important couple and whisked past us. I later found out it was a member of the Nepalese Royal family. Oops!

Transferring in Delhi where we had to claim our luggage, one lady in our party managed to become detached from us and for some reason known only to her, headed off to the connecting flight by going through the area for those just in transit, in spite of my instruction to come out and collect her bag. Anyway, the rest of us came out and collected our luggage ready to check it back in. We were asked in some cases to open our luggage. It was at this stage that two of the party, one single lady and one married gent, realised that the single lady had the glasses and the pyjamas of the married gentleman in her suitcase, and that he had one of her nightdresses in hers.

This led to a bit of eyebrow raising from the rest of the group, until in time we established that all our bags had been rifled through, probably in Nepal. We say that because we found the people to be rather morose and unhelpful as well as discontented and also rather intent on asking for tips for bag carrying, even when most of our group were wheeling their bags. In some suitcases, things were missing, evidently stolen.

A security official in Delhi said this was common as, in spite of locks, baggage handlers or anyone with access could get into zipped cases rather easily, and he showed us a trick with a ballpoint pen that even we could use to get into a case … another lesson learned!

CHAPTER TWENTY-TWO

———

More travel anecdotes: Cuba, Norway, Hawaii and Corsica

Cuba is such a wonderful destination, with surprises around every corner. We arrived into the heat of the Caribbean and our group seemed rather quiet as we headed from the airport into Havana, the capital and our first port of call. I thought they may just be tired from the nine-hour long flight but I think the nature of the scruffy, littered approach into the city and then the dilapidated buildings added up to them thinking, 'What have you brought us here for?'

We could not park the coach in front of the hotel where we were staying but we walked a short way around a corner and there it was, the classic Hotel Inglaterra. It's the oldest hotel in the city, founded in 1875, and was made a National Monument in 1981. It's located in Old Havana, looking over Central Park. The architecture, history, location and quaintness of this dated property made it a special place to stay. It is one of the most famous hotels in Cuba, boasting Sara Bernhart, Enrico Caruso and Anna Pavlova among its previous guests. The Colonial Restaurant, designed by the most famous Cuban painters, is in need of renovation but nevertheless wonderful to breakfast in.

After a good-night's sleep we started to engage with Havana and realised the rundown nature of the properties are actually part of the character of

Group picture with John Lennon Caption: Imagine - a farming group with John Lennon in a Havana Park

Dianne eyeing up a new Cadillac fifties style in Cuba.

Paladar La Guerida
Paladares a fabulous hidden
restaurant

the place. As you walk along the street music seems to be coming from every alley, street, bar and home. Everywhere and everybody seems to be so colourful, the streets full of old American Cadillacs, other sedans and open-top cars of yesteryear left by the Americans in the fifties and early sixties after the Revolution.

On the plane over we'd chatted to a British embassy official and were intrigued to hear about a secret restaurant in one of Havana's derelict buildings. We hired a cab and asked to be taken to this hidden restaurant, which seemed to be guarded like a state secret, so much so that after twenty minutes it was clear this cab driver was confused as to how to find it, stopping and asking people, before eventually heading to a taxi tank and asking fellow cabbies.

Finally, we arrived via a series of narrow side streets in front of the Paladar La Guarida. Paladares are invariably small family-run restaurants as opposed to state-run restaurants. This property was originally the setting for a film called *Strawberry and Chocolate*. After paying the taxi driver we walked through the grand façade into what appeared to be an unoccupied building and we were wondering if we had been dumped and ripped off. Then we heard people's voices, so walked up the shabby marble staircase and reached a floor where people seemed to be living in rather basic conditions with washing hanging out on balcony railings and who completely ignored the curious visitors passing by.

We kept climbing the stairs, asking the odd person in their bijou, shabby-chic homes for directions, who told us with hand signals to keep going up.

We reached the third floor and walked along to a closed door which seemed to be the only likely place. With trepidation and hope we knocked and the door opened to reveal a positively buzzing grotto of culinary delights, walls bedecked with photographs of famous visitors and old film paraphernalia, presumably referring to the Oscar-nominated film which was made here. There were three small rooms crammed with tables. Four of us huddled around a small table in one corner. There was also a balcony and a rooftop terrace with views of Havana up some narrow stairs, but we stayed in the less breezy interior. You paid in dollars for an assortment of good food such as fish tacos, tuna and other big fish steaks. We started with watermelon soup and finished with a blob of ice cream on a pineapple base. It was all very tasty, a great experience and well worth the slightly higher prices we paid here.

One of our party, Edgar, was a real character and took everything in his stride, always smoking a pipe. He was like our Welsh Hemingway and I am sure he could have written a book. Even when we left, we were all on the coach and wondered where he was then he appeared behind us being seen off by lots of lovely Cuban girls.

This place is magical, full of places which attract your attention as well as people, from once-regal building facades, ever-present music, to wrinkled-faced old women smoking huge Cuban cigars.

Cuba is just the ticket for Hemingway fans, perhaps those who have visited his old house in Key West would understand better the draw Cuba had for him! Hemingway's room on the fifth floor of the Hotel Ambos Mundos is set out as if he were still there and in a great position on the corner of the hotel. It is supposedly the room where he was inspired to write *For Whom the Bell Tolls*, which earned him enough royalties to buy his Cuban home, Finca La Vigía. Hotel Ambos Mundos was in good shape when we visited. The rooftop terrace was fabulous for a mojito sundowner!

Hemingway's favourite bar was the small Bodeguita del Medio, where he used to regale those who would listen with his fishing exploits and probably offer to take them on in a boxing match after a few too many drinks! His favourite restaurant in Havana was supposedly El Floridita, where, legend has it, they invented the daiquiri.

We were privileged to tour around his nineteenth-century former home, Finca La Vigía, which means 'lookout'. It's in a little town called San Francisco de Paula about ten miles to the east of Havana. He bought it in

1940 for $12,500 and apparently part of *For Whom the Bell Tolls* and *The Old Man and the Sea* were written here. After Hemingway's death in 1961, the Cuban government took it over without reference to his estate or his wife Mary. Needless to say it was left neglected, so had to be refurbished and recreated, but it is well worth a visit. You can feel the presence of the big man in more ways than one.

Hemingway had a boat called El Pilar, which is now on display at the Finca La Vigía. Apparently, the vessel inspired *The Old Man and the Sea*. It was kept at Cojimar where he loved to go fishing with the locals. He would have various libations at La Terraza restaurant where he hung out with his friends and fellow fishermen.

Viñales was an interesting excursion from Havana. It is a really verdant valley with huge foliage-covered peaks of limestone which give the topography real character. In between you will find tobacco plantations, where scenes of farmers in battered straw hats with their ox-drawn ploughs take you back to a time of past rural activity in the UK.

Back to Havana, there was plenty to explore especially some of the famous hotels from the twentieth century when Cuba was well visited by the 'mob', the American gangsters. The Hotel Nacional is famous for mafia guests pre-Castro when Cuba was run by Batista. It was a big gambling joint frequented by Meyer Lansky and 'Lucky' Luciano. Old classic taxis from these days in all shades, especially pink, line up outside.

Tropicana Cabaret with a greeting for Ron

Che Guevara Mausoleum Santa Clara

The Tropicana cabaret, which opened in 1931, has played host to Benny More, Nat King Cole, Jack Nicholson, 'Lucky' Luciano and many, many more famous names. This is a Vegas-style cabaret but with outdoor seating for up to 1,400 people and about 200 performers, gyrating to a pulsating Cuban dance routine. The dancers are wonderfully exotic in colourful costumes and the whole show is exhilarating. Food is served, which is not up to much, but you are there for the musical experience. We sang all the way home, probably because our entrance price included a bottle of Bacardi and can of coke each.

The Buena Vista Social Club is worth a visit, but it's not open all the time so you need to check this with your visit dates. We spent a lot of time trying to find a restaurant in Cathedral Square one day where they bring you a huge beer with a tap on it, like a glass barrel of beer. You can have such a wonderful time in Havana, don't miss it … nor those mojitos as sundowners.

The city of Trinidad on the south coast was a very surprising destination where we spent a couple of nights, a really wonderful place in a time warp. I's a perfectly preserved Spanish colonial settlement apparently built by money made from sugar cane, so magnificent colonial-style mansions abound on narrow cobbled streets which open into a beautiful plaza.

You cannot visit Cuba without hearing about Che Guevara, who was actually born in Argentina and died in Bolivia. The Che Guevara Mausoleum, a memorial in Santa Clara, houses the remains of the revolutionary Ernesto 'Che' Guevara and twenty-nine of his fellow combatants killed in 1967 during Guevara's attempt to spur on an armed uprising in Bolivia. It has a bronze 22-foot statue of Guevara. The Bay of Pigs was an historical turning point too, when the Castros became heroes with their lieutenants for defeating an American-backed rebel incursion.

We rounded off our tour in Varadero, a line of beach-front resort properties, which frankly is characterless and could be Miami beach! It's okay for some relaxation, but if you go to Cuba, you must visit Havana!

*

Norway is a stunningly beautiful country and in 1985 I made my first visit there escorting a group of farmers. We based ourselves in the lakeside town of Voss.

Norway and indeed all the Scandinavian countries have always been expensive to visit and alcohol, a prerequisite of some of our soirees when on vacation, is especially so! I had advised all the party if they wanted a drink at reasonable prices to take in their full duty free allowance and most did. On arrival at the hotel we had to ascend about six or seven steps and when I got to the top I looked around to see some of the farmers struggling with their luggage. I deposited my bag of duty-free on the top step and descended the stairway to give some assistance. It was then that I heard a crash and the people with me gasped. I turned to see a litre of whisky and a litre of brandy forming a small waterfall down the hotel steps. I just had to laugh.

As we expected the alcohol was very expensive even back then thirty-five years ago a bottle of beer was £2.50! We explored every avenue to find a cheaper solution, including speaking to a local farmer who offered us some locally distilled hooch which we tried as a prelude to mass purchase but a few sips told us this was not the way to go as eyes watered and smoke came out of our ears.

Back then supermarkets did not sell alcohol. When I enquired how the

Thomsons Spirit anchored in a Norwegian Fjord

alcohol arrived at hotels and restaurants and how the public bought it you apparently had to go to the post office and fill in a form.

With this cunning plan in mind I found a post office, where I was given a form to complete but before I started to fill it in I enquired how long it took for the alcohol to arrive. The post office clerk looked up over her glasses and with a wry smile said one week, so that little avenue of pleasure was scuppered and the week became a rather sober one after the initial supplies of duty-free had been consumed.

Food at the hotel was excellent, wholesome smorgasbords with a plentiful supply of very tasty food, not least of which was a huge salmon as the centrepiece of the display on the long table.

One morning at breakfast, a few days after arrival I was accosted by the Maître d. He took me aside, I confirmed I was the party leader and he said, 'I really must ask you to tell your group members to stop taking excess food from the breakfast table to make their lunches with. If this does not stop, we will have to impose a surcharge.'

Embarrassed, I started to apologise, but he had not finished. ' which leads me on to say the group will not be allowed to approach the evening dinner table until other staying guests have had an opportunity to have a plateful of food as some of your party appear to be taking two plates of food each not least of which the salmon, possibly again for a later meal!'

I was speechless and that day on the coach I had to diplomatically broach the subject with my fellow travellers. Fortunately, everyone took it in good humour and notice was taken.

We were enthralled with the majestic scenery of Norway, especially on the 'Norway in a Nutshell Tour' when we travelled by coach through small charming villages and dramatic landscapes, towards Gudvangen. The most hair-raising was going down the steep hairpin bends of Stalheimskleiva down the Stalheim gorge. From here we took a cruise on the narrow UNESCO-listed Nærøyfjord, surrounded by high, steep mountains, and the magnificent Aurlandsfjord, a picturesque fjord affording stunning views.

We arrived in the small village of Flåm, nestled on the banks of Aurlandsfjord, where we took the legendary Flåm Railway. The Flåm Railway offers spectacular panoramic views of the wild Norwegian landscape. We changed trains in Myrdal, to the Bergen Railway and travelled back to Voss. It had been a long day and we all slept well that night.

The next day we had a free morning so a few of us went for a swim followed by a sauna in the hotel facilities. Three or four of us fellas enjoyed the warm waters then the hotter rejuvenating temperature of the sauna. Sitting on the wooden slatted benches in the room perspiring like crazy the door to the other side of the sauna connecting the ladies changing room creaked open and in walked two young, attractive, totally naked Scandinavian ladies.

To say we all perspired all the more was an understatement. The ladies said hello but I think the best we could offer were wide eyes and a nod. The ladies chatted away and our conversation dried up as we all tried to look straight ahead at the pine wall. We became hotter and hotter until after ten minutes the ladies left, and we finally relaxed. All very British, we said little about it and showered and headed back to our rooms. But I bet that story has been told a few times at the livestock markets of Shropshire and Mid Wales.

Dianne and I later returned on a cruise along the Norwegian Coastline up to the Arctic Circle on the Thomson Spirit a enjoyable journey to the land of the Midnight

Dianne had always wanted to visit Hawaii as she had a small book of

Hawaii Collage!

postcards sent to her by a relative many years ago, so off we went to try and locate all these scenes. Flying via San Francisco into Oahu and stayed at Waikiki beach by Diamond Head its famous for the Hawaii 5 O theme tune opening vista in the series.

We took in Maui, Kauai, and Big Island too, the latter being our favourite as it was diverse with active volcanoes, beaches, waterfalls and so much more and yes we found all the scenes may have changed a little from the 1950's but still idyllic. One day we kayaked across a beautiful bay to Captain Cook monument which was constructed to commemorate the place where he made landfall. I had previously cut my leg so I was wary about going swimming as sharks had been seen that morning in the bay, but I did swim with trepidation. On other days we walked a few metres from our small but ideally situated beachfront apartment and climbed over a few rocks entering the ocean and swam with green turtles in the wild and the most amazing experience was snorkelling off the beach further up this same west coast above Kona and hearing whale sounds only to realise just a few metres away a huge whale and its calf had joined us, Dianne was oblivious until I pointed them out behind her that she was just a few metres away, absolutely amazing! A fascinating land and worth the visit.

Such wonderful memories of the sunset above the lava flow and the eerie glow of destruction tracking its way down the mountain side.

Aloha Theatre was built in 1932 and we found it by driving past it about 8 miles from our apartment on the way to the Volcanoes National Park. Set in the heart of Kainaliu is the home of the West Hawaii Performing art company. We managed to get some tickets for a performance of 'The King and I' a day or so later and what a delight it proved, from the welcome, the ambience of this fabulous old building on to the performance, absolutely wonderful don't miss it. A highlight was visiting the iconic Pearl Harbour a place we grew up learning about as the turning point of WWII. Apart from viewing the whole site including S.S. Arizona the opportunity to meet two veterans who survived the attack on the 9th December 1941 and were able to describe their haunting and frightening experiences to us was a real eye opener. Leaving Oahu airport for a journey home ukelele under the arm and many memories in our minds we were replete with happiness and contentment.

There are so many wonderful places in this world. One year we headed south to Corsica. We stayed in Calvi at a small family-run hotel on the north-west coast. Calvi is a fabulously relaxing and attractive resort with a beach of pure white sand and a very attractive medieval citadel which you can enter and wander through relaxed pedestrianised streets lined with restaurants and small shops. If you go in June there is a jazz festival, but we were there in September and witnessed a fabulous evening performance of polyphonic singing in the old Citadel at the Cathedral of St John the Baptist.

We saw a group called A Filetta, who have been part of the revival of the tradition of Corsican polyphonic singing, which had almost disappeared. It was a way of communicating in the narrow streets of medieval citadels and steep mountain villages. We stumbled on the performance by chance and it was wonderful! This polyphonic festival attracts singing groups from all over the world, from Africa, Asia, the Pacific and of course Europe.

CHAPTER TWENTY-THREE

———

Peru

Everyone's travel bucket list seems to include Peru. We had long since looked forward to heading to Peru and in particular exploring the iconic Machu Picchu.

Lima, the nation's capital, is situated on the Pacific coast and was a good base to start our tour of the country. Following our first breakfast we had a short city tour of the cathedral, catacombs, main square and so on. When a number of us stopped for some lunch, one of the guys with us on tour wandered along the road in search of some reading material and came back with a copy of the the *Daily Mail*. Such is the Brit abroad many still like to catch up with home affairs, even though nowadays most folk do it online.

Shortly afterwards we were all chatting whilst eating lunch and this same chap sighs and says 'Oh blimey, I've already read this. It's yesterday's.' After a short while he added, 'Crikey, someone's even done the crossword!' We all burst out laughing. I guess he bought a second-hand newspaper in more ways than one.

We headed down the dramatic Pacific coast to Paracas, a long journey and at times quite a hair-raising one on this beautiful wild coast as the elevated road, backed by rugged mountains, often exposed huge roadside drops to the ocean below with little space between the poorly protected roadside edge. At times the mountains gave way to sand dunes, including the highest sand dune in the world, Cerro Blanco

We stayed in Paracas at a lovely hotel called Double Tree El Prado. The *Daily Mail* reader and I went for a run, while the others relaxed by the pool drinking copious quantities of the local cocktail favourite, Pisco Sour. Pisco is the local strong liquor and it is made sour by adding a mixture of lime juice, a syrup, angostura bitters and ice. Sometimes egg white is added.

By the time we went on a boat cruise around the Ballestas Islands the next day the cocktail imbibers regretted it deeply. My good lady received the accolade of being the quietest, most refined 'vomiter' in the world. No one knew she had been ill, as she'd managed to hide her distress by using a plastic bag and disposing of it totally unobserved. Even so, Dianne could not really enjoy the plethora of wildlife. The cormorants, sealions, penguins and fur seals seemed like a mirage, I'm sure.

We had an unforgettable experience flying over the Nazca Lines, a series of ancient shapes up to 1,200 feet in size, drawn in the Nazca desert in southern Peru, and a UNESCO site since 1994. Taking a flight was certainly the best way to truly appreciate the seventy-plus designs of animals such as birds, fish, llamas, jaguars and monkeys, as well as human and tree-like shapes. I pondered why they were there, whether it was religious or spiritual in some way. Who knows?

Before we flew, to the consternation of some, especially the ladies in the party, we all had to be weighed to share us out equally amongst the small aircraft to make sure the planes were balanced and not overloaded. We were aware one or two planes had crashed in the not-too-distant past whilst taking tourist out over the Nazca lines, which naturally made some of our party reluctant flyers, despite reassurances.

My good lady, perhaps because of her slight build and possessing the dark hair and similar hairstyle of fellow Asian travellers, was conveniently put in a plane with a group of dinky Japanese ladies, while Jan, another diminutive lady and a great friend, was put on the other side of the plane to balance her out, such was the fine line on plane load. It was a great experience, but as the small planes were buffeted by a little turbulence most passengers were clearly happy to be back on terra firma when the excursion ended safely.

We then headed inland and to the higher altitude of the attractive 'white city' of Arequipa with its many churches, convents and monasteries. We enjoyed a drink at a rooftop café terrace above the Plaza de Armas with views of the Chachani volcano in the distance, before resting a night then heading on to the Colca Canyon (about a five-hour drive on some unpaved roads with stops along the way). One of our brief stops was the highest on our journey at 15,800 feet, where we alighted from the coach and found it difficult trying to suck air into our lungs due to the air pressure change at this altitude.

The scenery when you pass by Chuccura and Patapampa is breathtaking with snowy volcanic peaks, herds of llama and the cultivated terraces of the Colca Valley. We also visited the town of Chivay with its very busy and extensive market and stayed overnight here at 12,000 feet above sea level.

We had an early morning tour planned to Condors Cross, where we were hoping to see condors flying over the canyon. Condors and altitude sickness were in conflict for one of our party, however. She was feeling awful, had hardly slept and kept vomiting. Concerned, we called a doctor. We feared it might be cerebral oedema, where the altitude causes the brain to swell, but he reassured us. Di suggested he gave this lady shot of Stematil to help with the nausea and gradually she seemed to improve. We left her in the hotel with her husband and headed up to Colca Canyon.

On a very poor track we bounced around all over the place at times, wondering if the vehicle was going to plunge into the ravine. The canyon at its deepest is over 4,000 metres, over 13,000 feet deep.

We could only see a condor way down in the canyon, but a great experience nevertheless. Dianne took a photograph of a condor on a poster the day before and people were amazed at her photography … well, it did look rather real! We returned to the hotel to check on our patient who fortunately had recovered enough to rejoin us for the long journey to Puno and Lake Titicaca, approximately seven hours' drive away with stops at the likes of Patapampa where we had a good view of volcanoes.

Puno was the first place I started to feel a bit of nausea and a headache. Perhaps it was a culmination of recent days, a build-up. My appetite was curbed and I learned to drink more water and tea and even chew a coca leaf or two. Drinking coca tea was de rigueur.

Puno is located at over 12,500 feet above sea level, so altitude sickness is to be expected. It is situated on Lake Titicaca, the highest lake in the world, the legendary birthplace of the Inca civilisation and now home to the Uros Indians on the Uros Islands. Across this vast lake is Bolivia.

We headed out on the lake to the floating islands, which are made up of totora reeds roped together. They feel very weird when you first walk on them, like huge waterbeds, but you get used to it, and start to trust them not to sink. The islands need constant attention as the reeds rot though so new ones are regularly added on top.

A lot of the local Uru ladies wear traditional costume and are skilled in traditional handicrafts, but there is an odd juxtaposition with modern-day life, as there was even a satellite dish on one of the islands, which for me was a shame. But, hey, I'm not living there and the islanders' lives can't be played out simply for the benefit of tourists.

We took our boat further across Lake Titicaca onto the south side of Taquile Island with its unique culture. We enjoyed meeting the villagers and buying their handicrafts, and taking photos. The Taquileños still wear traditional clothing and are known for their high-quality handicrafts. While the women make yarn and weave, the knitting is actually done exclusively by men. Taquile has a small central square with a handicrafts market, small shops and a cafe or two. We were treated to a meal on a covered verandah with superb views. One or two of us felt the altitude more here today as we exerted ourselves walking upwards from the boat to gain the high ground for a terrific view. We were worried about the elevation of Machu Picchu but were already higher than that, so started to feel more comfortable about our upcoming trek.

A real highlight was the Lake Titicaca train to Cuzco, 'The Andean Explorer', which is run jointly by Peru Rail and Orient Express. It is spectacular, one of the world's great train journeys. We arrived at the small station and were checked in and greeted by a band as we boarded. It was all very special, a Peruvian version of the Orient Express.

The journey itself takes about ten to eleven hours as it wends the 384 km from Puno to Cuzco in a very slow snake of five or so carriages, often slowing and stopping. We passed though small towns and villages where there was often a market on the tracks, so we would slow and stop and watch as the sellers cleared their colourful array of goods off the line, then reassembled them after we passed through.

We reached over 4,000 metres en route. We followed rivers and continued onto plains between mountains, watching llamas and everyday Peruvians go about their working day. By now we had headed to the open-air observatory area at the rear of the train. Inside it had a bar, and during the journey some enthusiastic and brightly attired entertainers came along, playing an assortment of instruments from the traditional South American pipes to guitars.

Photographs showing life on Lake Titicaca

Ron waiting to travel on the luxury train Andean Explorer

We ate at tables adorned with table lamps, seated in comfortable armchairs, each one with a pillow. We were served lunch and afternoon tea during the journey. We were all so chilled, what a wonderful experience living in the lap of luxury. Appetiser, main course and dessert followed by petit fours and coffee for lunch. Wow! A real treat!

We made another stop at La Raya, the highest point on the journey at 4,319 metres, where some just stretched their legs and one or two checked out a little artisan market. All back on board, we started to descend and followed the Urubamba river, which starts right up in the Andes not far from La Raya. Afternoon tea was about 5 pm whilst we were slowly 'clicketty-clacking' down the railway line through the verdant valley. It was like a private club on wheels and a great way to see the scenery and arrive in

Cuzco. Cuzco was our base for heading towards the valley from which we would head up to Machu Picchu. Part of the reason for being here was to acclimatise before our trek up to Machu Picchu but, in reality, Puno had been higher than the 10,800 feet Cuzco sat at. Nevertheless, it was good to have some more time to get used to the high altitude.

Little girl street seller in Cuzco

Our hotel was not far off the main square, which is dominated by a cathedral. Next day after a relaxing breakfast and morning stroll around the shops near the square, we had an afternoon tour of the city, seeing the cathedral and the Koricancha temple. We then proceeded to visit the nearby ruins such as Sacsayhuaman, Qenqo, Puca Pucara and Tambomachay. We had the evening free and wandered along the small streets and alleyways, coming across a huge 'Amazonian'-sized guy in full costume who you just had to have a photo with, for a fee, of course.

Our next stay was in the sacred, very fertile valley of the Urubamba at a place called Yucay. Our hotel was old enough to be reputed as the place Pizarro, the Spanish conquistador, stayed with his men back in the sixteenth century. From here we visited the fantastic ancient city of Ollaytaytambo and marvelled at the skills of the engineers and architects and indeed the workers who created these phenomenal structures of Temple Hill. How could they construct such perfect walls of huge blocks of stone, all perfectly placed and all without modern techniques and machinery?

Food had become an interesting talking point, as all travellers will testify. Lomo saltado had become a favourite, which was a simple meal of beef, onions and tomatoes with chips. The food was often spicy so we learned to check before ordering. Empanadas, essentially small pasties, were a popular choice for many. As we walked around, we often saw in the smallholdings, pens of good-sized guinea pigs. It soon became clear that they were fattened to be eaten by the locals. We carefully studied the menu to avoid these Cuy (pronounced 'kwee'), but just occasionally we were not convinced we had not eaten this local delicacy, and who knew what was in the empanadas?

The real gem and the reason we were in Peru was still ahead. We transferred by bus to Ollanta train station. Here we joined the 'Vistadome' Train. As its name suggests, the train's roof was glazed to give a panoramic view of the passing scenery on the journey to Machu Picchu.

Some of the party were to take a morning tour of the ruins for around three hours, during which time their guide would explain the significance and technology behind the tight construction of the walls, the water channels, the staircases carved from the rock of the mountain, and the Intihuatana (or 'Hitching Post of the Sun'), which served as an astronomical calendar. All show technological advancement far beyond anything this area of the world was thought to have achieved. These, along with the Temple of the Sun, the Royal Tomb, the Priests' House, the Inca Baths and the Temple

Winay Wayna on the Machu Picchu trek route!

of the Three Windows, allow us to glimpse back into the past and, with the help of our guide, revisit the lives of these truly remarkable people. These lucky folk then had lunch at the luxury Sanctuary Lodge hotel.

But for about a dozen of us intrepid travellers we were offloaded with our backpacks at a halt on the track. We had opted to hop off the early-morning train at Kilometre 104 of the Cusco–Machu Picchu railroad. We had a spectacular 15-km (9-mile) guided day hike to the Lost City of the Incas to look forward to. This is one of the world's most famous treks and we could enjoy the satisfaction of puffing up the Monkey Steps to the Sun Gate to earn our first glimpse of Machu Picchu. The hike starts off steadily and is quite straightforward, although some parts of the path are very narrow, maybe only a foot or so wide at best, and if you suffer from vertigo, you don't want to look down into the valley far below. We leaned into the mountain without thinking about it. I would class the effort needed not as tough, but a fair challenge which takes its toll on the quads and calves.

On our journey we visited the former religious site of Chachabamba', before walking for the morning along the Peruvian mountain paths overlooking the Urubamba River Valley way below.

After stopping for a break and water stop beside a spectacular waterfall, we had lunch and explored the agricultural terraces of Winay Wayna. Then perhaps surprisingly we descended into the cloud forest and grew more and more excited until finally we were there at the Monkey Steps which led to Inti Punku, the Sun Gate, all by late afternoon.

Some ran up the last steps, others ambled up, and then we just stood there, most speechless to behold the wonderful vista that is Machu Picchu. It was a picture we had seen in magazines and on travel shows a thousand times but we were now there in person, we had earned the privilege by walking up to discover this ancient site. It was a very emotional moment.

Ron and Dianne at the Sun Gate looking down over Machu Picchu

We congratulated each other, took pictures, then descended the gradual track downwards into the ruins of Machu Picchu which itself is set at 8,038 feet above sea level. It is the largest Inca city ever found and unbelievably it lay hidden until 1911 when an American explorer, Hiram Bingham, stumbled across it.

Llamas wandered around the tiered ancient community. Tourist crowds had thinned out and it was almost as if we had the place to ourselves. We took lots of photos before descending by coach along the switchback road to the valley below Machu Picchu to complete our day with a celebration in the town square at a bar in Aguas Calientes.

We ordered beers all round. They were huge, but we were so elated and indeed thirsty. We drank and celebrated and ordered another round, but the second flagon defeated us. We needed to catch the train, so gave the rest of the beers to the band who'd been entertaining us and went on our way.

We took the train back to Cuzco, feeling quite smug at our achievement. It might only have been a trek of nine miles or so, but it was the higher altitude, the narrow paths with scary drops and rapid steep ascent that made it seem like much more. The cherry on the cake was the way we arrived in Machu Picchu and made the trek very special. Sure, the rest of the group had enjoyed a wonderful day, especially their lunch in the Sanctuary lodge but our picnic at Winay Wayna amongst the Inca ruins was atmospheric and memorable!

On the way back there was bizarrely a fashion parade by models walking up and down the aisles. I just cannot see it happening on the Shrewsbury to London service, let alone Aberystwyth.

Pen Y Gwyrd, famous for its mountaineer guests amongst them Sir Edmund Hilary. The Briggs family have run this hotel for many years

Whilst I have been to a huge number of places in the world thanks to my occupation as a travel agent, there are plenty of wonderful destinations in the United Kingdom we love, such as Pen-Y-Gwyrd, a famous hotel on the junction of the road up to Pen-y-Pass in the scenic idyll of Snowdonia, where Edmund Hillary and others stayed whilst preparing for their successful attempt on Everest in 1953. The residents' bar is festooned with mountaineering artefacts and pictures, some of the 1953 expedition. The food is fabulous here too. You are still summoned in good old-fashioned style with a dinner gong! For those who like the outdoors it is both perfect and quirky. We love the wild swimming lake behind the hotel complemented by the sauna, which is much needed after testing the very cool waters of the 'swimming pool'.

Whilst we were at the Himalayan Mountain Institute in Darjeeling recently, in a display case we noticed on documents on show the signatures of Chris and Jo Briggs, members of the family who have run the Pen-Y-Gwyrd for generations. It was Chris who awoke his guests at 1am on 1 June 1953 inviting them to share a glass of champagne in the bar when he heard that Hillary and Tenzing had summitted Everest.

CHAPTER TWENTY-FOUR

Vietnam, Cambodia, Thailand, Malaysia and an Alaskan cruise

Vietnam and Cambodia had loomed large in our consciousness during the sixties and seventies and were always featured on the news as we heard of the terrible bombing and atrocities in these countries as well as the senseless loss of life of all nationalities and Dylan's protest songs and anti-war rallies were prominent on the airwaves. So, over thirty years later, it was strange flying into Hanoi and visiting places synonymous with those times.

It was a fascinating trip. Hanoi was buzzing and the constant hum and whine of motorcycles and the hordes of people moving around the busy city were a real assault on your sensibilities. Our diminutive Vietnamese guide, ironically called 'Charlie', showed us the many stalls and spoke to the stallholders and explained produce to us before suggesting the unimaginable: to cross the road devoid of any break in the disorderly swarm of vehicles and cyclists. We feared certain death until miraculously we walked into the throng and the various vehicles managed to avoid us, allowing us to reach the other side. It was a 'Moses' moment and a leap of faith we have applied all over the world and thus far survived! Would I try it on the M25? Well, only when the traffic is at a standstill!

Motorcycles were often carrying a little family, sometimes four people including children and all sorts of goods. One had a stack of egg cartons all kept in place with strong elastic bungee cords, I think the bike was a 'scrambler'. Sorry, I couldn't resist it! Ho Kiem lake was a beautiful haven of tranquility in the city, which we very much appreciated.

We took a short drive to a village called Bat Trans where they make ceramic pottery. Here we hired some bicycles, so cycled in and around Bat Trans. It was interesting to see motorcycles passing us with huge vases strapped to either side. It doesn't bear thinking about! Joe Thuong, our guide, shared a milk apple with us, which was very tasty and looks all milky when cut, hence the name.

Ho Chi-Minh was the founder and first leader of Vietnam's nationalist movement. He sought support from the Soviet Union and Communist China. He became known as 'Uncle Ho', and was the catalyst for the liberation of Vietnam from first the French and then the United States during the Vietnam War. His mausoleum in Hanoi is a must, but beware there is likely to be a long queue, and you have to be very respectful. The outside of the mausoleum is monitored and controlled by soldiers who have no worries about telling you to shape up, no talking, stand straight, definitely do not smile and woe betide you if you laugh. I was told to take my hands out of my pockets. No water or cameras are allowed inside but strangely a phone is. The embalmed body of Ho Chi Minh is guarded by soldiers. We were in and out in a few minutes.

Danang, once the big U.S. Air Force base and also where troops used to go for R & R on the beaches, is almost like a modern-day Miami these days with large resort hotels lining the beaches, ironically many of them American with American guests aplenty!

Hoi An was a really pleasant surprise. We stayed in the delightful Ancient House Resort and the main reason for most people's visit is not the beach but walking through the city, visiting Hoi An's colourful market, then taking in the ancient houses, some dating from the sixteenth century. We visited the Chinese Assembly Halls, the Art and Craft Manufacturing Workshop and the 400-year-old Japanese Covered Bridge and we watched locals raising silkworms to produce silk. There are lots of cafes and restaurants to enjoy for some people-watching and the city really comes alive at night.

We travelled some days by bicycle, sometimes by rickshaw, occasionally by car, and walked some days. We visited the remains of ancient monuments and temples at My Son, with shell craters from the American B52 bombers still in evidence. Apparently, there were seventy-two ancient buildings here when the French were in Indochina, but you can barely see anything now.

One day we hired bicycles from the hotel and cycled to the beach. On the way a traffic policeman put his hand up. Two of our friends and I stopped, but Dianne sailed past, waving at him. Much whistle-blowing later we caught up with Dianne and asked her why she had not stopped for the traffic policeman. She just said, 'I thought he was waving at me!'

We flew down to Saigon, now Ho Chi Minh City and enjoyed a cocktail on top of the Rex Hotel, where journalists, broadcasters and expats used to meet in the sixties and seventies. We couldn't visit Vietnam without exploring the Chu Chi Tunnels, famous for the way the 'Vietcong' managed to hide and get around American positions undiscovered. They were so narrow and small, average Americans could not go down them and many were booby trapped. I am not a big chap but I could not even get past my waist to enter them. They have enlarged one section so that you can wriggle and shuffle through on all fours to get an idea of what life was like for the Vietcong. Some areas opened up so they could cook and sleep, and they even had field hospitals underground. Astonishing.

I have to say, when a chap in front of me stopped in the tunnel and tried to reverse that was enough claustrophobia for me! But it wasn't just the size, it was the smell of the damp earth and goodness knows how many tunnels would collapse when it rained heavily in the monsoon. Life must have been hell yet these people were determined to drive out the invaders, first the French, then the Americans. There was a lack of decent air to breathe and you soon became very hot down there in the tunnels.

We were able to have a go at firing an AK47, which twenty years ago cost about $20 for a clip of ammo. You fired at a target backed by an earth mound. If you squeezed the trigger the bullets were gone in a trice, so you put it on single shot and tried to hit the target but the sight was off and notoriously inaccurate, but nevertheless it was the weapon of choice for the Vietcong and was very effective when they sprayed the field with bullets.

You may recall seeing footage of a tank belonging to the North Vietnamese Army crashing through the main gate of Reunification Palace, ending the Vietnam War. The palace remains in a time warp and whilst Ho Chi Minh used it for meetings he never stayed here. The former US Embassy where that helicopter is seen landing and taking off, helping people escape, is now just an empty compound with plaques to American and Vietcong soldiers killed there. On that fateful evacuation on 29 and 30 April 1975 thousands of Vietnamese were filing into the embassy and then climbing into the compound hoping for salvation and fearing the purge from the Vietcong who were closing in. Many were saved but there were still marines there on 30 April. When they were finally helicoptered out onto USS Okinawa at 11.30am the Vietcong entered the embassy to the sight of looted offices and documents that had been shredded and mostly burned.

The embassy was demolished in 1998. During the demolition the ladder leading from the embassy rooftop to the helipad was removed and sent back to the United States, where it is now on display at the Gerald R Ford Presidential museum in Grand Rapids, Michigan.

The next leg was a flight to Siem Reap in Cambodia and the awesome Angkor Wat, a former twelfth-century Hindu temple used by Buddhists over time and taken over by the French. Beautiful temples, many restored, some in disrepair, are surrounded by a wall and accessed via a causeway over a pond-cum-moat. You are allowed to climb the temples, a great place to watch the sunset for sure. Some of the area has been used in films such as *Indiana Jones and the Temple of Doom*. Set just 5 km outside Siem Reap, it is a great short-stay destination and well worth the visit.

We came back home via Bangkok. I have never spent more than a day or two here but have enjoyed seeing the splendid Grand Palace and the floating market where narrow craft were bedecked with vegetables and all sorts of goods and people paddled in and did their shopping. The red-light district of Pat Pong, the oldest such area in Bangkok, had to be seen to be believed and what better way to see it than hiring a tuk-tuk, which vibrated us all the way there at great speed interweaving through the traffic whilst we ingested huge swathes of fumes from the other vehicles. We were dropped at the entrance to the night market and strolled down the parallel streets being beckoned into go-go bars by attractive night girls who were probably lady boys. We stopped for a drink and then having done the 'been-there-done-that' T-shirt thing took a tuk-tuk back to the Sukhumvit area where we were staying and visited a restaurant called Cabbages and Condoms, once a family-planning clinic which developed into a café then a restaurant. It has great food at a reasonable price and free swift broadband. When you finish your meal instead of an after-dinner mint, they give you a condom. Life is wacky sometimes!

One trip we enjoyed was the very sad visit to the Bridge over the River

Kwai. Ironically the film was mostly shot on location in Sri Lanka but we didn't let that take away from this nostalgic and hard-hitting journey into history. We were picked up by our driver from the hotel in Bangkok, and travelled by road to Kanchanaburi, a green verdant province 80 miles west of Bangkok. We started to get a feel for life in rural Thailand once we left the city limits, passing sugar cane fields, rice paddies and pineapple plantations. We reached the River Kwai and boarded a narrow dug-out canoe they called a 'longtail speed boat'. Well, it did have a whining outboard motor on the back which whisked the water very well as we zipped along, holding on to our sun hats, to the site of the bridge, an icon of the classic novel about wartime Thailand.

Paying our respects to the prisoners of war and other labourers at the Kanchanaburi cemetery was very poignant. The thing that sticks in my mind is how young many of them were. We continued to the nearby JEATH War Museum set out as thatched attap huts. Seeing the World War II artifacts and images we could imagine the prisoners living here. But we couldn't imagine the pain, hardship and despair many of these endured in captivity before dying and it must have been worse for those who came back with all the memories and lifelong afflictions their incarceration had left them with, not least the thoughts of their friends and comrades they left behind in the jungle.

We took the Death Railway through the jungle, passing through Hellfire Pass, which had been hewn out of the rock by these sick and emaciated men who had lived and died here trying to build this ambitious engineering project under the vicious supervision of the Japanese overseers racing to build a railway between Thailand and Burma. Even today the Japanese do not take ownership of the cruelty they dished out, saying it is exaggerated or in some cases did not happen.

Our friends headed on to Australia whilst Dianne and I caught the train from Bangkok down the length of Thailand into Malaysia, our target destination being Singapore for a flight home. The train journey was relaxing and we could not believe how fast it went considering it must have been about nineteen hours, but we met and chatted to people and took in the countryside enjoying an airline-type meal swilled down by a Singha beer. Watching the landscape of palm trees and paddy fields drift by was perfect before getting some sleep in our comfortable sleeper.

A few people boarded en-route and when we woke in the morning, we started chatting to a young English guy called Eugene Sidwell who had been teaching in Cambodia, and was now heading south to Singapore for a flight back home to the UK. Amazingly he was from Whitchurch Road in our home town of Shrewsbury!

Crossing the border, we reached Butterworth and walked the short distance to the ferry to take us over to Penang. As a child I thought of Malaya and rubber plantations as being synonymous but Malaysia is far more than rubber plantations. Penang itself was not as exciting as we had hoped, but we took in Chew Jetty, Fort Cornwallis and Cheong Fatt Tze, the nineteenth-century Blue Mansion. We did find the historic Eastern and Oriental Hotel rather beguiling, a little grander than the Mingood hotel we were billeted in. It's located in Georgetown, the UNESCO World Heritage Site of the capital's nineteenth-century heritage buildings. It featured classic wood furnishings and floral prints, spacious colonial-style suites and it takes you back in time to colonial British influence.

We were quite happy not to linger on Penang. The beaches didn't do much for us but we enjoyed the funicular train up the 833-metre-high Penang Hill, the oldest British hill station in South East Asia with splendid views overlooking Georgetown and Penang Island.

Happy to be back on the train, we were in our sleepers when some guys boarded who we got chatting to and it turned out that one chap was a DJ from, yet again, Shrewsbury. What a small world it is!

We arrived in Singapore late at night without any local currency and no way of changing money but such was the advanced technology of Singapore the taxi took a credit card and we were relieved to be in our hotel in Little India and the familiarity of a favourite city of ours. Visiting old favourites like Raffles Hotel and taking a bum boat on the river.

What a wonderful trip we had enjoyed but whilst in Penang we had taken the decision to extend our holiday and spend a few days in Borneo by flying on cheap Air Asia flights to Kota Kinabalu and then on to Australia to surprise our eldest son Peter, who was hosting our friends with whom we had shared our Vietnam tour.

*

We have been lucky enough to do quite a few cruises over the years. One year we went with friends on a cruise up the Alaskan coastline. It was

Vietnam and Siem Reap Hanoi, Halong Bay

Saigon where I tried unsuccessfully to enter the Cu Chi tunnels underlining how slim the Vietcong were

Angkor Wat

Elephant riding in Thailand

Raffles Hotel Singapore

fabulous, with lots of great experiences including flying from the port over the mountains to a glacier in the interior where we were met by some very excited huskies and their very experienced musher owners!

Our Juneau dog-sledding tour was via helicopter, very expensive, but a once-in-a-lifetime opportunity. Dog-sledding behind a team of huskies belting across the icy snow-covered and decidedly windswept plateau nestled between majestic mountains on the Juneau Icefield really was a breath-taking and thrilling adventure.

The journey began when we were picked up at the main cruise ship terminal in Juneau and transported to the helipad in the Mendenhall Valley. The helicopter took us over glaciers and in-between mountain peaks, giving us magnificent views of the Juneau Icefield. On arrival at the dog-sledding camp on Herbert Glacier, we met the excited huskies and enjoyed a tour of the kennels. Before heading off on the dog-sled trek we were given some instructions by our guide, who used to compete in the Iditarod, a very long cross-country dog-sled race held each year across the wastes of Alaska. It's been going since 1973 and takes well over a week to compete as sixteen dogs race the sled, encouraged by their musher. Because the race is so long the mushers often put booties on the dogs' paws to protect the pads.

We learned commands like 'hike', which means let's go, 'haw', turn left,

'gee', turn right, 'on by', pass another dog team. 'Easy' means slow down and, guess what, 'whoa!' means stop. Luckily, we didn't have to endure the 1,000-plus miles of the Iditarod with its sleeplessness, frostbite, ice, snow danger and mud. We were only gone from the ship for three hours in total, but what a fantastic experience. I was amazed we were allowed to mush ourselves and, critically, had to understand the calls and how to brake. Contrary to driving a car, you step on the brake to keep the sled moving; the moment you step off the brake it digs into the ice and snow to slow down the sled. About 160 Alaskan huskies live on the glacier. We heard personal accounts of the Iditarod competitors and their experiences racing and mushing with Alaska's most amazing form of transportation.

It was really exciting as the eager team of huskies barked and strained to get going as they were harnessed up, then we took off on a race over the snow. Driving a dog-sled team and taking part in a great Alaskan pursuit truly was thrilling, just tempered slightly, perhaps, when, all of a sudden, some dog poo would come speeding towards you! The dogs had the ability to defecate as they ran, a term our musher referred to as 'going on the fly'.

One thing that did tickle me and I thought of many times as cruise passengers queued up at the midnight buffet for their eighth meal of the day was a sign that said 'Those wishing to come on the helicopter tour to the glacier please note: Persons weighing 280lbs or more must pay an additional $100 surcharge for safety reasons.'

One day on board we headed to the Hubbard Glacier to witness the truly awesome sight of a glacier carving, in fact not just to watch it but to hear the eerie sound of climate change. Our party of four all went up on deck as we entered the sound and approached the glacier, but it was rather cold so Dianne, Cath and I headed off to the sauna which had a picture window where you could sit and watch this amazing sight whilst keeping very warm. Cath's husband Pete decided to stay on deck and when we returned to him, he was swaddled in blankets from head to toe sitting on the sundeck on a sunbed. He peeked out from the blanket wrapped around his head looking like ET and said, 'Where have you been? I've been keeping you some sunbeds.' There was no one else on deck and a plethora of empty deckchairs and sunbeds. We tried not to be too smug when we told him we'd enjoyed the view from the warmth of the sauna. I gazed down at the book nestled on Pete's lap to see that he'd been reading *Story of the Titanic.*

CHAPTER TWENTY-FIVE

———

China, Tibet and Nepal, including Everest Base Camp and the highest railway in the world!

Whilst the longest tour we ever put together was the Round the World trip in 1995, one of the most epic trips we travelled on with friends was to China, Tibet and Nepal in October 2014 for almost three weeks, when we were assaulted by so many experiences, sounds, smells and superb scenery.

I had been to China before, an interesting experience, especially with the first group I took, who were all farmers from Shropshire and Mid Wales. It was nowhere near as polluted as it had become on my second visit over a decade later. The millions of bicycles I witnessed on my visit with the farming group by 2014 had been replaced by cars and trucks churning out fumes to add to the industrial gases output which was literally choking the nation's people!

Eating with chopsticks was both a challenge and amusing, but by the time we took our first internal Chinese flight, frustration and the fear of starvation had kicked in and nearly all the group appropriated the plastic cutlery off the plane for use on the rest of their tour.

It's a wonderful country and as large as it is diverse. Wherever there is a tourist hotspot like the Great Wall of China or Tiananmen Square,

Street Haircut in 'the Hutongs' Beijing style!

lots of hawkers try and sell you things so, whilst not great linguists beyond Welsh and English, the whole group soon became very proficient in the use of the phrase 'Bú yòng xièxie', pronounced phonetically 'boo yow', which basically means not needed, thanks.

Whist Beijing came complete with a Peking Duck feast and an assault on our ears from the questionable delights of the Beijing Opera, Xian offered us its well-known ancient attractions in addition to its Grand Mosque and the intriguing Muslim Quarter. We also visited the southern capital of Nanjing, formerly Nanking, founded in 1403 during the Ming Dynasty, a large port on the Yangtze river. Two-thirds of the 40-foot-high ancient city wall still exists and is most impressive.

Outside the city walls we were intrigued by the blue-tiled roof of the white marble mausoleum of the Chinese National Party leader, Dr Sun Yat-sen, still thought of as the Father of Modern China. The 392 steps we endured on a rather warm day leaves the experience etched in our mind. Dr Sun Yat-sen had the ability through his beguiling personality to bind people's thinking, even from opposing factions, whilst seeing a pathway to power. His knowledge of the West, gained from the time he spent with his brother living in Hawaii, was heads above the understanding of his political rivals and proved a useful conduit to factor agreement between the Nationalist forces under Chiang Kai-shek and the Communists. The latter referred to him as the pioneer of the revolution which led to the overthrow of the Xing Dynasty and paved the way for modern China, but when he died in 1925 a power vacuum emerged.

From Nanjing we took the unusual route of a train to Wuxi before boarding a working cargo barge along the canal to Suzhou. We followed

Great laugh on the Great wall of China

a varied itinerary in Suzhou before a train ride to Shanghai, where we explored the Old Quarter of the city and the new shopping area where I bought a fabulous tailored jacket which I still wear today over twenty years later. In the evening we wandered the kilometre-long Bund embankment of the Huangpu river in the former International Settlement and the French Concession, taking in the rising towers of the modern Pudong across the

Farmers Group drinking Snake wine on the Li River near Guilin

300

Ron on boat and
Guilin Li River
Cormorant Fishing

river, a light cooling breeze caressing away our day before bedtime. We were yearning for the countryside, however, and were rewarded over the following days with a stay down south in Guilin.

We were blessed with a full-day cruise on the Li River to Yangshou. Our early morning start was well worth it, Guilin's wonderful karst limestone landscape unfolding before us, calming our souls as we glided away from Mopanshan Pier. The journey was about 50 miles but it didn't seem that far, enveloped as we were by the serenity of this mystical land and served with delightful food and drink. We watched the passing small craft as the river wound its way through the mountains with beautiful and magically named sights such as A Painting of Nine Horses, the well-known scene from the back of the 20 RMB Chinese banknote. We were busy taking photographs when we were plied with a rather interesting snake wine, complete with the snake still in the bottle, a somewhat powerful beverage not to everyone's taste.

The evening would not have been complete without a visit back to the Li River to experience the age-old practice of cormorant fishing. This traditional way of life now earns more money from tourists taking photographs than from the actual fishing. If you visit the riverside at Yangshou after sunset you will likely catch a glimpse of the cormorant fishermen complete with gas lamp-lit bamboo rafts. Whilst they fish during the day too there is something quite mesmerising about this evening display, where the lights cast their illumination into and over the water so that the cormorants can see their prey. As trained by the fishermen, they dive into the river, dispersing the shoal of fish. As they return to the raft with their catch in their long beak and gullet, the birds are deprived of all but the smallest fish by a ring tied around their neck. The larger fish are then retrieved by the fishermen and the birds are occasionally rewarded with a small fish before they make another plunge into the lamplit waters. Often the fishermen alight on the bank with their cormorants and nets for a photographic opportunity, but sadly this whole Li River experience from the daytime cruise to the evening cormorant fishing has developed into a far bigger tourist attraction than twenty years ago, with larger cruise ships on the river. I feel blessed we did it before this explosion in tourism. Nevertheless, the sight of the small fleet of lamplit rafts twinkling on the river stays with you for ever!

After a restful evening exploring Yangshou and visiting the antique and craft markets, the next day we took the morning cable car up Yaoshan Mountain to see the stunning views over Guilin. After a special longevity lunch in which peanuts, fruit, vegetables and noodles abound washed down with copious amounts of green tea, we climbed up Fubo Hill and visited the Reed Flute Caves, an illuminated underground cornucopia of stalagmites and stalactites.

Our epic journey was coming to an end as we headed back to Guilin airport with our last glimpses of paddy fields nestling under green hills and the reflections of farmers cultivating the lush soil with their water buffaloes.

However, as we were a little early for check-in we had one more unusual delight. We were taken to a massage centre of Chinese medicine for a 45-minute foot reflexology massage and a five-minute head massage. We were told they activate the circulation and lymph system and therefore eliminate toxins trapped in the body, a perfect way to prepare for a long flight. We were ushered in by smiling Chinese personnel who made up for their lack of English with their kind mannerisms and smiles and we were sat in very comfortable armchairs and our feet were then placed in a rather warm coloured liquid which we soon discovered was urine, but reassured this was a perfect cleansing agent and disinfectant for our feet. We reclined and enjoyed a very relaxing relief for our tour-weary feet.

Refreshed, we were soon on the road to Guilin airport for our flight home via Hong Kong, suitably informed about a very different culture and extremely interesting country.

*

On my second visit we flew out with Qatar Airways via Doha to Beijing, arriving in the relatively new airport terminal built for the Beijing Olympics in 2008. We stayed at the Jade Garden Hotel in the district known as Wangfujing, a decent well-situated modernised hotel with character and an unbeatable location close to the East Gate of the Forbidden City. Built in the 1930s, Cuimingzhuang, as it was known, then became in 1946 the residency of the Chinese Communist Party delegation. It was used as a villa-cum-country house for party officials until it returned to being a hotel in 1996.

During the hour or so's journey into Beijing we were shocked at how poor the visibility and air quality had become in Beijing since we were last there over ten years previously and whilst we loved and were beguiled by all the bicycles and rickshaws back then, they were rare to see now and this underlined how the traffic and industrial fumes led to poor air quality, a real threat to health here. The sun just could not burn through the depressingly thick grey blanket of gloom.

We took a brief evening stroll around the area close to the hotel, filled with western stores like Prada, Zara and Apple which made the whole area feel surreal to us. We came across a fabulous building called The Children's Theatre at the end of Wangfushing Street, meaning 'Prince's Mansion Well' and indicating the site of the old royal palace. In this street there is an inscribed metal plate the size of a drain cover, marking the location of an ancient royal well.

After a good night's sleep, we started our tour, visiting the Great Wall at Mutianyu and the Olympic Venues. The Great Wall is over 4,000 miles long, but we headed to Mutianyu, 70 km from Beijing, taking about two hours. The Mutianyu section dates back to the sixth century but was largely rebuilt in the sixteenth century during the Ming Dynasty. It is one of the best-preserved sections of the wall largely due to the high standard of construction.

Again, the haphazardness and quaintness had changed since our last visit. Now a slick Disney-like approach greeted us without so many street hawkers who had been replaced with concession stands. It left me yearning for the past harassment.

Views from the top of the wall can be spectacular, helped by Mutianyu having a high density of watchtowers which give a good idea of the scale of this section as it passes over the mountain ridges. There was a fairly arduous uphill walk from the coach drop-off point and then even on the wall it was surprisingly undulating. Unfortunately, the pollution was even affecting this area, the smog reducing the visibility more and more as the day wore on and thus in turn weakening the impact of this amazing engineering feat. It was such a shame as the views are impressive and you needed to appreciate the skill of those who built the wall.

Back in Beijing we drove to the 2008 Olympic sites such as the Birds Nest stadium, quite a sight and known as an Olympic wonder with 110,000 tons of steel lattice work and best seen from the outside. We also saw the nearby Aquatic Centre, known as the 'water cube' with its unusual bubble-like cladding and now used as a water park.

This tour was full on, as we packed in Tiananmen Square, the Forbidden City, the Temple of Heaven and Summer Palace before taking the Bullet train to Xian. Tiananmen Square, the largest city square in the world., is home to the Mao Mausoleum and monument to the People's Heroes. It was great to see the changing of the guard where Mao had raised

Ron, Kate and Dianne at Beijing Summer Palace on Kunming Lake

the Red Flag in 1949. This spot is flanked by the Great Hall of the People (China's parliament building) and the National Museum of China, both built in the early 1960s with help from the Soviet Union, which shows in the architecture. As we wandered around the square with our guide he told us about the various structures and the history of the place. We were occasionally approached by hawkers to whom we politely said 'bu yao xie xie' (pronounced 'boo yow shay shay') which roughly translated means 'no want, thank you', used as the time-honoured polite refusal, a phrase taught us by our guide. Amusingly some of the hawkers would say it to us before you had a chance to, breaking the ice and making us smile!

It soon became clear that our small group also had some Chinese people standing on the edge of the group as if listening to the guides' interesting facts. It transpired they were making sure the guides were not telling us anything detrimental to the Chinese people and government. In other words, they were minders and employees of the state, which was rather disconcerting for us all.

To the north of the square is the equally vast palace complex of the Forbidden City, reached by passing under the Gate of Heavenly Peace ('Tiananmen' in Chinese). Built from 1406 to 1420 by up to one million workers it contains 980 buildings with 8,707 rooms and was the former home of the emperors and their court. We walked up the middle axis from south to north, taking in the splendour of the most important central buildings before finishing in the relatively small garden by the north gate.

The Temple of Heaven was next, smaller than the palace but size doesn't matter as this building is considered by many (including us) to be the most beautiful building in China. Perfectly symmetrical, it was constructed at the same time as the Forbidden City and its primary purpose was as a prayer hall for the emperors to ask for good harvests. The temple is set inside a vast park giving a wonderful feeling of space in such a packed city.

We entered the harem adjacent to the temple. Apparently, of the 800 concubines kept here, those who were in favour were picked by spreading cards on a table. They then let a butterfly, a symbol of love, circulate in the harem and whichever card it settled on, the girl that card represented was the chosen one!

Around this neighbourhood were 'hutongs' (lanes) of old homes and businesses, which we explored by rickshaw. These old houses had high thresh-holds, a little like some British houses used to have. Our high thresholds came about to keep the threshed straw on our floors contained. In China the higher thresholds were also to stop ghosts and demons coming in.

The last stop of the day was at the Summer Palace to the west of Beijing, built in 1750 by Emperor Qianlong. It is dominated by the Kunming Lake which was hand dug. The lake is dotted with peaceful pavilions and gardens and is as popular an escape from the city for today's Beijingers as it was for the imperial court.

After the Summer Palace tour, we took a transfer to the station for the late afternoon Bullet train to Xian. All the time you are so aware that everything is in Chinese be it speech or writing, so unless you have a guide with you or speak Mandarin you are a little hamstrung.

The Bullet train reaches speeds of over 300 km per hour, shortening the normal rail journey time and on a par with flying to about five hours. It was fascinating to watch the digital display at the end of each carriage telling you the current speed. The highest I saw was 312 kph. The also showed the upcoming stations, most of which we noticed ended in 'Dong', such as Hai Bai Dong. All the seats were reclining and facing forward like on an aircraft in a configuration of three seats one side of the aisle, two the other and lots of leg room.

Before long we left the city behind and were passing paddy fields and the rural splendours of cows and sheep in the fields, the latter not something you immediately associate with China. I love the fact you can be on this train and a drink on the table barely moves and everyone watches the speed on the carriage monitor, excited as it gets faster and faster. During the journey we enjoyed one of the ubiquitous Bento boxes, a sort of Chinese lunch box, cheap and cheerful. We heard the words 'cau tau' being uttered by fellow passengers and asked if this was the same as our phrase the anglicised form 'kowtow'. We were told this was a Cantonese phrase and Chinese custom of kneeling and putting their forehead to the ground, considered as worship to a god or a form of submission, though in the West we look on it as a form of grovelling or very subservient manner.

Simple comforts such as a western-style toilet were available, as was the traditional hole-in-the-floor Asian staple defaecatory relief option, one of each at the end of each carriage.

Arriving in Xian (pronounced 'sheann') at around 10pm we were transferred to our hotel, the Grand Noble 5*. We immediately noticed the air seemed a little clearer even though it was dark and the ubiquitous neon signs did not have to be seen through a haze as in Beijing.

After a stroll around the Bell Tower monument, it truly felt less polluted here, your throat did not feel dry as you grasped for air and your chest, nose and throat immediately felt relieved. You can see why so many where masks. Up until then I'd always thought it was because they were trying not to spread viruses. We spent time in the old market area and were amused to see little children just squat in the streets, undoing a purpose-made flap in their trousers to facilitate the action. Quite astonishing, really, when we have problems with dogs doing the same.

Next day in Xian our plan was full on, to see The Terracotta Army, Big Wild Goose Pagoda and City Walls, take in a show and then head off on the overnight train west to Xining. We were so excited to be revisiting one of the highlights of China, the extraordinary ranks of the Terracotta Army. Created by China's first Emperor Qin Shi Huang – the fabled 'Yellow Emperor' – to provide an imperial court and, of course, an army for the afterlife. Work began in 246 BC and at its height it is said that 700,000 people worked on the project. At some point, however, this vast undertaking became lost to time and was only rediscovered in 1974 by a farmer digging a well. Excavations have continued since and the three pits unearthed are thought to contain 8,000 soldiers, 130 chariots with 520 horses and 150 cavalry horses. Several thousand of the soldiers have been pieced back together and make for a uniquely impressive sight today, even without their original paint. The Emperor specifically stated that no two soldiers should look alike and all are life sized.

We had driven into what appeared a very rural location where the farming was quite basic and feudal with ploughs pulled by bullocks working the land. This was the second time Dianne and I had been in Xian and it was a truly awesome experience. Before we left, we met the farmer who discovered the terracotta warriors. He signed a book for us on both occasions, but we couldn't help but feel it was two different people, which made us wonder on the authenticity of these meetings.

In the afternoon after a lovely lunch at a rural restaurant we visited a relic of the Tang Dynasty, the Big Wild Goose Pagoda, built in 652 AD and containing a large volume of Buddhist scriptures which were obtained from India by the eminent monk Xuanzang. After this we finished on Xian's largest construction, the 13.7 km City Walls, which are up to 18 metres thick at the base and 12 metres in height. The walls were constructed during the Ming Dynasty from 1370 AD and replaced an even larger wall. It is possible, if you have time, to hire bicycles and ride some sections of the wall.

In the early evening, we headed out on a short walk to 'The Tang Dynasty Dinner Show', a fabulous experience with traditional instruments and dance from a golden age, showcasing spectacular dancing, very colourful costumes and vibrant music, the highlight of which for me was the beautifully choreographed ribbon dance.

This was a full-on tour and after returning to the hotel we were picked up and transferred to the railway station for the overnight train to Xining along the world's highest railway! We were served meals en route whilst we took in the changing alpine landscape to Xining. It was amazing how much junk food we were eating as we took every opportunity to purchase familiar looking packets of peanuts, crisps, dried fruits and other snacks which served most of us as a dinner because you were never quite sure when the next opportunity would arise to find any. Added to these were beers aplenty. Logic told me we should be drinking bottled water as this helps immeasurably when rising in altitude, which this train was doing at speed, which affected our appetite adversely.

Nobody had slept much even though we had relatively comfortable if not salubrious sleeper compartments in groups of fours and sixes. Four hours' rest was about the norm as just before we arrived in Xining the sleeper compartment lights were remotely turned on at 6.30am as our cheeky wake-up call.

Our early morning arrival in Xining, capital of Qinghai province, home of the most devout members of Tibetan Buddhism, was greeted by a cold fog. We could just about see that it had been snowing on the surrounding hills. We were met after a slight delay and taken to a good local restaurant for an American-style breakfast. The town had lots of what appeared to be newly built but deserted apartment blocks. This turned out to be correct as they are trying to move swathes of the population from more rural areas to create a workforce here.

We drove to the Kumbum Jampaling Monastery, which is situated in Rushar Drongdal and founded in 1560 to commemorate the birthplace of Tsongkhapa, a famous teacher of Tibetan buddhism. It's built around a tree which marks the birthplace. Kumbum has been sacked and rebuilt several times, particularly in the Muslim rebellion of 1860 when hundreds of monks died protecting the chapels. It was quite ethereal as it was very cold, snow lay all around and then started to fall. It was wonderful to listen to the melodic chanting and the rhythmic drumbeats accompanying it, very mesmeric and enchanting as the incense smoke curled to the ancient wood of the ceiling and teased your senses alongside the burning essence of the yak butter lamps. This monastery held the remains of the third Dalai Lama

We were then taken for lunch, at which time our guide and escort ate with us and explained the various simple dishes, mostly vegetables like broccoli, beans and tofu, but also the one pot of soup-cum-stew full of chicken's feet, rice and pork dumplings. All very tasty!

In spite of us having confirmed reservations for the next part of our trip, this means nothing here. Our guide had been having trouble getting us tickets for the onward journey, as is often the case, Chinese military having taken many of the sleepers, but he jubilantly informed me we can have the sleepers we need in first class, which we had already paid for, but since arrival in Xining through negotiations between our guide and me we had paid the ticket office a supplement, better known to us as a bribe. Fortunately, I had read about these scenarios during my due diligence and factored in a slush fund for this. After a bit of negotiation, he had left to try and obtain the tickets we needed and succeeded!

After lunch we enjoyed a brief visit to the local Tibetan museum which held the 618 m-long 'Thangka' scroll in the south hall – the world's longest – which charts most of Tibetan history.

By now we were quite tired from the journey and fatigued by the bitter cold. We returned to the train station in the hope we truly did have the sleepers promised for the train to Lhasa. Our guide appeared all smiles as he had been successful, thank goodness, and reiterated all was confirmed. The only compromise was that I was separated from the party, sharing a sleeper for four with some French tourists and a Chinese girl who slept from the time she boarded, lucky her! My fellow French passengers were anti-social to the point of being rude, but fortunately the rest of our party were together in groups of three and one four, plus one two, so it worked fine. Again, we were served breakfast and lunch to keep our spirits and well-being in order.

Leaving on time, we headed up on to the Tibetan plateau passing a multitude of large paddocks and flocks of Tibetan sheep. Groups of herders appeared alongside yurts some of which were colourfully embroidered complete with chimneys funnelling smoke from fires and cooking pots, which no doubt warmed the tents in this bleak windswept landscape. The sun shone brightly from the clear blue sky. Herds of yaks dot the area too.

Gradually the landscape changed and, rather incongruously, sand dunes surprised us. Were we seeing a mirage? The sand was replaced quickly by more frozen ground. Everywhere is quite flat at this stage, quite featureless, until a lake appeared, then mountains covered by snow started to change the horizon and then swallow up the track, a very hostile environment. Any habitation appeared to be that of nomads.

The altitude was starting to get to a few of us and myself probably more than most, as I deteriorate so much. I started with a banging headache and I was becoming a little dizzy and found myself short of breath doing the slightest activity. I was drinking a lot of water as advised to do at altitude and tried to eat in the buffet car, green vegetables and mushroom with rice, served by a lady who was dressed as if from an Italian café in a fifties time warp and was very animated and verbose as we tried to order meals from a menu with no prices.

Sleeping was difficult and we arrived in the second largest city in the province called Golmud, which boasts a population of over 200,000 where the train waited for about 40 minutes I was given some Diamox, which we had used going up Mt Kilimanjaro, but this didn't really help so finally I had to access the train's plumbed 40% oxygen through a mask. You could monitor the altitude on displays in the train carriages. We reached 5,241 meters and most of the journey is above 4,000 metres on this 22-hour journey of almost 2,000 km. It was fairly desolate, featureless country with the odd yak herd and a few simple huts dotting the terrain. One strange occurrence, though: passing a small halt there were Chinese soldiers who stood saluting the train as it passed like a guard of honour.

During a full day aboard the train crossing the Tibet Plateau, at times we had seen pilgrims walking and prostrating themselves beside the track on the way to Lhasa. Finally, we arrived at the city's striking new station, greeted by brilliant sunshine and a clear blue sky as we took a short journey into the city, crossing the river Lhasa, and having our first sighting of the iconic Potala Palace.

It was exciting to be in Tibet, somewhere we had always wanted to visit. Lhasa sits at a quite manageable altitude of 3,650 metres (12,000 feet) and is a unique city. The setting, the history, architecture, people and contrasts all made for a very special travel experience. We soon discovered that the

Ron and Dianne - Potala Palace Lhasa

Chinese are really trying to put their mark on the community, creating lots of depressing, featureless buildings to bring in Chinese citizens and make the Tibetans a minority in their own country.

We drove through a rather modern area that even boasted a Zara store before entering the old part of the town. We stopped in a narrow street and walked a short distance to our hotel, the characterful, 3-star Dhood Gu Hotel. It's in a fantastic location close to Barkhor Square in the heart of the Old City, only 100 metres from the famous Jokhang Monastery. The quirkily set out and furnished building was lovely, it felt like you were entering a temple as it immerses you in the beautiful and colourful Tibetan style. From its roof-top terrace you can see over the Lhasa skyline with the iconic Potala Palace, named after Mount Potalaka, standing out and you have that little frisson of excitement and disbelief combined. The Indian/ Tibetan and Nepalese cuisine was tasty and fairly priced. Rooms were all unique but not very warm. They looked very much in a middle ages time warp except for the huge 1970s-style television replete with loads of Chinese television channels.

We couldn't wait to stroll out in the cool daytime temperatures and follow the flow of people, some strolling, most doing their laps walking the 'Kora' which is a pilgrim circuit. Many devotees go on the Barkhor Pilgrim Circuit around the ancient Jokhang Temple. On Barkhor Street, you can see thousands of people with prayer wheels, Buddhist prayer beads or incense walking clockwise around the Jokhang Temple, muttering their prayers as they go. As they walk around, they pass many Tibetan arts and handicraft stalls. This 'kora' is lined with traders selling everything, as well as small alleyways leading to temples, markets and homes. So pilgrims can do their shopping at the end of their kora around the monastery as part of their day. Tibetans, in general, consider this temple as the most sacred and important temple in Tibet. All around the square are shops mostly small and family run and had off shoots into back streets.

We climbed a series of stairs into a rooftop café called Makye Ame, which had views to the streets below and the constant flow of pilgrims, many of them prostrating themselves, which almost looks like a gym workout called a 'burpee' – they go down on their knees which invariably are padded and then with a piece of wood in their hands stretch out fully before returning to full height, whilst chanting all the time in worship of Buddha having travelled many miles on foot. Incense burners had been lit around the square and were constantly supplied by worshippers bringing juniper branches. They walked around each corner of the square, which was bedecked in prayer flags of various hues.

There is a very heavy Chinese police presence around the square with checkpoints including scanners for weapons and other items of concern. The Tibetans were resplendent in traditional costume, many stopping and asking us to take their photograph while some took pictures of us.

Next day the vast Potala Palace beckoned, the former chief residence of the Dalai Lama of Tibet until the 1959 Chinese occupation of the country, which the Chinese refer to as the 'Tibetan uprising'. The palace, reputedly the highest palace in the world, is today a museum and UNESCO Heritage Site. It dates from 1645 and has been added to over the centuries. It is thirteen storeys high and contains over 1,000 rooms, 10,000 shrines and 200,000 statues. Visitor numbers were restricted to about 700 a day when we visited, although this seems to vary into the thousands, dependent on a seemingly random decision by the authorities, although it always feels there is little time to have any solitude from other visitors.

We wandered about the delightful Zongjiao Lukang Park below the Potala Palace, giving us some picture opportunities looking up at the iconic red palace atop the surrounding and supporting white palaces with the Chakpori Hills behind. A huge queue encircled the palace but we were allowed priority access with our guide as overseas tourists. We had to produce our passport, which needed to correlate with the visitor permit we had pre-arranged. Now here is the thing: you have just one hour to explore, which does not seem enough time but, whilst it would have been great to have more time, this works and allows you time to see everything and take it all in.

The Potala dominates Lhasa not just because of its size, reaching 3,700 metres in height, but also its location on Mount Potala. The views you get from the palace stretch right across Lhasa to the mountains that surround the city. We had the chance to walk the devotional 'Potala Kora' that surrounds the palace, but the altitude and heat were getting to some of us. A bottle of water was essential and as you enter the main area of chapels there is a square where you can buy water and other refreshments as well as use the toilet facilities, probably better than the toilets in the quadrangle higher in the palace where the urinal was below a lattice glass window with a fabulous view, but the Tibetan toilet for ladies and gents was very interesting, where you could sit on a wooden seat which had an exit area as you sat on the edge and did what you needed to do which then descends in a 10-metre drop to rocks below.

A strong sun was beating down from the clear blue sky and felt surprisingly warm in the thin air at altitude. We slowly ascended the many steps to see some of the meeting rooms as well as the extraordinary chapels of this seventeenth-century building with vast gold and jewel-encrusted shrines to previous Dalai Lamas (the chapel of the fifth Dalai Lama has a stupa coated in 3,727 kg of gold and 18,680 precious stones). The scent of burning yak butter lamps pervade all the temples. A wealth of history fills these hallowed walls, such as the art wall with 698 murals telling the story of Tibet in religious and historical scenes. The Potala Palace also preserves the tangkas art, beautiful tapestries on ancient cloth and silk.

It's a really humbling experience seeing the former Dalai Lama quarters, his audience rooms and study area. We learnt that 'Dalai Lama' means knowledge and highest, so in essence the wisest one.

As we ascended through the immense structure it seemed to be drawing us in and enveloping us. It truly gives you a spiritual peace and assimilates into your soul making you feel enlivened and relaxed. The fabulous architecture and artefacts assault the senses and the spiralling scents of yak butter and burning incense along with the Buddhists chants just make you feel like you have had a life-changing experience in this majestic Palace, banishing your fears and life's woes. Anything now feels possible if you believe.

The Cultural Revolution between 1966 and 1977 saw monks removed and the abandoned palace was looted and damaged by Chinese soldiers. I cannot help but feel that the constantly creeping Chinese influence and control over the country will eventually do for the heart of Lhasa, probably starting with restricting numbers to Barkhor Square and the Jokhang Monastery. There are already armed checkpoints at every access point. You fear for Tibet and its people as they look and feel oppressed. Many checkpoints on the roads between provinces where you have to show your documents enforce the Chinese control.

As we returned to our hotel and then another stroll around Barkhor Square for some souvenir hunting we couldn't help but be affected by the awful habit of some people clearing their throats, hawking and spitting into the street, especially common in China. Little children are walking along with commonly split breeches showing their bare bottoms, and parents regularly held them over a gutter or latrine to use the toilet.

It was almost time to leave this fascinating and controlled city. We went for one more wander around the old city and its labyrinth of alleys and multitude of shops. Enchanting Barkhor Square is a real joy as we browsed all sorts of produce and local handicrafts. I didn't want to leave the place. We visited Dropeling Handicrafts Centre with its leather bags, rugs and other woven goods, all genuinely Tibetan made. As we strolled past all the small stall fronts, some open, some enclosed shops, we witnessed pasta making and yak butter being sold in large pats. Wool that had been collected off

the surrounding hills was being made into garments, silk goods were being made, butcher stalls creaked under the plethora of yak meat.

Finally, we reached a small milliner's stall. Those who know me know that I cannot resist buying a hat. I tried a few on and paraded outside the shop for all my fellow travellers and some locals before deciding on a very smart 'Derby' style in brown

Ron with our superb Tibetan driver Dawa!

314

100% wool and I was told it was foldable and packable, ideal for a travelling nomad like myself …

More shopping ensued, a number of souvenirs bought, including gilets filled with down and synthetics which were very cheap at just £15.

We found some decent restaurants in Lhasa, particularly the Lhasa Kitchen. On our final night we enjoyed a farewell to Lhasa meal at the delightful Dunya restaurant run by a Dutch couple This restaurant was such a breath of fresh air, with its bright, airy ambience and outside balcony for any time of the day. Brilliant and tasty food with a great selection of dishes local and Western and a really mean-tasting ginger, lemon and honey tea, perfect for the altitude and any nausea! Clean toilets too, always a plus, and very friendly and efficient service. We chatted about the journey so far and how we were ready for our adventure onwards towards Everest.

Next day we headed off on our expedition towards Everest Base Camp and the awesome Himalayas.

After breakfast our convoy of four Toyota land cruisers departed Lhasa setting out in front of the Potala Palace one last time before heading towards Gyantse which is 230 km south-west of Lhasa. Our driver was Dawa whose name means moon.

Dawa would chant his prayers throughout the day as he drove. The city was soon left behind once we'd crossed the Lhasa river and before long the road started to wind upwards through a quite barren landscape dotted with the sparse growth of hardy plants and bushes, which yaks wandered between, reducing the food supply even more. We ascended via tighter and tighter switchbacks in the road towards the 4,780-metre (15,680-foot) Kampa La pass.

We had stopped briefly beside the mighty Brahmaputra river for photos and a toilet break 'au naturel'. Kampa La, like many such passes in Tibet, with colourful prayer flags fluttering in the stiff breeze, offered spectacular views in all directions before the road descends slowly to the unbelievably turquoise Yamdrok Lake. This 72-km-long lake has a spectacular backdrop of snow-capped mountains.

We stopped to take photographs but a police vehicle arrived and moved us on. You could see the fear in our drivers' faces. We stopped at another viewpoint, where Dianne took advantage to have a ride and photo opportunity on a yak! There was a toilet block here, very cold and draughty but a fabulous open view of the lake and the surrounding mountains, a real 'loo with a view'. It smelt so bad, which is an understatement, it could be described a little more graphically!.

We followed the lake shore past a load of hawkers selling yak bone souvenirs and bracelets and necklaces made from polished Yamdrok lake stones. A guy with a large St Bernard offered photo opportunities for a fee. We then headed back into the hills passing the Norjing Gangsang Glacier, a good place to stop for photos at over 5,000 metres above sea level.

We continue along the road stopping at villages and scenic spots and a small place for lunch before rejoining the Friendship Highway. En route we encountered a few speed cameras and police checkpoints, which again made our drivers very nervous. It was a stunning day of clear blue skies and sunshine all the way, although the wind was very cool.

We passed Marivan Reservoir, which serves Gyantse and affords fabulous views of the many prayer flags up the mountain before we dropped down into Gyantse at 3,977 metres (13,050 feet) and Tibet's third-largest city. If I had the energy a walk up the Dzong (fortress) that dominates Gyantse would have been well worth it for both the views and a bit of acclimatisation. Instead, we visited the Pelkhor Choede Monastery, which was very interesting and set in the lee of the mountain wall. The monastery consists of a wide range of buildings and a stupa dating back to the ninth century. The most famous part is the Kumbum, a 32-metre-high nine-floored shrine which contains 108 unique chapels and is the largest such structure in Tibet. The main temple is called Tsulaklang and is well worth a walk around. These monasteries are far less busy than those around Lhasa. Unfortunately this monastery had been ransacked by the Japanese and a lot of the age-old prayer scrolls had been taken to Beijing and reportedly desecrated and used as toilet paper, such is the disdain and lack of respect towards the Tibetans held by the Chinese. Other ancient artefacts were used by the army as target practice.

After a long journey of about seven hours with a stop or two on the way for refreshments we overnighted at the Gyantse Hotel. This three-star hotel is about the best Gyantse has to offer and a bit of a shock after Lhasa, as the room facilities were very up to date including a flat-screen television, a brand called Panda. I wonder where that is from? It was here on BBC World News that we learned there had been an avalanche on the Nepalese side of Everest, one of the worst ever, with many casualties reported especially those doing the Annapurna Trek Trail.

We were not hungry, altitude does that to you, we were feeling a little cool as the temperature had dropped so we picked at some snack foods and had an early night.

There was little in the way of urbanisation or settlement on the way to our next destination of Shigatse. It was a further 100-km drive to Shigatse at 3,840 metres (12,600 feet). It was a cool and cloudy morning as we headed out on a straight road through a fertile valley. It was harvest time and most of the barley heads had been cut and hay had been stacked for silage.

It soon warmed up and before long we reached a small town and the inevitable police checkpoint. Agricultural equipment we saw seemed somewhat old, basic vintage tractors and more and more horse-drawn ploughs and man-held drills for planting. Juxtaposed with this almost ancient scene were speed cameras and signs besides the road threatening fines for travelling over 80 kph for which you could be punished with a 1,000-yen fine and points on your licence. This would be crippling financially to anyone with a typical Tibetan income, so the drivers had to make sure we reached each checkpoint no earlier than the speed limit would allow. Every now and again we pulled over to make sure we were safe from penalties. We reached Shigatse in the early afternoon .

We were impressed with the new railway station we passed that had just been completed to link Shigatse to Lhasa. We swung by the local police station to collect our Everest passes to allow us access to the region, then, all in order, we checked into our three-star Hotel Shigatse, about the best the small town has to offer, before exploring the town.

The Tashilhunpo Monastery and market were on our agenda today. The monastery was founded in 1447 by the first Dalai Lama and has since been the traditional seat of the 'Tashi' or Panchen Lama. The monastery is fascinating, as is the walk you can do around the Kora which stretches up the hillside behind the monastery, offering good views of the complex and Shigatse itself. Again, it was one of the monasteries looted by the Chinese in the so-called 'Cultural Revolution'. We visited the local market full of trinkets mixed with skinned animal carcasses, all displayed in the shadow of the impressive but replica dzong (fort), another building destroyed by the Chinese but rebuilt by local businessmen in the 1980s.

Shigatse feels more Chinese than Tibetan now, or so we felt on a post-sunset stroll around the featureless concrete shopping areas. The creeping degradation of society was highlighted with a lit shop window where there was a prostitute trying to look alluring in a mini skirt reclining on a couch as if we were in the red-light district of Amsterdam.

We found a rather dingy but intriguing restaurant called the Yak's Head. Entering through a clandestine heavy curtain we found ourselves in a Tibetan world of painted wooden beams, carpet-covered banquettes with elaborate decorated lighting above a prominently displayed model of the Potala Palace.

Two lovely Tibetan ladies served us and spoke no English so it was matter of pointing at things on the menu which had some pictures and we did a lot of hand interpretation with the occasional animal noises to work out what we might be ordering, true to form not all was what we were expecting when it arrived. An older Tibetan couple watched our performance and confusion and just smiled and nodded at us. We certainly kept them and the waiting-on staff amused.

We asked for some beers which did arrive but with shot glasses to drink out of as opposed to beer glasses. Our food was very yak-meat influenced, no surprise there, with spicy beans cooked in yak fat, sliced vegetables and luncheon meat, bacon fat soup, cold spicy chicken, yak ribs and rice. All very interesting, eclectic and indeed tasty. It was a great night in a wonderful space full of Tibetan ambience and nostalgia for what had disappeared in the town outside. It was like a room you would find in one of the monasteries and all this for about £3.50 each. Perfect!

It is so sad how the Chinese have tried to erode the Tibetan culture and desecrate ancient buildings. We were heading for late October now and our next leg was Shigatse to Shekar.

Altitude had been increasing and breathing was a little more challenging, making us out of breath with the smallest of efforts. It leaves your lips cracked and the body comes under a bit of pressure from the elements. Shekar, or New Tingri as it can be referred to, is over 14,000ft above sea level and just 60 km north-west of Everest.

After a pleasant rest we were all feeling brighter to prepare ourselves for the push on to Everest Base Camp. The long valleys are strewn with rocks reminiscent of the Welsh Valleys back home including stone-built sheep pens. Soon we headed up the Jugala Pass at 5,200 metres. We ascended the road to the top, en route coming across a British female cyclist who was cycling Lhasa to Kathmandu.

About thirty minutes later we were greeted with our first distant sighting of Everest. It was so clear with just a wisp of cloud over the south col.

In the afternoon we reached the small town of Shekar. We had some free time to walk around or rest up in preparation for tomorrow's big day to Everest Base Camp. We overnighted at the Qomolangma Hotel, the Tibetan name for Mount Everest.

After a restless and chilly night, we were ready for the big day, heading for Rongbuk Monastery and Everest Base Camp.

After leaving Shekar we drove through a long valley alongside the freezing-cold waters of a mountain river before we started to climb again. Before long we turned left off the main highway to reach Sakya Monastery. The seat of the Sakyapa school of Tibetan Buddhism, this monastery looks very different to others we had seen as it is of Mongolian architectural style. Much is in ruin or has been rebuilt, although some original buildings dating back to 1268 survive. It has an amazing Sutra Library. A number of us were under the weather, feeling the altitude effect, but we were given a simple but enjoyable meal by the monks, which made us all feel a little bit sharper.

We pushed on and climbed higher, reaching the 4,500-metre-high Gawu La pass, which gave us breathtaking panoramic views of the highest peaks of the Himalayas, all over 8,000 metres, the highest of course Mount Everest, then Mount Makalu, Mount Lhotse, Mount Cho Oyu and Mount Shishapangma. It was just awesome in the very real sense of the word.

We pressed on for Rongbuk. It was slow progress as the undulating landscape turned from barren brown foothills, to mountains and then to snow-capped peaks. Before long we were at the hallowed gate to Rongbuk Monastery, our destination for the next couple of days, where a passport check took place.

Rongbuk Monastery at the end of the Dzakar Chu Valley is the gateway to Everest Base Camp. It was early afternoon when we settled in to this bleak little grouping of buildings around the monastery, the cool wind, bearable in the sunshine, whipping around us. The restaurant with its central stove was a welcome, warm oasis.

After some sustenance it was time to explore the Everest Base Camp area. The hike to Everest Base Camp takes two hours from this accommodation at Rongbuk situated at 4,970 metres (16,300 feet), although there is often a national park bus in service to help you get to Base Camp if you don't want to walk. Rongbuk is often referred to as the highest monastery in the world but in reality Rongpo Monastery claims that crown set in Tingri County Tibet and located at 5150 metres above sea level.

Some took the short trek to the 5,200-metre-high Everest Base Camp,

Everest from the
Tibetan side

while a few took the Land Cruiser as they were very affected by the altitude. There was a basic toilet block beside a glacial river run-off and a mound dotted with many prayer flags fluttering in the bitter wind. We were all kitted out in many layers of top winter gear to keep out the bitter winds. Memorials to fallen climbers were in evidence.

Back at the monastery Pauline and I took a walk into the Rongbuk Monastery. Apparently, thirty-five monks resided and worshipped here, but they must have been having a siesta as they were nowhere to be seen, except one who wanted a fee for entering, which we refused to give him initially as we thought much to our shame he was an opportunist trying to sell us pre-used tickets, but it transpires the £4 each he wanted was the fee to explore on our own without a guide. He then said we could only go into the courtyard as the other area were off limits, so we didn't get to see too much.

To witness sunset over Everest was special. The real bonus was still to come when the truly awesome peak was seen under the stars which stood out in the deep blue canvas of the night sky. This had to be the icing on the cake. We had prayed for no cloud and we were blessed!

It was a rough and ready high-altitude night at Rongbuk Monastery, one of the highest in the world, its accommodation set in a quadrangle alongside the monastery. There were some rooms in the monastery block too and a row of chalets further down. The even more challenging tented accommodation at Base Camp itself had already been taken down because the season had ended for the mountaineering fraternity.

Our overnight stay was a real challenge with the combined effects of altitude, severe cold and the very basic toilet facilities, which comprised open stalls with no doors, what you might call well ventilated and they needed to be.

On the upside there was a great view down the valley from the toilet block's small window, but this only suited the blokes when standing, if you know what I mean. Another downside was the slits in the floor which were very difficult to target and, shall we say, many missed the bombing target, making it tricky for those following to resume the crouch position without putting your foot in it, literally.

The monastery served simple food for us tourists, soup and chips with fried egg being a big hit for us westerners. Ginger, lemon and honey was the drink of choice to try and combat the altitude sickness. Rooms are extremely basic and we slept four to a room which contained a small table with a flower on, some silver tin washing bowls, lots of blankets and covers and one pillow each.

It was time to embrace dinner and the night we all huddled together around the cosy stone stove with its silver metal smoke stack conveying the heat and smoke out of the building through the roof. This stove, fueled by Yak dung, was like a very old cooking range back in the UK, its top bedecked with huge heavy kettles and pans. It created an atmospheric dining room, warmed also by the wave of infectious laughter and excitement generated by being together in such a hostile climate so close to the iconic Mount Everest. Most had the ubiquitous bowl of chips accompanied by peppermint and nettle tea, trying to stay there as long as possible before braving our bitterly cold huts, but by 9pm we were being encouraged to vacate the room and get some rest.

It was one of the coldest and most uncomfortable nights of our lives, made even worse by drivers trying to run their vehicles to stop them freezing and fumes from the engines entering our rooms. A nightmare! We were wearing every item of clothing we could get on and wrapped ourselves in all the bedding provided and still we were frozen. My middle-of-the-night visit to the latrine was, shall we say, challenging, but the dark blue sky peppered with twinkling stars was a real highlight of this excursion, even if the facilities illuminated by my head torch were barely one-star quality!

After a very restless, and agitated sleepless night we tried to wash a little before going to the cafe again for hot jasmine tea, coffee and fruit and pancakes, which were most gratefully received. We were all quite exhausted by the combination of altitude sickness, headaches and nausea. Though fatigued and feeling a little unwashed, we didn't feel too disheartened especially when we were warmed by the beautiful sunrise in the clear blue sky over Qomalangma an abiding memory of this great mountain.

Just before 9am we headed away from this beautiful vista and descended down the road to Old Tingri. Our driver Dawa conceded he had slept little and had a dreadful headache as well as feeling very cold, but within the hour we entered a valley full of Himalayan sunshine. The scenery was wonderful as we bumped along the stony track, spotting nomadic yak herders and their herds with colts riding freely alongside horseback riders. Two and half hours later we entered Old Tingri. Although one of our Land Cruisers had a puncture on the way in, we had arrived safely.

We were pleased to arrive at the low-ceilinged An Do restaurant for some sustenance and a rest, and the obligatory honey, ginger and lemon tea. The antiquated stove in the middle of the room acted as a heating element for both the room and the kettle, a bit like a Tibetan Aga!

Relaxed and renewed we headed south to Nyalam. The wide valley was dotted with buildings, some as old as the eleventh century, some bearing the scars of Gurkha invasions from the past. We stopped to take our last distant photograph of Mount Everest and headed on past glacial streams edged with the melting snow and ice. We passed Gu Co Chong Oo, one of the many ruins, before ascending a pass where the great Himalayan range reaches the Shishpangma range. We then entered Nyalam county, where we felt as if we were flying, level with the high peaks. There was snow everywhere in this winter wonderland.

We stopped at a simple restaurant in Nyalam at 4,330 metres above sea level. We washed down vegetables, egg-fried rice and yak curry with either beer or green tea. We carried on along a yak-speckled road. It started to snow. A river ran below us, pounding down into the valley through the steep sides of the chasm of steep unforgiving sides of sheer rock.

Before long we came across a police checkpoint where it started to sleet heavily and a clap of thunder resounded around the heavens. Like a starter's pistol, it set us off on the steep, winding road down to Zhangmu, where we were greeted by a road congested by a long line of lorries, which were stuck as the road after the town into Nepal was impassable for these vehicles due to landslides.

Luckily, our plan was to stay overnight here in Zhangmu. It's a small town, its buildings clinging to the hillside, threaded with narrow roads barely able to take a truck's width in places. We reached our aptly named Zhangmu Hotel, about the best in town and only two stars, clean though, with large rooms. We had a fabulous view across the gorge to the far mountainside. We said goodbye to our guide Tsidor, who had been friendly enough but not the most informative. It was our devout Buddhist Dawa, our driver, who we would miss most. In fact, we learned more from him than the guide. He barely spoke any English but had a way of getting his message over to us and showed us interesting sights and pointing out people we would otherwise have missed.

At breakfast fellow travellers shared their night's adventure with a rat in their room. They managed to isolate it in the bathroom and reception changed their room for them. When they returned in the morning, full of trepidation, to collect their toiletries, they were relieved to find the rat gone, but not before it had chewed all their toothpaste and cosmetics. So we were looking for a made-up, sweet-smelling rat with fresh breath.

After our rather basic breakfast, we checked out and descended the twisting narrow road by Toyota Land Cruiser to the small, modern border terminal fully staffed with Chinese military personnel, only to find it didn't open until 10 am. We whiled away the time looking at the Friendship Bridge across the gorge which connects Tibet with Nepal.

Our passport control experience was quick and we soon crossed the bridge and met our Nepalese guide, Rasu, who took us through Nepalese immigration, which was rather more basic than the Chinese side but efficient in its own Nepalese way. As we entered Nepal a bevy of ladies and young girls enveloped us to act as porters. A pretty girl called Syreeta tagged on to us, Dianne protested she was fine, but the girl insisted and we couldn't resist her in the end. The bag was bigger than her and we felt sorry for her carrying this burden, pointing out that she could wheel it. We walked down the potholed narrow thoroughfare past stalls selling all manner of basic products as we tried to avoid mopeds and motorbikes. Finally we reached a widening of the road where a rather old and rickety-looking coach awaited us. It had a very large steering wheel and we soon found out that power steering was not one of its assets.

One of our passengers was surrounded by an amoeba of children as she made the mistake of handing out a few sweets from her bag. Eventually we were all emptying our small bags of goodies of every sweet, chocolate bar and even packs of dextrosol tablets to share with them.

Off we went in our rather ancient mode of transport on a narrow road which wended its way above a precipitous drop to a river gorge below, the vehicle barely clearing the edge of the road. A boy hung off the side of the coach guiding the driver as to where the wheels were, blowing a whistle every time the tyre was on the very edge of the abyss thus saving us from plunging into the ravine many metres below. On the way we saw lots of small suspension footbridges. After a while we reached a police checkpoint on the outskirts of Barabise on the east bank of the Bhote Koshi river, where a small girl sat with a little puppy garlanded by a chain of marigolds.

A police officer pointed out to our driver and guide that the road ahead was very badly damaged by the recent landslides and he was not sure we would get through. This now underlined the gravity of our group's decision the previous evening to risk taking the route south as opposed to the ironically named Intrepid Travel group who were told by the company they would not proceed and would return all the way to Lhasa by road and fly into Kathmandu because the route south was too dangerous.! It appears we were the adventurers and not Intrepid Travel, gallant or stupid we would soon find out!

Sure enough, a short way along this road towards Kathmandu we reached a point where a very severe landslide had occurred and we could see the remains of dwellings in the river far below. We were told over 250 people had died because the landslide took place at midnight when people were asleep. Huge boulders had come down the mountainside, breaking trees like matchsticks. The amount of debris and collapsed mountainside that had reached the river had now caused a dam, forming a small lake. Just a small amount of water was able to flow through, so the water was forcing its way higher up the valley side, creating more erosion.

We cautiously took the road onwards as narrow and rather basic restoration work to the road had been done. We were concerned that our bus might be too heavy for the hastily repaired road, and everyone wanted to be on the mountain side of the coach trying to act as a counterbalance against any shift that might take the bus over the edge of the gorge. A rather hairy journey, for sure, but we made it through.

A little further downstream we came across an electricity pylon beside a house, both almost completely submerged. Fortunately we managed to negotiate the rest of the road until we reached the valley floor and stopped at a pleasant location which was used for adventure rafting on the river called the Sanroshi Bar near the village of Balthali. We had a well-earned beer to steady the nerves and a bite to eat and sat in the sunshine amongst green fertile terraced slopes, pondering the disasters we had witnessed back upstream and lamenting the desperate loss of property and life the Nepalese had endured.

We were only about 40 kilometres south-east of Kathmandu but on this narrow road every time you met a vehicle coming the other way there was a lot of shunting and maneouvering, which left very frayed nerves for us passengers yet the driver would take it in his stride and sometimes be on his mobile at the same time whilst we were aghast, especially as every bend in the road was accompanied by the loud honking of horns, letting people know we were coming.

We ascended towards Dhulikel, sometimes passing vehicles, where it was a miracle we didn't hit one another but neither vehicle stopped. I'm not sure our driver closed his eyes, but many of our group did! Finally, we stopped for a pre-arranged meal at the Dhulikel Lodge and Resort, a fabulous spot set high up and with terrific views across the valley and hills to the Himalayan mountains from Mount Annapurna in the far west to Mount Karulung in the far east. The farthest reaches of the range were like a snow-topped mirage piercing the distant blue sky.

We sat on a lovely outdoor terrace, thought about the journey and sights we had seen en route and thanked our lucky stars we were safe. Relaxed, fed and watered, we were soon back on the road and one hour later we were descending into the capital of Nepal, Kathmandu.

We had been to Kathmandu a few times before and appreciate the way it feels like a crossroads and feel comfortable in the ramshackle labyrinth of alleys and narrow streets that make up the old part of the town known as Thamel. We loved spending time at Helen's Rooftop Restaurant, looking over Kathmandu, which bustles below. One of the bizarre places you see in the city is the Department of Money Laundering Investigation. Given the size of the building there must be a lot of it going on!

The city is the gateway to the Nepalese side of the Himalayas, and home to a number of World Heritage Sites. In addition to Swayambhunath temple, Boudhanath stupa and Pashupatinath temple, we of course love Durbar Square, where the chaos of a plethora of people and motorbikes flying around is part of life.

At the junction of Durbar and Basantapur squares, a red-brick, three-storey building is home to the Kumari, the girl who is selected to be the town's living goddess and a living symbol of *devi* – the Hindu concept of female spiritual energy. Inside the building is Kumari Chowk, a three-storey courtyard. It is enclosed by magnificently carved wooden balconies and windows, making it quite possibly the most beautiful courtyard in Nepal.

We were blessed in an audience with the Kumari of Bahal at her small palace just off Durbar Square, the residence of a living goddess! She was eleven years old when we met her, but she was selected as goddess at the age of five. When she reaches puberty, she will return to her family. She is looked after by a high priestess in the meantime, who accompanied her to a balcony above us from which she blessed us.

A kumari, or kumari devi, a living goddess, is traditionally worshipped in Nepal as a manifestation of divine female energy. In Hindu and Buddhist religious traditions this is common. Kumari means 'princess' in Sanskrit and the young girl is chosen from the Shakya caste or Bairacharya clan in the Nepalese Newari Buddhist community, yet also revered by some Hindus.

We revisited all the major sights, such as the Swayambhunath Buddhist Temple complex high above Kathmandu where we viewed the city below. There were lots of monkeys scooting around here so you had to have your wits about you. They will soon attack and steal anything from you, including your camera. We learned about the meanings of the five colours of the prayer flags which are ubiquitous in this Himalayan region and indeed throughout all Buddhism and its temples. The five colours consist of blue for the sky, white for air or wind, red for fire, green for water and yellow for earth. The centre of the flag often shows a 'lungta' (strong horse), seen as a symbol of speed and transformation of bad fortune to good, bearing three jewels on its back that represent the Buddha, Buddhist teachings and the Buddhist community. Images of four sacred animals – dragon, garuda, tiger and snow lion may appear in the corners. Mantras and prayers are usually written on the flags for the life and fortune of the person tying the flag. Himalayans believe that when the wind blows the flags, it spreads the blessings, goodwill and compassion embodied in the images and writings and distributes them across the land. Eventually the prints fade and the prayers become part of the universe, and the prayer flags are replaced anew.

I love strolling along the Pashupatinath area to the Ghats and witness the funeral pyres beside the river before they collapse into the waters, a very solemn and thought-provoking location. Hindu family members walked in a circle around their loved ones before the body, draped in an orange cloth, was set alight. This spot is for the common people, another location across the bridge is for the more privileged, which makes me smile. We all share the same fate, whichever way you gild the lily, it's nature's way of making sure we all have the ultimate equality Lots of Saddhus sit cross legged dotted around above the Bagmati River, Kathmandu's most important Hindu cremation site. Saddhus are India's wandering holy men. They have renounced their worldly life and all their comforts, so given up their possessions and their families, and now lead a life of celibacy and yoga in their search for enlightenment. I think some of them have given up on hygiene, too.

Everywhere people were creating 'rongali' patterns by their homes. These are very bright and create a trail, a sort of invitation and welcome to the god Laxshi and other dead spirits into their home for the festival of Diwali.

We enjoyed a delightful lunch at the atmospheric café in the Pathan museum's garden. Where an Everest beer or two was enjoyed, I was taken by the label on the beer showing the Tibetan-born Sherpa Nwang Gombu, who climbed Everest twelve times and his seven brothers also climbed Everest. Who can say that about their families? Quite an achievement. Gombu made his first ascent as a seventeen-year-old with Sir John Hunt and Sir Edmund Hillary when they were the first to conquer Mount Everest. Gombu was the youngest member of Tenzing's team of Sherpas.

We concluded our visits with time at the impressive Buddhist Boudanath stupa and enjoyed our last supper in our very comfortable Hotel Annapurna which we had been switched to as an upgrade. Last time we'd stayed at the Yak and Yeti. Don't you just love the names? We celebrated our wonderful adventure on the last night with a Nepalese meal and a few more Everest beers with their label of our now legendary pal Sherpa Nawang Gombu.

We had survived, endured and absolutely loved a wonderful adventure. It was time for home! It was not until we returned that we realised how close a shave we'd had as in 2014 a snowstorm disaster occurred in central Nepal during the month of October and resulted in the deaths of at least forty-three people of various nationalities, including at least twenty-one trekkers. Injuries and fatalities resulted from unusually severe avalanches and massive snowstorms which we were oblivious to on the Tibetan side of Everest on and around the mountains of Annapurna and Dhaulagiri. The incident was said to be Nepal's worst trekking disaster. Our family had been trying to contact us and the embassy authorities had been fearing the worst.

The landslides we witnessed on our precipitous and dangerous passage from Zhangmu on the Tibetan–Nepalese border along the Sankoshi river were the result of torrential rain in August before we arrived, stranding thousands of villagers and sparking fears of a cholera outbreak and destroying villages and roads. Things became worse when in October just before we passed through more heavy rain brought further devastation just as the Nepalese were trying to recover.

The main (and only) artery of goods and people to China, the Araniko Highway, was blocked by the landslide ripping out 5 km of highway and causing huge traffic jams, especially hundreds of trucks normally supplying goods to Nepal. Dozens of houses and community buildings had been swept away by the landslide. The landslide had a volume of 5.5 million cubic metres and had massive consequences far beyond, not evident from the many pictures you might see. The dammed river was threatening to unleash a torrent of water to hundreds of downstream villages that would ravage as far as Northern India once it flowed with full tempestuous force. Despite the use of dynamite, it took the Nepali Army forty-five days to dig a canal through the blockage to allow water in the lake to drain more gradually. The lake created was 47 metres deep and over 400 metres long. Even so, the emergency draining through the lake canal itself caused damage to houses downstream and threatened to destroy the Lamusanghu Hydropower Dam.

It left us with a very sobering view of life and counting our blessings!

CHAPTER TWENTY-SIX

———

Interesting people

I first became aware of a distant relative Gareth Jones through a distant cousin, Stella Mathias, who was living in Haverford West at the time. A film about him has been made called *Mr Jones*, starring James Norton. She recommended a book to me called *More than a Grain of Truth*, written by a mutual relative of us all, a very sharp lady called Margaret Siriol Colley, who was a niece of Gareth Jones.

Gareth Jones was a young and very brave Welsh journalist and foreign affairs adviser to Lloyd George, who died in mysterious circumstances on the eve of his thirtieth birthday in 1935 in Inner Mongolia, having been captured and held for ransom by bandits.

He flew with Hitler in February 1933 and a month later became the first journalist to expose the famine then raging across the Soviet Union. By telling the truth, instead of becoming acknowledged and applauded, he was denigrated and disparaged by other Moscow correspondents, shunned by the British establishment and of course became the subject of intolerance and viewed with suspicion as well as being blacklisted by the Soviet secret police.

This brilliant and illuminating biography by Margaret Siriol Colley uses her uncle's letters, articles and diaries to create a picture of a man who was not afraid to tell the truth whatever the cost and is especially fascinating as it is written by someone who knew him personally. It is also an enlightening view of social and political history in the early thirties.

Well, I have met loads of interesting people in over sixty years, lots of great footballers like Arthur Rowley and Harry Gregg, a surviving 'Busby babe' of the awful Munich air crash in 1958, Alan Durban and even Joe Hart, as I've already mentioned.

Meeting Stephen Venables, the great British mountaineer was special. It was after a talk he did at the local theatre that I got to know him better. We chatted over email and I read the book Stephen wrote about his son Ollie, a young boy who dealt very bravely with a number of illnesses, before succumbing very early in his life. It was then that I asked Steve to consider becoming patron of our charitable fund, Dreamcatcher Children, and he kindly agreed. More about Dreamcatcher later.

Dianne and I also met another intrepid mountaineer and explorer, Rebecca Stephens MBE, in 2016 when attending a talk at Ludlow. She was an inspiring speaker, drawing on her experience as the first British woman to conquer Mount Everest as well as the first British woman to climb the iconic 'Seven Summits', a mountain quest all climbers want on their résumé.

Eva Hart was a survivor of the sinking of the Titanic. We met her at a Titanic evening at Newtown in Powys, arranged by Harold Beadles, who was the President or Chairman of the Titanic Society at the time. Eva gave a great talk. She had the audience in the palm of her hand as she described that fateful night. Afterwards we had a brief chat with Eva and she signed my ticket for the evening. Eva was only seven when she lost her father in the disaster while she and her mother survived. It's her description of the screams that pierced the still, cold night air followed by an eerie silence beneath the star-filled sky that has stayed with me.

We used to do cricket tours to Barbados in the West Indies, courtesy of a great fellow called Basil Matthews. Basil was a character, in his seventies back

Ron with the legendary Sir Garfield Sobers sharing a rum and coke!

then, a really fit guy, who swam in the ocean every morning and played golf with Sir Garfield Sobers and others regularly. We had the pleasure of meeting and working with him in Barbados. We took a local club, Perkins Cricket Tours, in the nineties, and one evening as a special surprise Sir Garfield and Everton Weekes came by to a little soirée we were having, all thanks to Basil. Needless to say, most of the age group there related to Gary Sobers. Everton, also a real character, spent a long while chatting with John, one of the older chaps on tour. What a great experience. At the end of the evening Gary invited me to play some golf with him, but I declined, saying the team schedule we had was too busy … what a fool I was!

Di and I had actually met Everton Weekes in a bar earlier in the week on the way back from one of the cricket matches. Basil was driving, even though he had imbibed copiously of the rum and cokes on offer post-match, and said I need to stop off here, I have someone for you to meet. We entered this characterful, small watering hole frequented by locals and had a very interesting hour or two.

Everton Weekes was so funny, a master storyteller. He told the tale of why he was called Everton, explaining that about the time he was born his father was really keen on football and Everton were playing Wolverhampton Wanderers in a very important cup game. Everton won and that was the day his son was born, so his father decided to call him Everton. With a big grin he took a swig of his white rum and coke and said, 'Thank goodness Wolverhampton Wanderers lost!' and we all fell about laughing. Sadly, Everton passed away on 1 July 2020.

Whilst in Barbados we went to a horse racing event, the annual Sandy Lane Barbados Gold Cup Day, held in March at the Garrison Savannah.

Perkins C.C. Barbados Tour group at Dover C.C.

What a great day! We had a box between us, though mosquitoes were still a problem as the 'executive box' we'd been allocated was open to the air and the pesky flying biters. Frankly, the Barbadian idea of an executive box actually resembled a stable with breeze-block walls, and whilst I am sure facilities may have improved, the only air conditioning then was if the breeze blew. Good fun, though, and a number of the West Indies Test players like Tino Best joined us.

Tino had actually turned up at one of the Perkins games, set amongst a sugar plantation which had grown higher than us. Perkins were doing quite well and Dianne and I were regularly looking for cricket balls hit for six in the triffid-like jungle of sugar cane. The home side said to our captain, 'Do you mind if this chap joins in?' and Perkins had no problem with this until the wickets started to fall and they realised they had allowed Tino Best to join the bowling attack. Such is the informal nature of West Indian cricket. Can you imagine that at a local Shropshire ground and, say, Stuart Broad turning up for a game? As an aside and Shropshire link, Tino played half-a-dozen games for Madeley in Shropshire before he entered the Test match arena.

At the shindigs after the cricket matches copious amounts of rum and coke and beers were consumed and party games such as drinking glasses of Barbadian rum and then trying to run ten times around a cricket stump with your hand on it, before running back to a starting point and eating a dry Weetabix. Many did not get past the running around the stump.

The Barbadian post-match buffets were, shall we say different. The local cuisine, included sweet yams and fried plantain bananas and banana bread, fish cakes, rice and peas, chicken curry and pigs' tails. Yes, you heard that right!

Everyone was asked to bring something unusual to Barbados for a competition for the most outlandish. The winner was a lad who brought

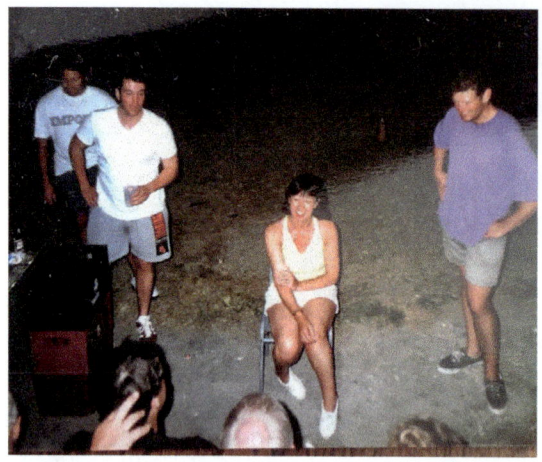

Barbados fines committee rum drinking penalty for Dianne

a 'men at work' road sign. How the heck he obtained it, let alone transported it and passed through customs and baggage control is to this day a mystery!

Preparation of cricket pitches was a whole new experience to me. I arrived early at the Dover cricket ground to make sure we knew the location and check out the usual basic facilities.

On arrival I saw the groundsman and a helper digging up the pitch with a fork. I wandered over to him. 'Hello, sir, is this where Dover are playing Perkins, a British touring team today?' I asked. 'For sure, man,' he replied and carried on with his work. He proceeded to water this ploughed patch of land which resembled a garden allotment which had just been forked. He then wheeled on the heavy roller and flattened it all. Proud of his work he sat back under the canopy of the small pavilion and supped a cold Banks beer whilst watching the sun bake the pitch hard, perfecting a good batting strip!

On one of the cruises with a farming group we were invited to the Captain's table and were seated with the well-known rugby union commentator Bill McClaren and his wife. Bill did all the talking, and was a real ladies' man, but very interesting when he told us his rugby stories.

Tania Szabo first came to my attention in the 1958 film *Carve Her Name with Pride*. She was the daughter of the Special Operations Executive heroine Violette Szabo, who, at the tender age of twenty-three underwent a gruelling training regime and was landed behind enemy lines in Nazi-occupied France.

We met Virginia Mckenna, the wonderful actress, wild animal conservation advocate and wildlife campaigner, at Wormelow in Herefordshire whilst attending the celebration of Violette Szabo's annual birthday remembrance. A lovely, gentle and gracious lady whose heart and soul has been put into the Born Free Foundation which she started with Bill Travers her husband and since his passing, she continues with her son Will.

Many of you will remember Virginia from the film *Born Free*, a 1966 British drama starring Virginia McKenna and Bill Travers as Joy and George Adamson, a true story about the couple who raised Elsa the Lioness, an orphaned lion cub, to adulthood and released her into the wilderness of Kenya. She also starred in wonderful films including *Ring of Bright Water* and *A Town Like Alice*.

Virginia very kindly shared experiences about the research she did to act the part of Violette Szabo in the film *Carve Her Name with Pride*, which helped my research for my second novel, *Etive*.

Unfortunately, although Violette was successful in a number of activities and very brave, she was caught and tortured and eventually died in Ravensbrück Concentration Camp along with three other female agents.

Whilst researching for my novel I made contact with Tania and arranged to meet her and we had lunch together, just Tania, her dog Mike and my wife Dianne. It was fascinating to meet this lady who had led an interesting life herself, spending a lot of time living in Jersey. She was full of fun and had a wicked sense of humour. Sadly, due to a house fire she had lost everything, but managed to salvage her mother's medals which she sold to a man who donated them to the Imperial War Museum.

One thing that sticks in my mind is what she remembered about her mother. She recalls being in her grandfather's arms and seeing her mother disappear down a big black hole. She was referring to the underground station and this was the last time she saw her mother.

When Ron Morgan Travel was in its original office in Princess Street, we met some very interesting characters and as always had wonderful clients who crossed the economic and class divide, from your ordinary man in the street to viscounts, to well-known sportspeople and television personalities. One of my well-known clients was Alan Whicker of *Whickers World* fame, who at the time lived on Jersey. For the most part I spoke to a personal assistant of Alan's but just occasionally it was rather strange to be talking to that familiar television show host's voice, almost surreal indeed, given my business. Alan was often referred to in the media as a 'travel journalist', which was understandable given the travelling he did time and again to far-flung destinations. He used to talk about being just a journalist to me and would add 'a journalist who travels'. I cannot dispute that fact.

Shropshire and the Marches has been the home to many musicians and associated producers and agents, like Ronnie Lane, Roy Wood and many more. I had been making travel arrangements for a lovely chap for quite a while whose name was Paul Samwell-Smith. One day he popped in to pick up some tickets and gave me a new CD album by Beverley Craven, who I was only vaguely familiar with at that stage. She had a lovely voice, and it turned out the single 'Promise me' from the album became a huge hit for her.

As we got talking it turned out he played bass for the Yardbirds. Not only that, he had formed the band with his school pal drummer Jim McCarty, and played with the likes of Eric Clapton, Jeff Beck and Jimmy Page before going on to be a producer for many records including tracks by Al Stewart, Paul Jones of Manfred Mann, Cat Stevens, Carly Simon, Jethro Tull, Chris de Burgh and many others. You see, you just never know who you might meet!

One day who should walk through the door but a familiar face to us in the eighties and nineties. 'Is Susie here?' he asked. No, she was at lunch was our reply. 'Oh, I'll come back in an hour. Tell her the fella she met on the ski educational called.' Susie returned and when advised immediately blushed. A few minutes later the character returned.

'Hello, Susie. Surprise!'

'Hello, Eddie,' Susie responded. Then the penny dropped: it was Eddie the Eagle, a nice chap and quite a surprise to us all, not least Susie.

My love of running means I am always interested in how athletes are doing, seeing the new ones achieve great things and wondering what older ones are up to. In 2016 I made contact with Zola Budd. It was by chance as my old contacts in North Carolina pointed out to me Zola, now Zola Pieterse, was at Coastal Carolina University as assistant track, field and cross-country coach.

Zola was a young girl I had great sympathy for as I felt at the time, she was much maligned and misunderstood by the media. It was kind and humble of Zola to take the time to engage. She was obviously enjoying her life now and seemed very happy to be more under the radar compared to her earlier life.

Over email we recalled the past, my links to North Carolina, how she'd ended up at the university, and the documentary which had been released about her rivalry with Mary Decker, all of which was fascinating.

To me she was the young girl who ran barefoot and broke world records in the seventies. Zola was very philosophical and ended our chat with the statement that after her injuries she was so happy to still be running and able to coach and run with the girls in her team, as well as compete in some running events.

It transpires Zola has three children. She has run a few marathons around the three-hour mark and done extremely well in ultra-marathons.

I am so pleased to know Zola is coaching and helping other athletes attain their goals, as of late she is concentrating on developing her prodigies in the cross-country discipline and hopes this will become an Olympic event sooner rather than later.

I find it hard to get my head around the fact she will be 55 years old this year 2021!

In 1985 Zola ran the mile in 4 minutes 15.57 seconds, still a British record! In a chequered career against the background of family disharmony and apartheid she was a 200m and 5000m track world record holder and won the world cross country championships twice, quite an amazing lady!

With my cancer diagnosis being almost identical to that of George Alagiah, the newsreader, journalist and novelist, after meeting him at the Hay Festival and Dianne sharing experiences with his wife, I have tried to keep in touch with George as he is an inspiration to me having fought off cancer for around six years and even beaten Covid-19. He said he didn't know whether he had beaten the disease for good but for now was counting his blessings. A lovely chap!

One of my inspirational heroes has been Vera Lynn. Her songs sustained my parents and grandparents and were shared and sung by us as children and indeed came to the fore yet again in the recent Covid crisis with the reference in the Queen's speech to 'We'll meet again'.

I had shared correspondence with Dame Vera Lynn when researching my first novel as she had flown with the Entertainments National Service Association to entertain the troops in Burma during World War II. Whilst some stories and newsreel exist of that time, she shared with me in 2014 that she was writing about her experiences and hoped to publish that year. In reality publication wasn't until 2018. The book was co-written with her daughter, Virginia Lewis-Jones, and called *Keep Smiling Through – My Wartime Story*. What a wonderful lady, loved and feted by many from a range of generations to the day she died in 2020.

About 5 years back we had the privilege of visiting Yr Esgwrn the home of Ellis Humphrey Evans better known as the renowned Welsh poet Hedd Wyn.

We were blessed to meet his nephew Gerald Williams who lived in a home just down the track not far from Trawsfynydd village where a bronze statue of Hedd Wyn commemorates the famous bard.

Ron and Gerald Williams at Hedd Wyn's home 'Yr Esgwrn'

We spent a pleasant and informative couple of hours with Gerald, a good deal of the time sat in front of the fireside at Yr Esgwrn where he boiled the kettle and shared a cup of tea with us. We were enthralled to see and appreciate the famous 'Black Chair' known as such because Hedd Wyn was posthumously awarded the Bard's chair at the 1917 National Eisteddfod which took place at Birkenhead Park in September 1917. Hedd Wyn had been killed in World War one, so the empty chair was draped in a black sheet. It was delivered to Evans' parents in the same condition, "the festival in tears and the poet in his grave", as Archdruid Dyfed said. The festival is now referred to as "*Eisteddfod y Gadair Ddu*" ("The Eisteddfod of the Black Chair").

Hedd Wyn was killed on the second day of the Battle of Passchendaele on the 31st July 1917.

Yr Esgwrn was sold to the Snowdonia National Park Authority and opened its doors to visitors on the 6th June 2017

On the statue of Hedd Wyn in Trawsfynydd village, Hedd Wyn had written in memory of his friend Tommy Morris who was also killed in WWI.

Ei aberth nid â heibio – ei wyneb	His sacrifice was not in vain, his face
Annwyl nid â'n ango	In our minds will remain,
Er i'r Almaen ystaenio	Although he left a bloodstain
Ei dwrn dur yn ei waed o.	On Germany's iron fist of pain.

CHAPTER TWENTY-SEVEN

Losing people in your life

We all lose people we love and care for in our lives. Some are taken far too young but losing my father in 2010 shook my world. He had been in poor health since his early sixties, precipitated by a bladder growth problem in his fifties. He lost a lot of weight and no specialist knew why all his tests came back okay. He did suffer from legionnaires disease at one time and was very ill in hospital. He never really recovered properly, he dropped to five stone five pounds and consolidated at about six stone. It transpired that my eldest sister Issy had encouraged him to eat more healthy foods. Sadly, he took this too much to heart and became somewhat obsessional, resulting in the weight loss. Losing his mum, my nan, was another big contributory reason his health suffered. This perfect storm of events somehow meant he was never the same after that. The stress and loss were unbearable for him.

Dad seemed to chug along and, whilst he continued with habitual routines like getting up at 5am as he did in the army and when he worked, he was now retired so he would go to bed late afternoon just after 4pm to listen to his radio and generally rest. His meal schedule resulted in him eating less too.

We felt he may not have long to live, such was his deterioration, so I decided to take him to Germany while he could cope with the journey, back to where he served just after the Second World War. With big excitement we embarked on a tour of memories, especially as it was the first time Dad had flown.

It was a lovely small but comfortable prop jet airplane complete with leather seats. Dad was in awe of it, he sat looking out of the windows at the clouds and looked like a kid in a sweetshop. He wished he had flown years ago. We landed in Hamburg and drove to Celle and Lüneburg where Dad was based in the post-war years and where our small hotel for the night was located.

When Dad got up to shave, rather early as usual, I got up with him made sure he was okay. I left him in the bathroom and lay back in bed. I dozed a little as I was bushed, but was startled awake by a bang and wondered what it was. I shouted, 'Dad! Are you alright?' I had no response. I tried to get into the bathroom but could only just open the door it was being blocked by something. Unfortunately, Dad had fallen. I managed to force the door open, pushing Dad's body aside and recognised he was bleeding badly from the side of his head. He was unconscious, as when he fell he'd banged his head on the sink.

I tried to bring him around by talking to him, and after a few minutes he responded, but he was very woozy. I wrapped his head in a towel to stem the blood.

I called the internal phone to reception but there was no answer. I ran downstairs and found no one. I called the emergency services myself. It seemed to take an eternity for them to arrive. I had gone back to Dad meanwhile and talked to him trying to keep him conscious.

Finally, two grubby-looking ambulance paramedics arrived. Poor Dad, they handled him roughly as they picked him up and put him on the stretcher. I could see he was in pain. Somehow, they managed to get Dad down the awkward winding narrow staircase to the ambulance. I travelled with Dad in the ambulance. The short journey to the hospital was one thing but we had to wait for over an hour in a corridor at the hospital.

I kept asking hospital staff what was happening, using my basic schoolboy German. No one spoke English. No triage took place and no pain relief. Once more it was left to me to try and keep Dad conscious.

Eventually a doctor came on duty. Fortunately, she was very good, efficient, friendly and spoke English. She examined Dad, put some stitches in his ear and head and said, 'I think your father should stay in for observation.' The x-ray was clear, and Dad didn't want to stay, he wanted to get on with his travels, so against the doctor's will, we signed Dad out with a bandaged ear and head.

Dad joked he had been attacked last time he was in Lüneburg in the army. He then told me a tale of being out at night walking back to barracks when some people jumped off a lorry and attacked him, knocking him out. He was not badly injured and filed a report on his return to barracks.

Next day, his commanding officer took troops into Lüneburg as a show of strength and searched house to house and by chance found a former SS officer hiding in the attic of one of the properties. Unfortunately, Dad could not remember the officer's name, but what a story!

Just south of the town of Lüneburg, near the village of Wendisch Evern, the Germans signed the instrument of surrender ending the Second World War in Europe. There is just a small monument remembering that historic occasion. It was in Lüneburg on 23 May 1945 where Himmler killed himself. He had a cyanide capsule hidden in his teeth and avoided interrogation and no doubt execution. He is buried in an unmarked grave in a nearby forest.

We started our tour of Celle and Lüneburg. We found a crane by the Ilmenau River in Lüneburg that was made in Coalbrookdale, near Ironbridge, Shropshire where we used to live, dated from the eighteenth century. Dad remembered it from his time spent their post-war in the forties.

We visited Lüneburg railway station, which had changed a little but that was where he worked in the Regional Transport Organisation office, invariably arranging the logistics of troop and equipment movements, as well as some time in Celle railway station. We went to the barracks where Dad was attached to the Scots Greys, then headed to Belsen. It was a pleasant drive and Dad seemed comfortable considering his accident. His head hurt but he wouldn't take any painkillers.

We arrived at Belsen. There was a fairly new museum and information centre for us to look around. But I was very surprised after his insistence on visiting that Dad wouldn't get out of the car, he was just content to be there, to say he had returned. It bothered him but he would not open up about it.

I said, 'Dad, I will go and have a brief look around and report back.' Dad just nodded and closed his eyes.

I walked up to the memorial monument which had been erected there, remembering the souls who had lost their lives. There were none of the original buildings. The wooden huts had been burned by the liberating British troops as they were riddled with lice, spreading typhus. It was just open land surrounded by woods, some of which would have had clearings with huts in. It seemed very quiet, no birdsong while I was there, quite eerie.

An army base was next door and Dad said that was where he'd stayed when he came to supervise German troops who were re-interring corpses of the former camp prisoners and covering them with lime because of the smell.

He talked little of it, but instead chose to go on to other stories. Such as being best man for a mate of his called Mick Sturdy, who married a German girl in St Nicholas Church, the garrison Church at Lüneburg when many of the allies were garrisoned around the Lüneburg Heath.

He commented, 'I liked and got on well with the German people. I was unhappy that locals would be expected to get off the pavement if allied troops were walking the other way. I felt it was demeaning and uncalled for.'

He told of the time they took a train of supplies up to the Russian border. Dad had a cigarette and some schnapps with the Russian border guards and waited for the train to return but it never did. This apparently happened often.

Another time, whilst shipping all sorts of supplies to the USAF base at Fassberg, he would stop and help to load the transport planes taking much-needed food and other goods into Berlin during the airlift blockade busting.

Dad had watched the fully loaded planes vibrating as they struggled to take off from Fassberg airfield. So, when he was asked, 'Hey, bud, you coming with us for the ride?' Dad apparently chortled and said, 'Not blinking likely'. It was probably where he developed his reluctance to fly.

We managed to return home safely and check Dad out at the hospital in Shrewsbury. He was fine.

He reached about 78 years old, then one day he fell in the house, this time breaking his hip. In hospital he developed pneumonia, spending time in intensive care. He was very weak and we didn't expect him to survive. Fortunately, he did and for 18 months he put on weight and was better than he had been in a long time.

He and I went down to his old home at Penwyllt in the Upper Swansea valley and caught up with relatives. On the journey Dad would be telling childhood tales, and singing along with a Welsh choir tape. These were very happy days. We used to stay in an old farmhouse where Dad used to go to as a child, as it was owned back then by a relative of the Watkins family and he used to get some farm produce from them to take home. We saw our cousin Janet Watkins with her family and had a Welsh tea with her, and Beryl Jones, another cousin who used to keep the pub in Pen-y-Cae.

Alas, when dad was eighty, just before Christmas, which ironically was his favourite time of year, he fell again and broke a hip once more. So back to hospital he went and predictably because of his immobility he developed pneumonia again.

Dad was deteriorating and inexplicably I decided to carry on with a college day trip to London we had planned. My stepdaughter Kate, who also worked at the college, and I were taking students on a field trip. We had a lovely day. I was also escorting a good Japanese friend of ours called Miho to London for her continued travels after staying with us in Shrewsbury. I knew it was a risk to be away from home with Dad being so ill, and just before we headed home on the train, I had the call I dreaded from my sister Sue.

Dad was asking for me and the doctors didn't think he would last long now. It was the longest train journey ever, I felt like everything was in slow motion. I finally reached the hospital and rushed onto the ward. Dad whose lungs were so full of bile and congested so badly he could barely draw breath, tried to say something to me.

Sue was on the other side of the bed holding his left hand whilst I held his right hand. I said, 'Have no fear, Dad. I'll look after Mum and all of us. I love you, Dad!' and with that tears came out onto his cheek. I too was crying. I brushed away his tears and mingled them with mine, he closed his eyes, coughed a little and died. I had lost my dear dad!

A song/poem remembering my Dad

Take me by the hand Dad
Walk me down the lane
Take me by the hand Dad
Stroll with me again

Tell me all those stories
Of when you were a boy
Remind me of the happiness
And your childhood joy

Dad, you know I miss you
Do you miss me too?

No one could ever have a Dad
half as good as you

Remember we'd play cricket
You'd bowl to get me out
I'd miss the ball, the bails would fall
But you'd say I was not out

You taught me how to handle
The ups and downs in life
You would get mad, because you're my dad
If I would cause some strife

But though you would be angry, you did not hit me at all
But I would be scared
when your temper flared
And dash off down the hall

When it came to working
You gave sound advice to me
Work hard son, get the work done
You'll get on you'll see

Always be kind and gentle
Never do any man wrong
Be kind to folk, he was that kind of bloke
And for him I sing this song

It's times like this at Christmas
When it is plain to see
All sitting by the fireside
Or around the Christmas tree

We miss your decorations
Your thoughtful presents too
But most of all we miss your smile
Oh Dad, we do love you!

CHAPTER TWENTY-EIGHT

———

Eulogies and letting go!

I have had the honour of being asked to write and deliver a few eulogies over the years, the first being my dear friend Helen. I was tearful when writing it, but it was only when I stood up in front of a grieving chapel full of people, many of whom I knew, that the gravity of the occasion really hit me.

I cannot say I enjoy writing and delivering eulogies but I do feel it a real privilege to be asked, and it's fair to say you become more adept with practice.

Ken Morris, a good friend and client over thirty years, was a tough one to do, but it's strange, when I get into the pulpit, I feel a different persona seems to take over, perhaps inspired by the genes from my grandfather Tom Morgan, who was an ordained minister. He was a reader in the Church of Wales, which is the equivalent of a curate in the C of E. I was also the eulogist for two of Di's uncles, Uncle Michael and Uncle Norman.

Of course, the toughest to date was my own Dad's eulogy in 2010. This was the basis of it.

Thomas Joseph Dennis Morgan was born in the small mountain-top village of Penwyllt on 20 September 1930. Growing up he was known as Tom Bach (Little Tom) as his father was also a Thomas and therefore was Tom Mawr (big Tom) and indeed he was, at well over 6 feet tall he was commanding in every respect and, as Granddad was an ordained minister, moral responsibility and good living were an important core to dad's values.

My dad had a lovely family childhood in Penwyllt, although the winters were a lot tougher than anything we have experienced, the name 'Penwyllt' meaning 'wild head'. Dad often spoke of his dad carrying him on his back in snowstorms home from school and this involved a climb up a very steep gradient aptly called 'the Task'. I think it was this upbringing and love of mountains that rubbed off on the rest of us and indeed my brother Chris and I were always off walking hills at every opportunity throughout our life, so it must be in the genes.

During the war Dad moved to Ironbridge with his brother Roy and his mum Daisy, to avoid any fall-out from the bombing of Swansea. After returning to the Upper Swansea valley, it was clear Shropshire had made an impression on Nan and Dad, and they persuaded my grandfather to move all the family to Ironbridge. Like the proverbial sliding doors if they had not, Dad and Mum would not have met ... but meet they did and fell head over heels in love, so thank goodness that some good came out of the war.

Dad had done his national service at the end of the war serving in Germany and returned to the UK to take up a job on the railways and this is where he eventually met Mum and the rest, as they say, is history.

Dad and Mum, like many in the post-war decades, had a pretty tough early life and they were so excited when they had their first real home together when we moved into Mary Webb Road in the late fifties. Indeed, we all were and we were joined a year or two later by my brother Chris and the family was complete.

As much as Dad always loved the railways, perhaps because a railway and tramway ran by his house as a child, he decided to take up a new profession in plastering as he could see the burgeoning economy needed many properties building and there was plenty of work ... although he would have bemoaned weather conditions like this December as it may have meant no work and no work meant no money. Dad went onto be a master at his craft, in demand by many, and specialised in restoring the decorative ceiling crowns in older properties.

In later years Dad worked at the Barracks as a maintenance craftsman and this allowed him a steadier, more reliable pattern to his work. He used to spend his weekends travelling up and down the country as he took an interest in collecting all sorts of things form cigarette cards to commemorative memorabilia.

Dad loved being out on the hills or walking the fields around our home at Meole, he just loved the outdoors and nature and enjoyed his garden, especially growing roses, 'Papa Meilland', a fragrant rose, being one of his favourites.

Dad loved music, all sorts of genres. I remember him when I was young, always slinking into the kitchen to put the portable Decca transistor radio on to listen to the Top Twenty records on a Sunday afternoon and singing along with them, especially if it was Elvis. There is no doubt he bestowed his musical talents on my sisters and brother, perhaps by them hearing him singing and encouraging music occasionally by the use of our piano. I don't know what happened to me, Dad, I like music but whilst my siblings can sing, I am almost tone deaf.

Still, we are all different and perhaps it was the joy of sport Dad and Mum gave me, as I played football and cricket. I always remember Dad playing cricket with me on the recreation ground just outside Meole Brace Church, which helped me along no end. Dad was a decent cricketer and sportsman in general and, surprising to some, in his army days he was a good featherweight boxer. His love of boxing would mean him staying up late to listen to a fight in America or somewhere else on the radio. I would stealthily position myself and listen at the door even though I should have

A young aspiring cricketer Ron is waiting for his Dad to bowl

Dad as a background artist in 1984 film Christmas Carol!

been in bed.

In 1984 Dad was an extra in the filming of *A Christmas Carol* in Shrewsbury. He loved this role and being on set and we have always been very proud of his involvement. In fact, I am lucky enough to take *Christmas Carol*-themed tours around the town each Christmas.

On one of the days' filming, Dad was having a break when Susannah York (Mrs Cratchit) came over to him and asked, 'Are those real sideburns?' Before he could reply, she pulled them hard. His eyes watered and he made a sharp exhale of breath through a pained visage and she had her answer. Smiling, she apologised and said he really looked the part. They chatted for a few moments which for my dear, now departed Dad, was a wonderful experience.

I always say the film starred George C Scott, Susannah York and Dennis Morgan. Indeed for me he is my star and when I look to the heavens and the night sky, I know his star will continue to guide all our family. Dad was the best role model you could ever wish for. He taught us right from wrong and showed us the path we should take. A gentle, kind, honest, loving, interesting man, always welcoming and interested in what you were doing and would do anything for you.

Dad related well to all his grandchildren and indeed great-grandchildren and they love him dearly. I feel sad they never knew him as a fit young man when he used to take us with Mum for a well-earned holiday at Monkstone Bay in South Wales for a week. He would hold two of us, one in each arm and walk us out into the sea and as the waves hit Dad's strong, fit body it took our breath away and we would laugh and want more. We would trawl the rockpools for starfish and other fine sea creatures and climb the rocks and explore caves and build sandcastles. We were all in our element.

We all have many memories of Dad, through childhood right up to the present day, but of course in life the impression left on you by Dad when you met him is a personal one, so how do you remember him? Maybe it's remembering him walking down to the Brooklands at the bottom of Meole Estate to get a lift to work in the depths of winter with his shirt open to his waist, his shirt off in summer, of course, wherever he was taking in the sun rays, in fact many in his younger years thought he was Italian, given his dark skin.

I have many memories but the time I spent with Dad revisiting his childhood village at Penwyllt and its environs stands out. Both dad and I as we neared the Upper Swansea valley used to get more and more excited and full of anticipation, as the Welsh say there was a certain 'hiraeth', a sort of belonging and desire to be there, which you could understand from Dad, but as I have never lived there it was strange I felt exactly the same, as though we were both going home! He revelled in meeting cousins and friends whom he reminisced with and his Welshness returned, helped by lots of Welsh cakes, bara brith and lava bread washed down with a ginger beer, perhaps, and greetings of 'Sutmae Tom Bach'.

Dad beguiled me with his stories, many of which ended up as the basis of tales I told to my children at bedtime. I used the characters Dennis and Roy (my Dad and Uncle) along with a dog called Lucky. I could tell it was a tough, simple, yet happy childhood that gave Dad the foundation and knowledge to provide us with a wonderful family childhood too.

So here we were in the twelve days of Christmas, a time Dad revelled in, from the time he put up the concertina paper Christmas decorations and the tree carried home from town and had the usual trouble balancing on a base and securing so we children did not knock it over when we helped dress the tree.

On Christmas Day, whilst we were opening our Christmas stockings in our bedrooms early in the morning, Dad used to light the coal fire. We of course first checked if Santa had been to eat his mince pie and drank his sherry (milk if it was a Sunday) in the kitchen, which amazingly he always had, then through to the warm living room where our eyes widened as we viewed the bounty before us. What special times!

Dad instilled in us that Christmas was not just about presents but was a celebration of the birth of Jesus Christ and indeed it was his faith and prayers that pulled him though many tough times in his life, not least of which was his many years of failing health and it is that faith that he knows will pull us through this difficult time now. Early Sunday evening was always about watching *Songs of Praise*, he never wavered in his faith.

Well, Dad, we miss you more than words can say this Christmas and every Christmas but we are warmed by the thought that at long last you are reunited with your mum and dad Tom and Daisy and your brothers David and Roy and your childhood family is finally together again.

We are heartened by the memories of all the wonderful times we were privileged to share with you and Mum here on earth and look forward to spending Christmas with you again one day when our family is reunited

Nadolig Llawen, Nos da, Dad and a very merry Christmas to you all
Hwyl fawr am nawr, Goodbye for now, Dad.

My brother, sister and I all stood together while I delivered the eulogy. Dad's coffin came into the church to the tune 'Fields of Gold' and left to 'Amazing Grace'. These songs will always be associated with Dad. But this Piece de resistance was when my sister Isobel, a gifted singer, sang 'Danny Boy' at Dad's graveside. I used to call him 'Denny boy' as a play on words. I struggle to hear any of those three songs now without having a tear in my eye.

*

Dad worshipped Mum and she always said he made her happy and content, but now Mum was alone, the love of her life departed. She was bereft, as were we all.

Mum started to deteriorate, repeating herself then staying in bed all day, not wanting us to leave and taking far too many codeine and phosphate. Then she fell and broke her hip, but recovered well and seemed to be okay for a while living at home but then she began to struggle again. It was clear she really had no will to go on. We would spend a couple of hours each day with her, plus I still cooked her lunch as I had for Mum and Dad for a couple of years, helped by my sister Susie. We took it in turns where we could and made sure we were not both away at the same time. I would leave and I would just arrive home when she would call, wanting me to come back because she was frightened or needed something. I would sleep there some nights to keep her as calm as possible.

One day I had returned home after cooking lunch and spending time with Mum when her next-door neighbour called, saying she was hanging out of the window saying she was locked in and couldn't escape. It was clear the loneliness was affecting her mental health.

For a while Mum spent respite time in a care home, then came home. However, before long it was clear she still could not cope and went to another care home. Many meetings were held with social services before Mum was moved permanently to Holy Cross residential home in Abbey Foregate.

Mum resisted the idea at first and refused to settle, not wanting to leave her bed. It was a natural reaction, but we felt she was safer and better looked after by a staff who were used to dealing with these situations. As time went by, thanks to the understanding and cajoling staff, Mum grew to accept her situation. There had always been a danger of her falling at home and not being found for a while, or overdosing on her medication, because she could not remember if she had taken her tablets or not. Her cleanliness was suffering too as her personal care was suffering if we were not there. It was a shame as Mum had always taken such pride in her appearance and the presentation of her home.

We had tried carers at home but they would often just come for a short time, do nothing much at all and say Mum had advised she didn't require help, saying they could not force her to do things like prepare for bed. They signed their sheet and went, still being paid.

Five years on, Mum was now a matriarch at the home and as happy as she could be. Still missing Dad, until recently mobile with a frame, eating well, and still going to the occasional football game with me. As time has passed by, a little dementia has set in, she was coming up ninety-five years of age after all. Now bed-bound, her spine crumbling, it was hard to witness, but her smile kept us going.

Mum has been the most wonderful friend, confidante, supporter and human being any child could ever wish for in a mother. She endured so much hardship and ill health during her 95 years on earth, surviving peritonitis as a child, she lived a full life having been born in Pontesbury at White Horse Cottages in 1925, she had many strings to her bow, she was a fantastic bookkeeper for Ron Morgan Travel well into her late seventies but prior to that had been a shorthand typist at the Sentinel on Whitchurch Road, a caterer at Shrewsbury Golf club before being a self-employed caterer servicing many events from weddings to bowling club championship days. Mum was well known as the Manager at Send A Card on Pride Hill and much more.

Mum and Dad Christmas
Carol 1984 Premiere

Our parents were absolutely the bees knees we were so lucky to have them and could never thank them enough for the life they have given us.

Mum passed away Sunday 21st February 2021 aged 95. I miss them both so much the best Mum and Dad anyone could wish for!

For me one of the most heart-wrenching things in life was clearing our family home of all the familiar chattels of a wonderful life together with my parents and brothers and sisters. We had been there a lifetime as brothers and sisters, but now the familiar high thick privet hedge protecting our sheltered front lawn and the path accessed by our painted wooden gate, all of which created by Dad and our fifties home was going to be gone.

It is so difficult sorting through other people's possessions and many of them made us weep. It was amazing what my parents had squirrelled away. We found old payment cards for hire purchase, every TV licence ever bought and of course lots of photographs. My dad was a collector! Fortunately, after his passing we had sorted a lot of these items and shared them, as per his wishes. There was lots more to do, including many cameras and watches. None of them were expensive or particularly special, but all collected by him as being of interest.

We had to arrange for furniture we had used for years to be collected and given away, an old pine chest falling apart in the shed was collected and restored to its former glory and sold on. All Dad's tools from his plastering trade we kept. I had Dad's old Royal Engineers army beret, disintegrating badly, but sentimental to me, as was the cap he wore up to his death. I could not let them go.

Finally, we had the home clear of all my parent's earthly belongings. It was a rental property, so we had to hand it back to the council. Sue and I locked the door for the last time, took a final look at the fishpond in the back garden, which Dad had sunk and cemented himself, envisaged where the sunflowers had once grown tall, viewed the vegetable patch, once abundant with potatoes, carrots and more. There was still rhubarb at the top of the garden close to where we buried a plethora of goldfish and hamsters. The centrepiece, though, was Mum and Dad's beloved rose garden.

We stood at the gate, recalled the games of football on the small front lawn, looked back at the red brick house with its large hollyhock by the drainpipe and through misty eyes we bid it adieu.

GOODBYE MUM

There she was the pillows cradling her sweet head
Her mouth agape nothing to be said
Her last breath her spirit gone
To the next life she has travelled on
Her hands clasped around a flower
I envisage her walking around the bower
Back with her sweetheart both smiling serene
Happier than they have ever been
Of course, we cry, our hearts are broken
We hold her hand no words are spoken
The loss immense, the emptiness real
The hurtful chasm, the way we feel
No words can ever salve our pain
Until we are reunited again
But we have our memories of a wonderful Mum
They will keep us warm until kingdom come
Her lovely smile, her loving arms
Her wicked humour, her many charms

Now the joy floods our souls
As we recall Mum's many roles
Her smile filled us with happiness,
her manner slow to chide and swift to bless
Always there to comfort us,
through troubled times she caused no fuss

She merely made everything right
and saw us through the fear of night
She helped us to reach the other side
and always knew we could confide
A shoulder to lean on, an open heart,
Arms to fall into until death we part
So now we don't feel so alone,
Mum's just journeyed on ahead
Tonight, she's an angel around the throne,
not in this body on the bed.
One day we will join her
we know not when
For now we are happy,
she is back with Den
Ronald D Morgan

CHAPTER TWENTY-NINE

Dreamcatcher Children, founded 1995

We were in Sydney, Australia, when Dianne and I had first discussed the possibility of helping children and their families who she nursed on the paediatric ward at Royal Shrewsbury Hospital. Dianne had travelled along with me as an assistant tour manager on a Round the World Tour and was very good at taking care of the pastoral needs of the guests we had travelling with us.

I had been very ill with ulcerative colitis the previous year and been in hospital for a while. I had recuperated but was quite weak, as proved when we arrived in Hong Kong, I barely made the flight to Sydney, and was still struggling when I arrived in Auckland but pepped up after seeing a physician and taking a strong steroid dose regularly. Dianne fortunately was on hand to advise and support the group on the pre-arranged daily itinerary.

As time went by, Dianne and I talked more about Dreamcatcher Children and decided to get this project off the ground. Initially the charitable fund's name was 'Sydney House' to give a nod to the location where the idea was conceived.

On arrival back in the UK we then thought we needed a more dynamic and all-encompassing name, so came up with Dreamcatcher Children. 'Dreamcatchers', we had learned, were used historically by certain North American Indian tribes. These sacred hoops, invariably made from the red willow tree, had a net set inside the frame to catch bad dreams when people were sleeping. The belief is they let the good dreams through to slide off the attached feathers to the sleeping person, bestowing them only with good dreams.

Dreamcatcher Children is now a well-known children's charitable fund in the Shropshire and Mid Wales area with a view to giving memorable experiences to children who are suffering from life-threatening and chronic illnesses. Over the last twenty-five years we have fulfilled the dreams of a good number of children and their families by providing holidays to many places including Disney World in Florida, Disneyland Paris, Legoland, Peppa Pig Land, caravan holidays, Center Parcs, cruises and lots more. We have also provided other children with great experiences such as helicopter rides, pop concerts, theatre trips, special nights out, stretch limousine rides and much more.

Every Christmas we run the Dreamcatcher Santa Special where, in conjunction with the Welshpool and Llanfair Railway and their fantastic volunteers and the understanding and patience of Kevin and John there, we have grown our event to over three hundred children and family members over a weekend in December. The Dreamcatcher Santa Special is a steam train that leaves from Llanfair Caereinion Station and takes its special

passengers for a ride beside the River Banwy. Whilst the train makes its merry journey, which takes about forty-five minutes, each child receives a present from Santa, who gets on with all his elf helpers about half-way along. The

Kerry and Simon treated Dreamcatcher's young ladies to a performance of Cirque Du Soleil at Theatre Severn then a limousine transfer to the Armoury restaurant for a meal!

looks on the children's faces are priceless.Back at Llanfair, all are served a complimentary drink and mince pie in the tearoom. It's a wonderful family festive fun day out and we always have so many wonderful photographs of precious memories.

Most of the children we help are from the Shropshire and Mid Wales area and most have been hospitalised at some time or other in the Royal Shrewsbury Hospital or Princess Royal Telford.

We are supported in many ways, through fundraising evenings such as discos or casino nights and through donations, small or large, they all help. Some folk refuse gifts for special celebrations such as wedding anniversaries and instead ask for donations to Dreamcatcher, some funeral donation bequests follow the same lead. Esme, one of the children Dreamcatcher has helped in the past, has overcome a brain tumour and cancer, and is now in her twenties and raises money for Dreamcatcher Children.

Some of the children may not recover from illnesses, but we want to give them some magic while we can. Latterly we have also endeavoured to help children less fortunate than our own in the UK, who do not get support from their governments and social systems and rely totally on charitable donations. We are proud to have helped Orchaa school in India, which we visited with our tour guide PK when we were escorting a tour in India. It was run by a German lady who also ran a restaurant and used the proceeds of her restaurant to educate the children. A proportion was spent paying the parents to allow the children to be excused from work on the land to go to school. The rest was spent on books, uniforms, teachers and so on. Sadly, local people and authorities stopped the school operating in the end because the parents wanted their children to work on the land.

We are also helping Ananda Myitta Agape Childrens Home Grace and Love in Yangon, Burma. Again we visited the orphanage, which has about a dozen children at any one time, and left a donation of pens and pencils and other small items and some dollars and have since sent money to have shower and toilet facilities mended and a dining room table recovered, buy a guitar and a bicycle to go and get groceries from the market and treat the children to a meal each Christmas as it is a Christian orphanage supported by the local church elders too.

The Two Sisters Orphanage in South Africa has been our latest overseas project and latterly we have joined forces with Manna Church Fellowship to

make sure Two Sisters children and other vulnerable children in the Greater White River district are cared for.

The very first child we helped was Jamie, who was very ill with leukemia. With a view to giving him something to look forward to we booked Jamie and his family a two-week holiday in Florida. The day before travelling he had a gastric food tube removed in hospital and we all hoped he would cope with the journey and with being able to eat. We needn't have worried, he was soon on hamburgers, had a fabulous time at Disney and other Florida attractions and to our amazement and joy regained his appetite for life. He flourished and started to test clear of leukemia.

A few years ago, Jamie's family contacted us, sending us a picture of Jamie as he is now in his twenties. We were astonished to see this fine young man in a uniform. He had become a Royal Marine. Thanks to the fabulous support of Pete and Cathy Ashby who had donated generously the family made a dream come true. Since then, many more kind and regular donors have been making dreams come true for Dreamcatcher Children and brightening the lives of the children and their families. It is such a joy to watch the families creating magical memories

CHAPTER THIRTY

———

Agricultural tours

There are many travel anecdotes you collect over fifty years, some I have probably forgotten I'm sure. For over thirty years I travelled regularly with a farming group co-ordinated and put together by a chap called Ken Morris who sold animal feeds sales for Criddle Billington Feeds, once known as Criddle Burgess Feeds.

He was a real character and used to go around the farms selling feed. At the end of the seventies/early eighties, his company gave the farmers an incentive to buy feed. Depending on the tonnage they bought they were given vouchers as discount off an annual overseas trip with agricultural visits built in as part of the tour. The more feed they bought the more discount off the holiday.

Now this holiday/business trip had to include agricultural visits, whether to an exhibition, farming institute or simply a farm. One way or another it had to be agriculture related because apparently there was a tax benefit for part of the trip. It was not long before this little avenue of fiscal gain was cut off by the Inland Revenue, but the tours continued.

I first met Ken in 1980 and arranged the first tour with him not long after that. It was a cruise down the Rhine using Cosmos tours. Ken travelled with his wife Mair, a lovely lady.

Previous to these dates the tours had been arranged by a company from Stoke. The first trip went okay but Ken had a few gripes and was looking for some compensation, which I successfully took up with the company on his behalf. I figured these tours would be better with a tour manager travelling with them and offered my services. So, in the early eighties my new additional role as tour manager came into full swing.

Now I have mentioned that I had taken some short breaks of no more than four days escorting people, but now this was a full-blown tour for a week at least with excursions every day. To say it was intense, exhausting and stressful would be underplaying it. To say it was enjoyable would in a rather masochistic sense be true as the people were wonderful and many are still friends today. Many are now departed and the youngsters of the tours, then in their forties, are now in their seventies and eighties but, then again, I'm now in my sixties. Father Time, eh!

We travelled together on shorter adventures too, such as weekends to Paris. On one such trip in the early eighties we stayed at the Hotel Bergère on Rue Bergère in arrondissement 9. It was a great position and we did all the usual tourist spots, Notre Dame, Sacre Coeur, Arc de Triomphe, a boat trip on the Seine and so on.

I had walked up to the Folies Bergères and managed to get group tickets for what proved a rather eventful evening, with rather more nudity than we were perhaps expecting, entrancing many of the male members of our party whilst the ladies seemed less than enamoured.

I had warned the folks that as we exited this fabulously ornate theatre, that it is quite common to be besieged by groups of North African children who come distracting people, asking for a handout but using the fracas to steal purses and wallets. Sure enough, right on cue we were deluged and surrounded when we came out, but thanks to a contingent of local gendarmes and our own diligence nothing was stolen.

We took in other great shows like the one at the Moulin Rouge. We loved walking around the artists' quarter behind the Sacre Coeur called Place du Tertre, trying to spot another Toulouse Lautrec.

One evening we came back from an excursion and a guy in full running kit swanned into reception and as I enjoyed my running I said hello and asked if he had enjoyed his run. I arranged to meet him before breakfast to go for a run together the following morning.

Next morning, we met in the lobby bright and early and set off up to Boulevard Haussman and dropped onto the footpath by the River Seine. It soon became clear to me that this chap, who was fairly slim and of small, muscular build, was a decent runner. In fact, I was having to go full tilt to stay with him whilst I felt he was coasting. He insisted on running up the steps off the riverside footpath, crossing a busy road and going back down to continue our run along the Seine. Exhausted but exhilarated we returned to the hotel and I thanked him for the run. I commented he was going at a good pace and, no surprise, he said he felt it was a good loosener and thanked me for the company.

I had established his name was Ian and said if he fancied another run to let me know, but he said he was leaving to do a race somewhere. I asked him if he ran regularly; it turned out he certainly did, the chap I was running with was a fellow called Ian Stewart, he was about five years older than me and more to the point was one of the world's leading distance runners.

Between the late 1960s and mid-1970s. Ian won the bronze medal in the Men's 5,000 metres at the Munich Olympics in 1972 (a race won by the amazing Lasse Viren). He also won the following championships: European 5,000 metres (1969), Commonwealth 5,000 metres (1970), European Championships (1969 and 1975) and World Cross Country Championships (1975). He had run a ten-mile road race in 45 minutes 13 secs and set other British records: 1,500 metres 3:39.12 (1969), 2,000 metres 5:02.98 (1975), 2 miles 8:22.0 (1972). Other personal bests were a mile in 3:57.3 (1969) and 10,000 metres in 27:43.03 (1977). Just before I met him, he had been awarded an MBE.

None of the above had he mentioned to me, he was certainly not a boastful sort, merely mentioning his brothers and sisters were keen runners too and he was evidently Scottish yet he told me he was born in Handsworth in Birmingham and knew of Shrewsbury, my home town. He was a Birchfield Harrier. We didn't talk a lot, actually I could barely talk at all the speed we were running at, but I feel privileged to have trained with such a fabulous athlete.

*

In the nineties I took a group to Canada, visiting Toronto and Niagara Falls and over to Calgary and of course across the majestic Rockies. A fine time was had and on the last morning we said goodbye to our Hotel Georgia, Vancouver, a comfortable, characterful stately old lady, but which had given us an unplanned early morning call when a fire alarm went off and we had to muster on the pavement outside in our nightwear.

Farming group at Niagara Falls

We headed to the airport for our flight home. As usual, before we left the hotel I stepped on to the coach and asked the departing group questions like 'have you paid your room bill, checked your room safe to make sure no valuables had been left?' and so on.

I walked up the coach aisle and retrieved a few room keys, which from experience I found was nearly always the case. Strangely, even when the keys are attached to huge pieces of wood or plastic people still have them on their person. How folk do not know they have them, I have no idea. I of course say to make sure they have their passports to hand and not packed in their suitcases, giving them an opportunity to check and in Vancouver's case at the time, a reminder to have some departure tax ready in Canadian Dollars.

All went well, and we were in the line queueing to check in at Vancouver airport. I walked up the line of chattering passengers reminding them to have their tickets available for travel (which I had handed out on the coach) and their passports ready for inspection. About half-way along a rather meek and pleasant chap leaned over and whispered to me: 'Excuse me, Ron. I don't seem to have my passport.'

I replied 'Oh dear! Do you think you left it on the coach?'

363

With a reddening face he said 'No.'

Getting a little worried and perplexed I quizzed him further 'Where do you think you've mislaid it?' expecting him to say what he had done between the coach and the check-in queue.

With a nervous chuckle he informed me: 'Well, we were taking a taxi ride yesterday and I think I may have left it in the taxi.'

You can imagine my concern and no matter how incredulous the predicament after all my earlier reminders and warnings, I knew I had a dilemma, my first thought being I may have to leave my clients here or remain in Vancouver with them and try and get to a consulate to replace the lost passport and of course travel home on new flights at extra cost.

Thinking quickly, I went to the desk and said to the desk attendant 'I'm very sorry to bother you, but one of my group has lost their passport and with the flight departing in two hours, we have no chance to get to a consulate for a new passport.'

You could see it was not what the attendant wanted to hear but in fairness to her she was very understanding. To cut a long story short, we debated the problem and negotiated that as the group were changing planes in Toronto and would all need to transfer anyway, she would allow us all to travel to Toronto as it was a domestic flight, and the matter could be dealt with there.

This was a partial result as it at least reduced the cost of replacement tickets if needed, but it did not solve the problem. My fellow passport-less passenger and myself spent a very worried flight considering our options.

On arrival in Toronto, I thought we would enquire if they would let us on the flight to the UK and allow the problem to be resolved by UK passport control. Now in this day and age there would be no chance and, in fairness, even if the airline allowed a passenger to travel without the prerequisite correct documentation, they would be fined so we knew we were on a sticky wicket, in cricket parlance.

Our group lined up for the connecting flight, my offending couple just in front of me, ready for me to try and plead their case, although I was rather forlorn about their chances. Just in front of us were a couple of ladies not from our group, who seemed animated and a little agitated. I thought they may just be excited and nervous about travelling.

I do not know to this day what the problem was, but they started speaking to the staff at the desk and all of a sudden, all hell broke loose, a lot of shouting and posturing took place and before long security were called and they were taken away, one lady very irate, the other trying to calm her down.

In the furore I said to the couple with me just present your tickets say nothing about your passport until asked. This they did and the by now flummoxed staff thanked them and said next. I was astonished, so said nothing. We all went to the gate and boarded the plane.

We could not believe our luck we were on the flight home. Of course, the next problem was going to be UK passport control, but now we were going to be back in the UK. We knew we could get information brought to the airport or satisfy an interview or at least have a good chance of such, especially as this couple were in their sixties and didn't look like butter would melt in their mouth.

On arrival in Manchester, again I stood in the passport line behind my clients. They went as a couple to the desk, and explained their predicament. I was ready to be called to the desk to explain as their group leader and tour manager. But incredulously I witnessed the gentleman in border control smile and then heard him say, 'You look an honest fellow. I'm letting you in, make sure you contact the passport office and tell them you've lost your passport.' I was dumbfounded, it just would not happen today!

A similar story with a plane ticket was witnessed by my good lady Dianne in the nineties when she was returning with a group from Cyprus. As usual all were asked to have their tickets and passports ready as we approached the airport, they all disembarked the coach and entered the terminal building and lined up at the check-in desk for Cyprus Airways.

As Dianne went along the line making sure all tickets and passports were to hand, one sweet diminutive old lady said in a broad Mid-Walian brogue. 'I am so sorry, Dianne, I don't have my ticket.'

Dianne asked where she thought the ticket was. The reply came: 'Well, a couple of days ago, I spilt some sun tan oil on the ticket and put it on the balcony to dry, and it must have blown away.'

'Oh no!' said Di 'Why didn't you tell me?'

'Well, I didn't want to bother you,' came the reply. A frantic completion and signing of a lost airline document form, to confirm this is what had happened, and the reissue of the ticket, was just completed in time for her to board!

I took many farmers on holiday over the years but one guy sticks in my mind, for the sheer fact that whenever he wanted to find his wife in the resort or hotel, he would walk around with his walking staff as if looking for his sheep and let out a loud wolf whistle followed by a call of cwm bye.

I specifically recall this in the foyer of a grand hotel in Interlaken and sure enough within a minute his wife appeared.

*

One tour to South Africa started with a new member of the farming group joining us and he was a sight to behold, travelling with a bag the size of a lunchbox, which we later learned had a change of underwear, socks and shirt. That was his entire travelling wardrobe in addition to the suit, shirt and tie he travelled in, the top pocket of his jacket sporting a tube of toothpaste and a toothbrush. Before we had even gone through passport control he had lost his passport, but fortunately we found it in the toilet block he had visited.

This chap was a lovely guy but had barely travelled out of Mid Wales, so this was quite an adventure for him. We took him on a Caribbean cruise the next year and he decided to go on shore and do his own thing even though his cousin, his travelling companion, thought he was still on the ship. He was still A.W.O.L. when the ship was due to weigh anchor. There was still no sign of this splendid fellow and we feared we were going to leave without him which I just could not contemplate, then all of a sudden he appeared in the distance strolling down the jetty, his image and gait unmistakable. He only just made departure, explaining that he'd spent most of the day with some local Jamaicans who he said were very nice but had obviously helped him spend the money he had gone ashore with.

However, my abiding memory of him is on Blouberg beach near Capetown, wearing his suit and tie in the midday heat with his trousers rolled up above his knees whilst paddling in the ocean surf with a knotted handkerchief on head! Whilst we had walked up towards the coach with the rest of the party, we turned to witness him staggering backwards on the shelving surf-covered sand before landing unceremoniously in the water. We retrieved him, a little shaken up and sodden, and escorted him, dripping a trail of briny water, to the waiting transport.

Quite apart from gentlemen in knotted hankies, Blouberg beach offers fabulous views of Table Mountain!

On the same tour we had a few mishaps. One day we were driving up

Farming group paddling on Blouberg Beach

a steep hill in our tour bus only to look back and see all the luggage strewn along the road. Somehow the luggage boot locker had come open!

One of the funniest occurrences was at Cape Point, a tourist hotspot where the Indian and Atlantic oceans meets. As some of us returned to the coach, some of those already on board were having a snack and a drink when all of a sudden, a troop of baboons decided to come down off the surrounding land and invade the coach. One lady who had a chocolate bar half eaten in her hand was suddenly dispossessed by a teeth-baring baboon. Screams and whoops of fear and confusion filled the coach as another baboon disappeared with someone's hat. It was scary and intimidating but in hindsight quite a funny interlude as no one was hurt.

You really must visit the wine district in the Cape which has its best-known wine-growing area around Stellenbosch. The initial impression as you drive in is not too exciting but as you pass the Legends pub, cross a couple more traffic islands and turn left into Dorp Street you find yourself in an historic and characterful town. The Dutch-influenced façade of the thatch-roofed Stellenbosch Hotel dating from 1876 is fabulous.

Our accommodation for the night was in the small town of Franschhoek which has a population of about 15,000 and is one of the oldest towns in South Africa. Translated, it means French Corner and with all the wines produced in this valley it is very apt as the Huguenots settled here escaping persecution in Europe and developed wine and perfume industries. The Cape Dutch architecture is unspoilt, it gives a lovely old-fashioned feel to the place from the Cape Farms to the Dutch Reformed Church.

Franschhoek has a great reputation for wines and top restaurants and

Ron at Robben Island by the garden
Mandela tended

the Franschhoek Country House
Hotel, once a perfumery, where we
stayed did not let us down. Set in
the Berg River Valley this town is
very attractive. Our hotel only had
about thirteen rooms and dated back
to 1890. It was full of history and
the furniture and decor felt very in
keeping. A big plus for Dianne was
that it had a swimming pool. The
Monneaux restaurant is outstanding. We celebrated an anniversary here
and the food and wine were excellent. We particularly loved the Chamonix
Rouge.

On leaving we went to pay the bill and had to speak to the hotel about
the total, it just didn't seem correct. Concerned, the manager came over and
checked the bill with the restaurant and confirmed it was accurate. He was
astonished to hear our reply when we said but it cannot be enough for a
night's stay and a fabulous breakfast and dinner. He started to smile and we
were very happy to pay the £43. Bear in mind this was twenty-two years ago!

We took a journey up onto the Franschhoek Pass, climbing steeply
from the Berg Valley and heading into the Little Karoo, which is a 290 km
strip of semi-desert that skirts the edge of the vast wilderness of the Great
Karoo. Before long we were greeted by a large troop of baboons on the
road. We slowed and then stopped, mistakenly, it seemed, as the baboons,
led by an extremely honorary alpha male baring his teeth, approached our
vehicle with menace in their eyes. It was clear they were not frightened at
all and started to climb on to the car. We had heard of baboons ripping off
windscreen wipers and worse, so we made sure the windows were wound
up and just accelerated, to scare them more than anything. Fortunately it
did the trick and they scarpered pretty quickly without injury to life or limb.

Relieved but amused, we continued our journey to the small western
little Karoo town of Montagu, an attractive place, in a farming area full of
fruit trees to the north and well worth a stop for coffee. We wended our way

through the narrow valley at the base of the Langeberg mountains boasting the fruit growing, dairy farming district of Barrydale on Route 62 before we came across a small tin-roofed breeze-block construction in the middle of nowhere with 'Ronnie's Sex Shop' daubed roughly in paint on the walls. I naturally denied any knowledge of ownership. A couple of hours later we arrived in Oudtshoorn on the Western Cape.

Oudtshoorn is billed as the ostrich capital of the world. Great fortunes were made here by the ostrich feather barons in two big booms when it was fashionable as an accessory in Europe, first in the 1860s and then again in the early twentieth century, peaking just before World War I. At these times the value of ostrich feathers was only just behind gold, diamonds and wool on the export market.

The town has enjoyed a renaissance since because of ostrich meat and leather. As well as checking out some of the splendid mansions once owned by so-called 'ostrich barons' we visited an ostrich farm.

This was a really fun experience as you learn about the ostrich and the breeding, get to feed an ostrich and see first-hand the huge eggs they lay and the wonderful feathers and what they are used for today as well as their historical uses.

One thing many of us were already aware of was that an ostrich can run pretty quickly, up to 43 mph, but never did I think I would get to see an ostrich race. It was fascinating to see them strutting around the paddock with their long necks and strong legs. They were really inquisitive and very attracted to anything sparkling. The ladies in the party were told to watch out for the ostriches trying to peck at their earrings.

Then came the race. Two local guys mounted their ostrich, which left a spare ostrich and people were invited from the throng of tourists to become the third jockey in the derby. Before I knew it my group had volunteered me to be their nominated jockey.

Ron with statue of his childhood
hero David Livingstone

Somewhat daunted, I stepped into the arena and weighed up this bird, which was not a natural weight-bearing creature. How was I going to get on it? These huge birds were pecking at anything and everything.

Fortunately, I was assisted to mount by some handlers. The next problem once on was what do you hold on to? The initial reaction is to grab its neck. Not a good idea, you would throttle it and no doubt be thrown off, you had to make sure you mounted the ostrich as centrally as possible, then I learned you held on by putting your legs under the large wings of this flightless bird and grasped as tightly as possible the front of the wings. No saddle and no safety net.

All the above achieved, there was no time for calculation or concern as we were off, the ostriches sped along this course like racehorses and I held on for dear life. I had been told to lean back, which feels totally unnatural, but I did my best to follow these instructions, and as the ostrich bounded down the course for about 20 metres we were in contention. Then, bang! I couldn't hold on and I came down with a bump, but fortunately all that was hurt was my pride, one at not staying on

Breakfast beside the mighty
Zambezi

and two because I didn't win! Great fun, though, how many people can say they have ridden and raced on an ostrich?

Later in the trip we visited the amazing Victoria Falls then a little later we were on the mighty river Zambezi in Zimbabwe, staying in a hotel bordering the river. It had a signboard in the grounds that stated: 'Beware of crocodiles and hippos.'

One of our buffets held in the grounds on the banks of the Zambezi was full of local produce such as hippo meat, elephant tails and crocodile steaks. I guess the other side of the hotel's sign which faced the river should have said beware of humans!

Some of the party were having a cup of tea in their room one morning with the door open to the wilderness when they had opened a pack of biscuits and before they knew it again the ubiquitous baboons arrived and tore into the room stealing the biscuits and screeching around the room before departing back to their own environment. What tickled me was the room occupants, after they recovered their composure, contacted reception to complain about the unruly baboons and ask who was going to replace the biscuits!

We found Zimbabwe at this juncture in time very intimidating especially when we left through Harare airport where the troops made their presence known armed to the teeth.

*

Sad times come too. I was party to a young man who was holidaying in Canada with his sister, and hearing that he had passed away and having to speak to the family in the aftermath. A lovely family who I knew as clients. It was heartbreaking as I endeavoured to support them as arrangements were made to repatriate the body of the young man.

*

Japan was very much a culture shock to our small party of eighteen farmers from Shropshire and Mid Wales.

In Tokyo the first surprise was in the hotel rooms of this technologically advanced, neon-lit city. Toilets were a revelation, indeed an excursion in itself. You could higher the seats, warm the toilet seat, have music playing, seats would light up in neon, press buttons and water squirts to cleanse the nether regions. You had your own sound and light show to accompany your ablutions!

We enjoyed a restaurant where they brought your food and you cooked it yourself almost like a fondue in the seventies. You went to a large buffet, helped yourself to a plate of whatever you fancied such as beef, shrimp or vegetables, then used a central little grill in the middle of the table to cook your food. You were limited by time rather than by the amount of food you could take from the buffet. We then went on to a karaoke booth, which we hired by the hour and all had a go in our own private session of singing along with a playlist we chose, a bit like a living jukebox. This impromptu evening proved a real winner; we were only passing when we decided to have a go. Ladies brought us in a tray of liquid refreshments and we all had a whale of a time (or in my case a wail of a time) singing along with favourites like 'Green Green Grass of Home' and 'Danny Boy'.

We visited the delightful resort of Hakone and were able to view Mount Fuji. One of our favourite destinations was the island of Miyajima, not far from Hiroshima. Our accommodation was a 'Ryokan', a traditional Japanese inn. Low futon beds were positioned centrally within our sleeping areas surrounded by traditional screens. Bossy housekeepers wrapped each of us in a thick cotton liner inside a bedroll to go to bed. If you wanted to go to the bathroom in the middle of the night you had to extract yourself from the imprisonment of your 'pancake roll' bedding, roll off the bed onto your knees and try and stand up to find the toilet, before doing the whole lot in reverse. Not so easy if you are in your seventies.

We were given lessons in origami, making paper cranes. One evening we dressed in kimonos we had been given, completing the local attire with complementary stockings. These socks had a separate part to encase your big toe and the rest of the foot was wedged into a pocket attached almost like a mitten, quite strange to wear and not unlike a hospital stocking you are given to stop you getting deep vein thrombosis when bedridden.

We went out looking for a bar and eventually found a cosy little establishment. As we entered in what we felt was fancy dress the locals looked at us quite strangely but were nevertheless welcoming. One of our party could not believe his luck when he saw a beer tap with Guinness on it and promptly ordered one, whilst others had a bottle of the local beer or the Japanese spirit 'Sake' made from fermenting rice. The young barman went into the back room and brought out a can of Guinness, put the glass under the Guinness pump and pulled the pump whilst emptying the can into the glass and simultaneously and nonchalantly making the sound of the pump, a little like the spoof coffee commercial simulating a coffee machine's noises. We all thought it hilarious.

Of course, we couldn't wait to go to Hiroshima and the Peace Park. We met an older Japanese lady on the train journey into the city. I gave up my seat for her but she didn't want to have my seat as it was accepted in Japanese society then that a lady should let the man sit first. On my insistence she took the seat, every time I had eye contact with her, she kept bowing in thanks.

On arrival at the Peace Park there were lots of school groups, mostly young children as it is part of the school curriculum to show the children what happened and to make sure they want this and war in general never to happen again. The devastation as shown in the museums here is horrendous. We were approached by two little girls and given a flower as a token of welcome and peace.

On 6 August 1945, a US B-29 bomber called Enola Gay dropped the atomic bomb nicknamed by the crew as 'Little boy', which exploded around 1,800 feet above the ground. Around 80,000 people were killed as a direct result of the bomb, and another 35,000 were injured with thousands more dying from radiation poisoning over the subsequent days, weeks and months. The attack flattened most of the city, but a few buildings survived, at least partly because they were made from reinforced concrete. Some bomb damage to the metal windows and doors is still visible.

One picture from the museum that sticks in my mind is what looks like the shadow of a person on the steps of a bank. Such was the ferocity of the blast, eerie shadows of incinerated humans were left imprinted on steps, pavements and walls. The haunting impressions were caused by the heat of the explosion, which changed the colour of surfaces and thus the outline of bodies and objects that absorbed some of the blast.

Hiroshima's most famous ruin from the bomb, of course, is the dome in the city's Peace Memorial Park, now an iconic structure. A UNESCO World Heritage site, it has undergone reinforcement work to make it more earthquake resistant.

This was a really sobering day and left an indelible mark on all of us. We headed to a restaurant for lunch where we had to climb down into pits to eat western style, either that or the traditional pose of sitting cross-legged at a table, both of which were tricky for our aging group. It was common in most restaurants to see a display of what food was served in the restaurant with a number to select by, a bit like a Chinese restaurant back home, which was helpful as everything was in Japanese.

In Tokyo we enjoyed an early morning visit to the Tsukiji Fish Market and were impressed by the huge specimens that had been landed, but appalled to see whale meat.

Dianne always greeted everyone, no matter the situation, with 'Arigato gozai mass', which translated means: 'Thank you, good morning'! This caused amusement and confusion in equal measure, especially in the evening.

We attended an early morning Sumo wrestling exhibition, where we ended up being very close to the action, as we arrived a little late and the bout had started. We quietly tried to shuffle onto surrounding benches in a rather small room given the size of the sumo wrestlers, it felt like we were in someone's front room.

Brian, who was one of our group, almost knocked over a tall Japanese vase. We all stopped and watched with baited breath as it swayed but fortunately sat back on its base and saved our faces. Sumo is a competitive full body contact wrestling match in which a wrestler, known in Japan as a 'rikishi', tries to push his opponent out of the circular ring called a dohyo or force his opponent to the ground. Sumo wrestling originated in Japan, and rikishi are as revered as film stars and other sporting heroes, possibly more so. They are larger than life and even more so in this limited-sized arena. It is quite a spectacle. These huge male specimens are clad in just an oversized adult nappy tied around the waist and wedged between the buttocks of each participant, and often used by the opponent to manipulate and throw their target wrestler to the ground. We were worried they were going to throw each other out of the ring, as if they landed on us, we could be rather … well, squashed!

Overall, we loved Japan. A guide is very important, not just to give you the history and explain local customs and foods, but there is so little signage in English, if you cannot speak and read Japanese you are at a distinct disadvantage! Of course, when people think of Japan often the bullet train comes to mind and we were fortunate enough to travel on a couple of these. We were amazed how fast they were but also how precise they are in coming to a particular stopping place. In each station people quickly board as the train does not dally long before moving on. Then you sit and watch the countryside whizz by as you look at the speed indicator above the

carriage door, amazed you are going so fast when the coffee in your cup on the table does not have a ripple in it, a big difference to the railway experience in the UK. Apparently, the maximum speed is 320 kph and we witnessed the train frequently over 300 kph, an exhilarating experience.

We visited some ancient temples and castles, not least the Emperor's Palace in Tokyo but were not allowed inside this one so wandered around the gardens taking photographs from the outside.

We had a splendid evening hosted by Maiko's, trainee Geishas. We had to be invited to join them as part of protocol. We had witnessed the geishas and Maiko's all beautifully attired and made up shuffling along in their high clog-like footwear through the alleyways of the wonderfully historic district of Gion. It's a long and enjoyable evening, a minimum of two hours, offering dinner, a good quality kaiseki meal (including drinks) and entertainment by Maiko's

and retired geishas throughout. They play musical instruments such as the traditional shamisen three-stringed instrument, accompanied by singing and dancing, and they also involve the entire audience in amusing drinking games like what's under the tumblers, a bit like magicians. We were encouraged to converse with the maiko who offered plenty of photographic opportunities, as well as to appreciate a wonderful insight into this ancient Japanese tradition. We enjoyed a mix of interesting delicacies such as shark fin soup and of course other fish dishes as well as others. A really delightful, quite authentic evening as we were in a traditionally screened room with the floors covered in tatami mats.

CHAPTER THIRTY-ONE

———

More agricultural tour tales

Russia was a fascinating country. Arriving in Moscow we stayed quite near the Kremlin and couldn't wait to explore Red Square, so familiar to all who remember the May Day Soviet parades shown on BBC News, dominated by the walled Kremlin on one side and, of course, the iconic St Basil's Cathedral with the multi-coloured swirls on its onion shaped dome.

The square is also flanked by Lenin's Mausoleum. We filed past his well-preserved body, but one thing stuck in my mind: he had a plaster on his finger. Strange, I thought. What some folk don't know is that Stalin used to be in this mausoleum too, but he was removed and placed in the Kremlin necropolis around 1961 under Khruschev's leadership.

Also, in Red Square is the GUM Store, which in Russian stands for 'Glávnyj Universálnyj Magazín' meaning 'main universal store', although a fine elegant building, its architecture dating back to the 1890s, it struck me that it was full of western franchises like Guess, Gucchi, Givenchy and Clinique and reminded me when lit up at night of Harrod's.

We visited the Kremlin, too, and this was fascinating. It has some very attractive and historic buildings, not least of which are the churches within the walls. The day we visited as we walked through Alexander Gardens, we were aware of a lot of men in sunglasses and long black coats. All the men had ear pieces in and as we entered Red Square it was clear that security was more than obvious in and around the Kremlin.

377

Ron in Red Square Moscow!

Two guys seemed to follow us, then tailed off, got in a large black SUV and sped off.

Inside the Kremlin's fortified walls were buildings containing offices and residences, all very splendid, and the stunning Cathedral of the Dormition where all the Tsars were crowned. It is fairly small inside the cathedral, almost intimate compared to the Cathedral of the Archangel. The Grand Kremlin Palace is worth visiting, but it did seem bizarre walking within the portals of the west's former Cold War enemy.

Moscow Metro is a sight to behold with fabulous architecture, especially the stations constructed under Stalin's regime in the style of what were described as underground 'palaces of the people'. Stations such as Komsomolskaya on the Circle Line, and at least twenty others built after 1935 are tourist honeypots, their surprisingly large chandeliers and ornate decoration not what you'd expect. It's almost like entering a palace from the Tsar's days. Novoslobodskaya with its stained-glass windows, Kievskaya, Tetralnaya and Maya are all worth a visit.

One day we had the privilege of visiting the Moscow State Circus and seeing a performance. Confusingly there are two circus buildings, one where the Circus Nikulin, the old circus featuring animal acts is performed, and the Bolshoi, the newer incarnation where trapeze and acrobatic acts take place. The latter is where we were enthralled by a scintillating and precision performance!

The real highlight, though, was our journey east on the Trans-Siberian express from Moscow to Ekaterinburg. It was rather daunting just checking in as all the signs were in the Cyrillic alphabet. There was a kerfuffle as we tried to board as they felt our documentation was not in order but after a lot of hand signals and document waving the exasperated attendants allowed us to board.

Our Trans-Siberian express was run by Providnitsas (females). Providniks are the male equivalent. The Providnitsa's compartment is equipped with a small kitchen, and you can buy food from them during the journey. Their duties include keeping the compartments clean, washing the floors twice a day and offering the travellers tea or coffee by keeping the samovar (large tea urn) full. At times our Providnitsas were a bit officious It was difficult to get a smile out of them even when they came to the carriage to sell you something like a 'pirozhki', which is a bun stuffed with … something, origins unknown. What tickled us more than anything was the fact that nearly all the providnitsas sported sixties beehive hair styles and wore bright red lipstick!

There were fresh sheets and blankets in our sleeper compartments for four as we overnighted on the 28-hour long train voyage. We had a small table on which we could stack our supplies – bread and cheese, snacks like pot noodles or daily hot meals that were provided in polystyrene packaging.

Miles of pine forests and the occasional lake passed by. We stopped at some stations and hawkers and elderly women complete with 'babushkas' tied under their chin (like a headscarf) with their goods laden in shawls, sometimes with children in baskets on their backs, tried to sell us food and souvenirs through the window. Sometimes we disembarked and engaged with them but needed to keep a lookout, as we feared missing the train and being stuck in the middle of nowhere, and even ending up in a gulag!

Yekaterinburg, sometimes called Ekaterinburg, is the fourth-largest city in Russia, located on the Iset River east of the Ural Mountains, in the middle of the Eurasian continent, on the Asian side of the boundary between Asia and Europe and was out of bounds for tourist until not that long ago. It is the main cultural and industrial centre of the region and named after Peter the Great's wife Catherine. We had actually entered Asia! This city when we visited was still coming to terms with how Russia had changed and were still cautious about speaking to westerners like us. There were very few tourists around.

We experienced people walking past us, then turning around and following us, saying 'hello' in a clandestine manner as they drew alongside us, or 'where are you from?' squeezed out of the corner of their mouths. They would then walk on and do the same as they shuffled back past the other way, always worried about being followed or fraternising with foreigners.

The main reason we wanted to come here was because Yekaterinburg is perhaps most famous for the killing of the Romanov family. On 16 July 1918, Tsar Nicholas II, his wife Alexandra, their five children and four of their employees were assassinated by Bolshevik troops in the basement of Ipatiev House, where the 'Church upon the Blood in honour of All Saints resplendent in the Russian Land' now stands. Bizarrely, a statue of the Beatles has been erected nearby.

The official story is that the Romanov's cremated remains were thrown down a mine shaft about 15 km out of town, the site now known as Ganina Yama. Today there's a monastery there made up of seven wooden chapels (one in honour of each of the murdered family members), called the Monastery of the Holy Martyrs.

But having done some research I asked our guide about the reality and asked whether the bodies were actually retrieved and buried at an isolated location in the forest. She initially stuck to the official story, but under pressure she agreed we could go to the actual site I referred to, just refusing to escort us there. She was obviously scared about losing her job, distinctly unhappy and uncomfortable and gave instructions to the driver rather abruptly to take us there.

We followed a track and sure enough came across the marked burial site in a forest clearing where we witnessed a couple of gravestones. This burial site was discovered in 1979 but the existence of the remains was not made public until 1989, during the glasnost era. The identity of the remains was confirmed by forensic and DNA investigation.

The Romanovs were reburied in the Peter and Paul Cathedral, St Petersburg in 1998, eighty years after they were killed, in a funeral that was not attended by key members of the Russian Orthodox Church, who disputed the authenticity of the remains. A second, smaller grave containing the remains of two Romanov children missing from the larger grave was discovered by amateur archaeologists in 2007. However, these remains are kept in a state-run centre pending further DNA tests.

We continued our tour with a visit to St Petersburg, once known as

Leningrad, Russia's second city and by far my favourite Russian city. We were staying in a fabulous boutique hotel, the Alexander House, once a grand residential house, about ten minutes easy walk along the canal past Kazan cathedral to the Mariinsky Theatre in Mariinsky Square and just 1.5 miles from the Hermitage. The building dates from 1826, and opened as accommodation in 2003. The rooms, each unique and quirky, are named after cities of the world, such as Rome, Venice, Marrakech, Mexico and Barcelona.

There is so much to see in this wondrous city: the Peter and Paul fortress, Senate Square and the monument to Peter the Great, the Academy of Fine Arts and 3,000-year-old sphinxes, the Church of the Saviour on Spilled Blood, St. Isaac's Cathedral, Nevsky Prospekt, the Kazan Cathedral/St. Nicholas Naval Cathedral, Palace Square with the Winter Palace adjacent to the stunning Hermitage museum, Catherine Palace.

The Grand Palace at Peterhof is the real wow-factor former royal residence, incredibly ornate with art filled galleries, exquisite period furniture and chandeliers everywhere. Gardens and fountains extended to the Baltic Sea.

The Pushkin area named after the famous Russian poet Alexander Pushkin is located outside of St Petersburg and previously known as Tsarskoye Selo (meaning village of the Tsars) is another local destination to swoon over. There is just so much to do you need a few days here. Just one of the fabulous sights is the summer residence of the ruling Romanov dynasty, the sumptuous Catherine Palace building with its wonderful grounds.

Yet one of my favourite visits was to the Yusopov Palace. This spectacular palace on the Moyka River has outstanding nineteenth-century interiors fitted out with gilded chandeliers, silks, frescoes, tapestries and some fantastic furniture, in elegant rooms. The palace's last owner was Prince Felix Yusupov. The real attraction for me was the fact that this is the palace where Grigory Rasputin was murdered in 1916, and the basement where this took place is part of a guided tour. Rasputin was poisoned and shot but, still not dead, he was finished off by throwing him in the river.

No visit to St Petersburg is complete without a visit to see ballet or opera at the exquisite Mariinsky, for the period 1919–1992 known as the Kirov. The theatre itself dates back to the 1860s and was named after Marina, the wife of Tsar Alexander II. This theatre boasts one of the best ballet and opera companies in the world. The very splendid decorations and elegant theatre is worth a visit in itself. This stage has been graced by Anna Pavlova, and

Rudolf Nureyev to mention just a couple of major artists. We witnessed a world class performance of Eugene Onegin made even more special by our prestigious seat allocation.

We had left it late to buy tickets deciding on the day to go to a performance. Needless to say, most tickets had been sold and trying to get tickets for a group of us almost impossible until the reservations lady came up with an amazing solution for the group of eighteen to be together.

We were allocated what the clerk described as 'special tickets'. They were quite expensive, so we thought she was taking advantage of our tourist status but we snapped them up as they were still a lot cheaper than London theatre prices. That evening we attended in our finery and, having walked up a fabulously grand staircase, we were shown to some splendidly ornate doors on production of our tickets. As the doors opened a sight to behold greeted us, we were blessed with the amazing panorama of the opulent and very ornate interior of this historical theatre with its chandeliers, blue seating and fabulous sculptures. What a special place to watch classical productions of opera and ballet.

Open mouthed, by our spectacular location directly facing the stage in the main Royal Box. We were rather embarrassed as all the people who came into the theatre turned their heads to look up at us, wondering which dignatories or royalty we were. A terrific evening and a crowning finale to our Russian holiday!

*

One year we travelled to South America and toured Argentina and Brazil. We landed in Buenos Aires, where we stayed in the classic Claridge Hotel. We felt it had British private club overtones and was in a superb situation to explore the city. From the colourful La Boca district to Buenos Aires central square, everything was intriguing. The show at night in the Señor Tango Club located in the Barajas district was superb, so evocative and atmospheric.

For me exploring the places Eva Perón frequented was magical. The fabulously ornate Teatro Colón was a sight to behold. We did a backstage tour of it to explore its amazing structure. It is thought very highly of in the arts and considered third in the world after La Scala in Milan and the Opéra in Paris.

Casa Rosada, the Presidential Palace, on Plaza de Mayo, was the scene of many of Eva Perón's evocative speeches. We just had to visit the Recoleta Cemetery, her final resting place being in the Duarte family mausoleum. Evita rests here, together with her mother, her brother Juan, and her sisters Blanca and Elisa. What I didn't know was

La Boca District Buenos Aires

that she was in a cemetery in Milan until she was repatriated in 1974. Messages and floral tributes are often left even now by the door of the mausoleum.

We loved the visit to the Santa Susana Estancia, about 80 minutes' drive out of the capital. We followed the huge girth of the River Plate for a while, its brown earthy ribbon famous to me for the wartime Battle of the River Plate, which resulted in the demise of the German Navy heavy cruiser, the *Admiral Graf Spee*. We drove past the River Plate football stadium used at times for international games and soon arrived at the Estancia.

We were welcomed with a glass of wine or a beer and a very tasty 'empanada', a pastry casing filled with meat or vegetables very much like a small Cornish pasty. Feeling relaxed and replete we were then given a little tour of the estancia in a horse and cart, clip-clopping and bouncing along the tracks of this extensive cattle ranch.

We were then given a fabulous meal with wine and entertained by a very characterful group, men with large moustaches and ruddy complexions, one with a large nose and

Dianne's dancing partner the Toothless dancer on the Estancia Argentina

383

cheeky grin, one chap playing accordion, another using 'bolas' in rhythm to the music. Normally these two heavy balls connected by a cord were used to snare the legs of cattle or horses to bring them down and control them. Then came a demonstration of the tango.

All this was followed by a display of horsemanship in a nearby corral. Free-running horses were kept in groups by a gaucho whistling and calling much like a sheepdog display, all followed by a gaucho racing on horseback at speed to target a small ring hung from a cross bar. He would put a stake through the ring and come back to the crowds, choosing a lady in the crowd and giving it to her, like a knight presenting a favour in medieval times.

Before we left, we toured the ranch house, a very grand building full of dark antique furniture and elegant portraits. A smell of burning hickory emanated from the kitchen stove giving the property a homely feel of baking. This was a fabulous day's experience!

One of my highlights was having coffee in one of Eva Perón's favourite haunts, the Cafe Tortoni, which dates back to 1858 and was once a haunt of many Bohemian characters. It has high ceilings and the feel of a European classic coffee house. The cost of drinks and food is surprisingly reasonable.

Riders displaying horsemanship on the Santa Susana Estancia

Eventually we flew down to Trelew from Buenos Aires Jorge Newbury domestic airport, named after an Englishman and known as the father of Argentinian aviation. Our target destination was to experience a little bit of Wales in Argentina. We stayed at Hotel Rayentray, a property that was stuck in a time warp. It was basic but the best available in Trelew at the time and it worked fine for us.

The Welsh people first arrived in Patagonia in 1865. They had migrated to protect their native Welsh culture and language, which they considered to be threatened in their native Wales by the anglicising drive from UK central government. Over the years the use of the language started to decrease and there was relatively little contact between the homeland of Wales and the distant community of hardy Welsh settlers in the Chubut valley in Patagonia.

About 150 Welsh adventurers sailed from Liverpool in late May 1865 aboard the tea-clipper *Mimosa*. Passengers had paid £12 per adult, or £6 per child for the journey, the voyage took around eight weeks, and the *Mimosa* eventually arrived at what is now called Puerto Madryn on 27 July.

Unfortunately, the settlers found that Patagonia was not like the green and fertile lowlands of Wales. It was a barren and inhospitable land, with little fresh water. The first settlers' homes were hewn out of the cliffs.

Sheila, Beryl, Nancy and Gwyneth at home with the Welsh community
at Casa Te in Gaiman

Farming group at Iguaçu Falls

In 1875 the Argentine government granted close to 300 Welsh settlers official title to the land, and this encouraged many more people to join the colony, with more than 500 people arriving from Wales. A major new irrigation system in the Lower Chubut valley helped them to thrive. Others came in the next 30-odd years and Welsh-speaking schools and chapels abounded.

We had not been in Trelew long when a lady came into the reception and asked for me. I was summoned to reception and greeted by an enthusiastic lady called Anita Lewis, who tried to speak to me in Spanish and a little Welsh, neither working very well for me. Through an interpreter

and some pidgin English we managed to talk for a while.

I smiled a lot and shook her hand and we had some photographs taken. All the time I was thinking how did

This rib boat ride under the Devils Throat of Iguaçu Falls literally took our breath away

she know I was here? How am I related to her as she claims? Whilst it was of course possible, it was perhaps wishful thinking on her part and indeed mine, as I was unaware of any connection to the original Welsh settlers from my family. On our return back to the UK we established Anita's claim that she was related to me was absolutely true, astonishing and wonderful in equal measure.

We had agreed to meet again at the 'Noson Lawen', a Merry Evening in English, which was going to be held in St David's Hall in Trelew, to meet other descendants of the Welsh settlers. Fortunately, we had a number of Welsh speakers in our group, but hardly any Spanish speakers!

Meanwhile we did a little wandering around Trelew and visited 'Ysgol Feithrin', the nursery school. Here Welsh was still the primary language. We then took a road trip to a little town called Gaiman.

Gaiman is full of little Welsh tea houses, and we were due to have tea together later but first of all had time to go walkabout. We visited the local cemetery to see familiar Welsh names and then strolled along residential streets with Spanish and Welsh names. We were commenting on the net curtains in a house on one of these streets when they started to twitch and then the door opened.

A lady addressed us in perfect Welsh. She greeted us, then asked where we were from. Four of our party of six were fluent Welsh speakers, the lady was elated and we were invited in. We were given tea on a large old oak dining table made from the timbers of one of the original ships which brought the settlers from Wales in the nineteenth century. It transpired this old lady was the granddaughter of the original settlers and she told us how her grandparents had told her that they wished they had never made the journey.

After tea and Welsh cakes, we were saying goodbye when in the hallway we noticed a picture of Prince Charles and Princess Diana. It was clear this lady and indeed many of the local inhabitants still had a very close tie to the old homeland. What was strange was there was a sticking plaster over the face of Prince Charles. We asked what had happened. She shook her head and said, 'Charles is out of favour because of the way he treated Diana and I don't want to see his face!'

We had our Welsh tea as planned and took lots of photographs. I was later to find out one of the cafes was still run by a relative of mine and returned there to knock on the door as Marta Rees aged 81 at the time still lived there. Unfortunately, no one responded and I later learned she was ill. I also tried to make contact with Ririd Williams but sadly our paths did not cross. I later sent him pictures of the Blaenau Farm which my great-grandfather lived in with his great-grandmother as brother and sister.

I had bought a book of historic photographs chronicling the past of Trelew, Gaiman and the surrounding area in a local bookshop. In there I found pictures of relatives like Daniel Rees and his family at his home Plas Y Coed, which is a café nowadays, the one I had called at.

That evening some of us went to chapel for a service. It was all in Welsh and very stirring. As the locals knew we were in the chapel I was invited to speak on behalf of the party, which was very daunting as my Welsh is so weak, but I managed it and threw in some English too. We all joined them afterwards for a cup of tea and some Welsh pastries.

On our last evening we joined in with the locals at the St David's Hall and had a wonderful time singing and talking, all arranged like the rest of our visit by Eluned Gonzalez, a strong character and very welcoming. We could not have been made more welcome, and when I returned home, I found out a lot about my family history as purported by Anita and now confirmed. I met other relatives like Elsi Evans and Rebecca and it turned out our guide's wife was the granddaughter of Marta Rees, a relative of mine too! Life is full of surprises.

We flew out of Trelew to the Iguaçu Falls at the confluence of three countries, Paraguay, Argentina and Brazil. It was an awfully challenging flight getting there, flying over dense jungle. Just as we were about to land and descending through the mist, we started to feel a lot of turbulence, then without warning we ascended again as the plane was tossed like a feather in the wind.

Our pilot told us to prepare for landing. We tried to land three times, each time circling around, buffeted by winds and lashed by torrential rain. The storm we were flying through was causing the pilots a lot of problems. On our final approach we came zooming out of the clouds once more, but to our horror we saw that we were almost on top of the terminal building. Our pilot pulled the aircraft up sharply just in time.

You could hear a pin drop on the plane, the previous nervous chatter had stopped and there was silence, not even crying or screaming. We headed off and the pilot came over the intercom telling us we were returning to a rather modern Posadas airport as the weather was too bad to land. We landed at a very modern airport about one hour away, they must have refuelled and we tried again a few hours later, this time successfully. It was some experience.

Foz do Iguaçu in Brazil was an interesting place. It means Iguaçu river mouth and we were here to witness the largest waterfall system in the world. The falls divide the muddy brown river expanse into the upper and lower Iguaçu, and to experience the force of the water is breathtaking. Iguaçu was named by the indigenous Indian tribe meaning 'big waters'.

We stayed at an old colonial-type property called Das Cataratas, which has no doubt been refurbished by now, but though a wonderfully characterful property when we stayed, it was in need of some upgrading. The area was alive with sounds from the likes of the colourful toucan. Lizards almost 2 feet long were occasionally spotted and we were told the coral snake could be in the undergrowth and was certainly so if we ventured into the surrounding jungle, which was not too reassuring.

We were coached to an area where we could take the wooden walkways to get up close and personal with the Devils Throat and witness how truly awesome this sight was. On arrival one of my party said to me, 'My word, Ron, these are fantastic! Did you know they were here?' Well, I'm not dumbstruck too often, but having put the tour together and standing next to one of the most famous sites in South America, this was one of those speechless times.

A number of us took a boat to the bottom of the falls. The small craft was nowhere near the size of the 'Maid of the Mist' at Niagara Falls but was more a RIB speedboat. The sheer force of the water coming down literally took our breath away, the noise and magnitude of the cascade of thundering water meant it was so hard to draw in air, truly awesome.

We finished our South American tour in the iconic city of festivals, Rio de Janeiro, meaning January river, framed by Sugar Loaf Mountain and Corcovado Mountain (meaning hunchback) with the Christ the Redeemer statue atop, looking down over song-inspiring beaches such as Copacabana and Ipanema. We stayed at the Rio Pestana on Copacabana beach, where we watched games of volleyball from early morning each day.

The beaches and sumptuous apartments overlook the sparkling bays against a backdrop of half-built slum dwellings, favelas, which cling precariously to the hillsides.

We of course visited Sugarloaf Mountain, between the city centre and Copacabana beach in the picturesque arty quarter of Urca. Its peak is reached

Ron and Dianne at the Christ the Redeemer statue in Rio on a misty day

by a spectacular cable-car journey done in two legs, and has views over Copacabana beach and the bluey green waters of Guanabara bay with its many islands, as well as forested mountains and across from us the pinnacle of Corcovado mountain with the illuminated Christ figure on top.

We then headed for the Christ the Redeemer statue which is amazing to be up alongside with the equally stunning view down over the Maracana football stadium and the Atlantic Ocean beyond. The statue is 30 metres high, perched on top of an 8-metre pedestal, and was completed in 1931. Maracana stadium once held a crowd of 200,000 people.

The poverty in the favelas meant that crime was high on the streets. Young children especially often acted as pickpockets while others use distraction tactics. Muggings are still common and you have to have your wits about you. It shouldn't stop you going, just take obvious measures especially after sundown, such as stay in busy areas, do not carry much in the way of valuables and leave your passport in the hotel safe.

Prostitution is a thriving business too, an example being when we arrived at our hotel, most of our party in their sixties and seventies or even older, we checked there were a few attractive ladies in the lobby. Not long after we checked in two single gentlemen in our travelling band, both cousins, were sharing a room and heard a knock at their door. One of the chaps was about to have a shower and opened the door wearing just his vest and pants, and two attractive Latin beauties attempted to persuade the occupants of the room that they had some services the men would enjoy... Well, for two aged farmers from Mid Wales confusion led to astonishment and then a very definite 'no, thank you very much' but it was a great story to take back to the Welsh Hills.

Our tour ended. Rio had shaken and gyrated its hips to the throbbing rhythm of samba music pouring out of bars, but I never met 'the girl from Ipanema'!

All in all, a fabulous tour of discovery.

It just goes to show it's not always the heralded tourist attractions themselves that are the most memorable but experiences in everyday life on those adventures that shape you and your future and the way you think, indeed it's not always the destination you head for but the journey you embark on to get there that gives you the greatest experiences.

CHAPTER THIRTY-TWO

Myanmar (Burma)

Burma, now Myanmar, was a joy to visit in 2012! We had deliberated long and hard about whether we should visit as a lot of political and indeed media coverage said if we travelled there we would be supporting a corrupt and vicious regime. But there had been a little softening of the rhetoric and Aung San Suu Kyi, the youngest daughter of Aung San, the so-called 'father of the nation' of modern Myanmar, had been released from house arrest. Born in June 1945 she was an attractive, charismatic and well-educated woman having graduated from Delhi University and then Oxford University before working with the United Nations for three years and now running for election to the governing body. Our thoughts were that the winds of change might well blow in our favour.

Dianne and I travelled in February 2012 with a view to getting there in the dry season and had a stopover in Singapore for a couple of nights to break the journey. Our flight on Silk Air to Yangon (formerly Rangoon) took just over two hours and we breezed through passport and customs control much to our relief, as we had an awful lot of crisp American dollars, which was and still is the currency of preference and we had to pay for our whole holiday on arrival as well as have spending money for the journey.

We were met by Judy who represented a local company called Journeys and proved to be such a delight. She and our driver Zaw (pronounced Zo) drove the twenty minutes in his battered but very clean Toyota Camry to their office to take care of financial matters. I felt like some sort of gangster producing incredibly pristine $100 bills, but they had to be this way because if you try and give them any with tears or marks or anything that is not in perfect order they will not take them! At this time credit cards were in their infancy in Myanmar. There was a chance of using them in a bank in Yangon but even that was debatable.

We were then driven with a couple of welcome gifts to our hotel. We checked in and relaxed ready for our tour the next day. But we first had a bite to eat and exchanged some money. For $300 we were given a little over 24,000 Kyat, the local currency. It was such a huge wad you couldn't put it in your pocket, so we hired a safety deposit box.

After a restless sleep we were met by our city guide Joseph, the first of an assortment we enjoyed travelling with throughout Myanmar. We took in all the colonial buildings, many of which were falling into disrepair not unlike many in the former British empire and soaked up the Burmese history. The jewel in the crown was undoubtably Shwedagon Paya Pagoda with its immense gold-leafed stupa and very dramatic architecture on entry. For many Buddhists especially those from Myanmar this is a must-do pilgrimage.

Most of the tourists were from the various Myanmar tribes but you could detect voices from North America, other Asian countries, Europeans including the Netherlands, Germany and Scandinavia, but there were no other British tourists.

We took in the huge reclining Buddha and other sights before a pleasant lunch at the Golden Duck (Be le) down by the river, set in a former ferry terminal-cum-warehouse which had seen many famous people pass through it such as George Orwell and Rudyard Kipling and of course the country's very own royalty, King Thibaw and Queen Supalayat.

The restaurant was a cavernous affair with the echoes of activity and chatter you might expect. A dumb waiter produced food from a higher level in one corner of the building, noisily cranking and labouring to let you know another meal was on the way.

We took our lives in our hands crossing the street to see the Sule Pagoda near the gardens that boast the Independence Monument. Nearby were the High Court and Secretariat building, the latter of which was dilapidated and awaiting designation of use. This, we learned, was where Aung San was assassinated not long after independence in the forties.

Transporting chickens Burma style

People bathing in the river in Burma

*Sedan ride we arranged
for Princess Dianne*

*Golden Rock Temple
the destination of many
Buddhist pilgrims*

As cities go, it was interesting without being scintillating and soon Scott's Market was calling us to do some shopping, as it had lots of stalls and sold just about everything.

On day two we were joined by two elders of a church who were involved with a Christian church foundation called Ananda Myitta. We were escorted by Judy Zamtei, a fellow parishioner who had met us at Yangon airport. We travelled about 45 minutes from our central hotel to the outskirts of Yangon to visit Ananda Myitta, which was set in a rural area. The Grace and Love Home, as it is also known, is a small orphanage run by the Christian church with a house mother and father to take care of the welfare of the nine boys and two girls, who at the time varied from about five to thirteen years old. No funding is received from the state as it does not recognise the Christian church and supports only Buddhist institutions. Some of the children had parents but because of their circumstances they could not be at home, others were orphans. As time has progressed some have gone to university, others have finished school and taken jobs.

The building itself is relatively austere with two floors, a couple of showers and toilets and dormitories, a small kitchen and a living room where they congregate. It also has a small vegetable patch. The orphanage pays for them to go to school and for their health and welfare including medicines and food. It is very basic and following our first visit we were able to give them some writing materials but also paid for their battered old kitchen table to be recovered and one of the toilets and showers to be repaired so they could finally use it again. We bought a bicycle so the house father could cycle to market to buy food and other necessities. It was only a small expenditure for us but it made a huge improvement to their lives. We still keep in touch and help when we can, always remembering this humbling experience of the small joy we brought to children who had very little. They sang for us and we tried to reciprocate with a little song of our own. A very special time.

Day three we were met by a young chap called Khin Naw, a cousin apparently of our guide Joseph from the first day's tour. Khin Naw's English was reasonable and we headed off with Zaw our driver towards Bago once known as Pegu and home to the immense Shwemawdaw Pagoda, where we took a break, then headed south towards a place called Kiprun at the foot of Mount Kyaiktoyo.

This was a bit more off the beaten track for overseas visitors but very important to the Buddhists as we transferred to a 'Pilgrims' truck' which wound its way up the mountain, throwing us back and forth at the switchbacks with our fellow travellers all in cosy proximity to a staging area where we offloaded to walk the rest of the way.

It was a hot day and the heat and steep ascent started to take its toll on Di who was feeling a little faint, so we flagged down two guys who were carrying people up and down in a litter like a sedan chair with a canopy. Di was glad of the ride and the relief from the heat of the unrelenting sun. We paid the men and I walked beside with our guide Khin Naw.

We reached the top and could not believe the amount of accommodation and visitors circulating at this important Buddhist site called Golden Rock. The rock is precariously perched according to legend by a strand of Buddha's hair. The rock, covered in gold leaf, defies gravity with a pagoda set on top of it. Women are currently not allowed into the inner sanctuary but there are some breathtaking views from this point.

This whole journey down to almost the Thai border at Thanbyuzayat was fascinating, and we felt like locals as we trundled through Mt Zwegabin set in dramatic outcrops of karst limestone. We stopped at Hp'an and the Thanlwin riverfront Shweyinhmyaw Pagoda, we witnessed local folk below us at the river's edge washing clothes and men in longtail boats bringing in produce such as watermelons from upriver. Some families were washing together and fishing boats headed out as the sun started to set. The rich orange sun created an imaginary bridge of golden light as it sunk below the horizon to signal the end of another day in this quiet land of smiling faces.

Another hot day greeted us. As the temperature rose towards 30°C we visited a large monastery at Thaton and we called into a little primary school and listened quietly to the lyrical language before engaging the young children with a little English and giving them some pens, pencils and books and amusing them with finger puppets we then left as a gift too.

Each day we saw many more interesting sights on the way to Mawlaymine (more familiar to us and Kipling as Moulmein). Kipling's famous poem, 'The Road to Mandalay', published in 1890, was apparently inspired by the pagoda here overlooking the river. Kipling made a stop here on his travels and was taken by the beauty of a Burmese woman. On our journey there were many guard posts at strategic junctions and over river courses, most with no one in them. We constantly had to carefully pass cyclists and heavily ladened pick-up trucks and even moped riders who had dozens (yes, dozens) of chickens in cages balanced around them. I can only think that the rider got on first and associates then added the various chicken carriers around them. How they kept their balance, who knows.

We made a beeline for the temple Kyaukthanlan Paya and were amazed at the vantage point it has over the Thanlwin River and surrounding countryside. There are many steps up and it must have tired Kipling, but for us the modern lift was a blessing in the heat. We were blessed to watch the sun go down over the Thanlwin River, a wonderful experience. The Seindon Mbaya Kyaung is by the side of the monastery in the lee of this structure and once the home of King Mindon and his wife, the last King and Queen of Burma. This wonderfully evocative structure should not be missed, the teak floorboards felt as though they were the original flooring and could tell many a story from over 100 years ago.

Next day, more than an hour further south, Thanbyuzayat was the northern terminus of the infamous 'death railway' built by prisoners of war of the Japanese from Thailand to Burma during World War II. It was a very poignant visit to the memorial and graves in the Commonwealth graves cemetery, the air was hot but the stillness and quiet made it a perfect place to remember the sad loss of these brave men, over 8,000 Asians and 6,500 Allied troops, mostly British, but Dutch, Australian, American, Gurkhas and New Zealand men too. In the cemetery itself there are 3.371 graves. Some inscriptions stick in your mind like one: 'I waited but you didn't come – life was cruel to us "Dorothy"'. We were not far from the coast so went to the Paya (temple) there and walked along the beach at Setse.

On our journey back to the capital, we witnessed schoolchildren all very well dressed, often in uniform, carrying little tin lunch pails. From time to time we saw long lines of shaven-headed novice monks and sometimes nuns, the latter in distinctive pink robes, also with shaven heads. Many people, especially ladies and girls wear 'thanaka', light markings of a ground paste made from the shavings

'Girls in Pink' – novice nuns, Burma

of a wood that is mixed with water and spread on their faces to act as a sunscreen.

Lots of folk cycled and were on motorbikes and mopeds, as well as trucks, pick-up trucks and three-wheeler vehicles, but oxen and cart were still in evidence in the more rural areas. Many people wore the traditional pointed wicker hat to keep the sun off. We saw lots of young children but observed they seemed to be carried, with no pushchairs or prams in evidence, which to our western minds was strange. You rarely saw people especially men in western dress. The men wore a wraparound long skirt-type garment with a shirt tucked in, called a longyi, while the ladies' equivalent was a 'htamein'. Lots of stray dogs roamed the streets, no matter where we were. We noted that a lot of the porters carrying suitcases and heavy loads on their heads were women.

We returned to Yangon and flew from Mingladon airport up to Bagan, pronounced 'Bahgunn', which is situated on the mighty Irrawaddy river, nowadays host to a number of cruise boats. The sunset over the mountains in the distance is fantastic as you gaze across the river.

This is a very famous historical site with hundreds of temples, many over 1,100 years old. It is magical to climb a pagoda and watch the sunset over the Irrawaddy. We visited the fabulous Ananda Pahto temple, where the four 9-foot-high buddhas shaped in teak and adorned by gold leaf stuck in our mind. Many of the temples were damaged in the World War II conflict and others in earthquakes.

Even more outstanding is to get up at dawn to be transferred by 1940s Chevrolet bus left by the Canadians from World War II and take a balloon ride over these more than 2,000 historical monuments and mystical temples.

Our balloon captain was Graham Luckett, from Oxfordshire originally, now based in Colorado, who travels the world dependent on the season ballooning for tourists. We joined six other visitors, four from Austria, a German and a Finn, stepping in foot holes set into the basket to climb into our own compartments. We felt like bottles of wine in a wine rack but the champagne popping feeling exhilarated us as we soared into the air. The sturdy rope tethers released, the balloon, charged by the hot air, ascended gracefully to a wonderful elevation allowing us to be surveyors of the waking land below.

We glided silently above this ethereal mist-covered landscape and the mighty Irrawaddy river, watching people go about their daily lives as brick

Bagan Balloon flight

makers and peanut farmers. As the sun came up and illuminated the day the mist was burnt off, peeling away from the land to let us see the fruits of history. In eerie silence, punctuated now and again by the blasts of hot air keeping us airborne, we used the winds to drift aloft like silent spies, a wonderful and unforgettable experience. We landed 45 minutes later in our brace position as we awaited the thud confirming our safe return to terra firma.

Chatting with nervous excitement about our shared joy that morning we were all presented with a 'Balloons over Bagan' certificate and a baseball cap as well as some banana bread, a croissant and a glass of champagne before returning to our hotel for a hearty breakfast, full of the amazing experience.

We stayed at the Bagan Thande hotel, which is located next to the Irrawaddy river within the Bagan archaeological park with its plethora of monuments and temples. This resort is a real haven to relax in, even Edward VII stayed here. We sipped a sundowner watching the sun set over the hills beyond the famous Irrawaddy then enjoyed a meal on the terrace and watched the excellent marionettes show or simply strolled along the river, which was so calming it was as if you were taken back hundreds of years as boats silently drifted past beneath the pastels of the evening sky.

A car journey through a somewhat biblical landscape, very feudal in its appearance to walk up Mount Popa 40 miles away was interesting especially stopping en route to meet a chap and his family who run a palm sugar business. It's a bit more than that as they make a 40% proof toddy from the tree and sugar oil, and the leaves are put to various uses from roofing to making bags, whistles and goodness knows what. A family member shinned up the palm tree to get sap and coconuts. Another chap behind the building was using a wooden grinding wheel on a stone plinth to make peanut oil. We bought some tamarind and coconut sweets they had made. Their ingenuity was quite astonishing.

Continuing our journey, we switch-backed up the road into the mountains, finally arriving at Popa village. This area was a popular destination for the Nat religion pilgrims as they, like us, would visit the Nat shrine in the village, then climb the auspicious number of 777 steps to the monastery on top of Mount Popa, an extinct volcano higher than Ben Nevis at 1,518 metres. The Nat religion is a widespread belief that there are forest guardian spirits who can act against environmental destruction. If nothing else, it deters people from felling trees, as they fear it would bring lots of bad luck and terrible things would befall the perpetrator. I'm not sure the Green Party in the UK could include this in their manifesto, though.

Fortunately, for much of the route up to the monastery you can stay in the shade under a covered walkway, although just when you thought it safe to relax you had to deal with groups of at times vicious and bold monkeys who tried to steal anything from you that was not held on to. We had been forewarned and the guide and ourselves were armed with sticks to deter their approach. Others did not fare so well. As tiring as the climb was it was well worth it to take in the fabulous views.

U Bein Bridge in the dry season

The next stop the following day was Mandalay, a very romantic name linked to the Kipling poem and it made us think of the film *Rebecca*, whose leading lady lived in a similarly named house, Manderley. The modern-day city of Mandalay was somewhat disappointing and not as romantic a destination as the notion conjured up in my mind from various books and articles I had read over the years. True, there were some interesting monasteries and other places and the iconic U Bein Bridge was outstanding and evocative, whether you saw it at sunrise, sunset or during the day. It is best seen out of the dry season, when you might see fishermen and people herding ducks on Taungthaman Lake. This is the world's longest teak wood bridge at 1,300 yards. Many of the 1,060 supporting poles were not in good condition and the teak planks were uneven with no side rails. Some of the bridge had covered areas, under one of which we bought a small painting of the bridge from a local man as we sheltered out of the 31°C heat.

It is well worth a visit over to the ancient island of Ava, where you board a pony and trap to explore, visiting some attractive monasteries and other buildings especially the stupas. Finally, while taking lunch on the tour we met a couple of Brits from 35 miles down the road from our home town, Dorita and David Hughes from Newtown!

Another little highlight on a blisteringly hot day was a small private boat ride up the Irrawaddy river. We sat in a couple of planter chairs under the tapestried canopy of the vessel as the rhythmic 'chug, chug, chug' of the engine lulled and relaxed us and the refreshing breeze cooled us. We cruised by bamboo and rattan shelters of the native fishermen and farmers looking after a few head of cattle.

We soon arrived in a sun-drenched Mingun, met by a bullock-drawn cart with a rattan cover like a western wagon, the word 'Taxi' stencilled on the side. Mingun was a historic settlement with, although cracked, the second-largest bell on the planet, beaten only by a bell in Moscow. A small boy banged the side of the bell with a block of wood almost as big as himself.

Over the coming days we explored the Mandalay Palace, which was a mostly restored building after it took a battering in the Second World War. The Golden Palace, the former home of King Mindon who was the father of the last king of Burma, King Thibaw, was a very impressive building of teak and gold leaf. The Kuthodaw is home to the largest book in the world of the writings of Buddha, hand-carved on 729 marble slabs each housed in its own small stupa, all fifteen books of the Tripitaika. Finally, we headed up Mandalay Hill to watch the sunset, taking an escalator up and a lift down. We could make out the Shan Ridge, the old racecourse, polo fields, a golf course and the amazing old British-built prison which was very large and shaped like a wheel, a fantastic vista as the sun set over the Irrawaddy.

Next day we headed to a summer retreat of the British, called Pyoo Oo Lwin nowadays but once known as Maymyo. The route we took was also the main road into China. The air became thinner and cooler as the road climbed into the hills. It was an enjoyable visit and exploration of colonial history. The 1936 Purcell Clock Tower in the centre of the town with traffic travelling around it was named after an American. An old British private club, now a hotel, called Candacraig was nostalgic, as was the garrison church of All Saints with standards flying inside of British regiments. I gave a small sermon from the pulpit. Kandawgyi Gardens were a kaleidoscope of horticultural colour and made our day.

We drove on, passing fields ploughed by Brahman cattle and small private fuel stations with yellow canisters of fuel and litre bottles for motorbikes. Regular teahouses dotted the road. Stands of teak trees gave way to open land dotted with scrub-like trees, then we climbed into the hills and finally arrived at the hill station of Kalaw and the delightful Pine Hill resort with its colonial-like reception, enhanced by the bungalows typical of colonial hill station life, one of which was our accommodation.

Kalaw was a quaint town and the one building that stands out for me was surrounded by a very English-looking garden full of typical English flora such as nasturtiums and sweet peas. Down in the market area a mother and daughter had a little stall and we asked if we could take their photograph, which they agreed to and loved seeing the image, albeit embarrassed by it. A small game of 'chinlone' with a handwoven rattan ball was taking place outside a shop between about six young fellas, the usual team size. 'Chinlone' is also known as Caneball and is the national sport. It is similar to western efforts with a hacky sack. 'Chinlone' is played by team members passing the ball among each other within a circle without using their hands whilst they are walking round, with one player in the centre of the circle.

We loved our stay here. Two of the girls working at the hotel were in the garden one day doing chores and singing like nightingales. One morning at breakfast when they had served us, they sang for us, which was so special.

We were sad to leave this relaxed town, but headed for the small railway station which felt as though it was stuck in a time capsule from a bygone age. A number of locals were waiting for the late afternoon train with their assortment of bags and sacks of rice. A tea room idled in the slow hot afternoon and a fruit, water and snacks stall at the other end of the platform sold their wares between fighting off the flies.

A few other tourists arrived and we purchased our $3 tickets down the line to Shweyaung. Nothing much had changed here since the 1930s. We had been told the train was an hour late before we left the hotel and, sure enough, in line with its revised arrival time, the train pulled in. We were fortunate to have tickets for the first class coaches with reclining seats and soft upholstery.

The train rocked its way along the rickety tracks and gradually made its way downhill towards the plains, rhythmically coasting past rolling hills and through fertile valleys littered with workers sporting pointed bamboo hats. A light breeze from the open windows eased the humidity of the day. Small stations en route had orange sellers and other traders. Soon the lowlands with fields of barley and wheat appeared as we neared the lakes.

Arriving in Shweyaung we picked up our car and driver who took us to our boatman on Inle lake. It seemed strange loading our cases into a longtailed motorboat, but off we putt-putted down the narrow man-made waterways passing all sorts of humanity with their huts built on stilts, lots of birdlife and people working pigs wallowing in mud. After about half an hour we were joined by an 'Intha' tribe guy, the engine had been cut and this wonderful native of the area leg-rowed us the final 1,000 metres to the fabulous hotel, the Inle Princess Resort Hotel which was our home for the next three nights in a lakeside bungalow with a delightful deck from which we could watch the world go by and the fantastic sunsets over the distant mountain range.

We had arrived here to celebrate Dianne's birthday and the next morning after breakfast we set off just after 7am with a youth leg-rowing us to a junction onto the main lake, where he disembarked and our guide and

Schoolgirls in Myanmar classroom

pilot struck up the outboard motor and progressed rapidly through the reed channels into the open lake. We were in shorts and T-shirt, which was cool at 7am as the air rushed past us but by 8am the sun was much higher and the day was a hot one once more, although we still needed the blanket provided and a fleece whilst the boat whizzed along in pursuit of the unique lake inhabitants.

We witnessed fisherman with their unfamiliar large cone-like baskets they were pulling up from the lake depths with great hope of a decent catch. As they used these upside-down funnel-like creations to corral a fish and then spear it, it was frustrating and labour-intensive work, but fascinating to observe. Other fishermen would slap the water's surface with their paddles to scare fish towards the nets of their fellow fishermen, a bit like the beating principal in grouse shooting, but instead of the air the lake is the killing ground.

All through the day we witnessed fascinating sights and events such as a four-year-old boy being taught to leg-row. How he kept his balance, let alone learned to row, amazed us.

We were heading to the south of the lake and encountered breakaway islands of water hyacinth. We saw lots of them in the canals around the entrance to the lakes and villages. We made our first call at the five-day market at Inn Thar, close to the location of the famous temple called Phaung Daw Oo Paya. This Buddhist monastery and pagoda is surrounded by a plethora of stalls and small shops selling handicrafts to tourists, vegetables, fruits, meat and fish with a variety of fresh carp, catfish and eel as well as dried fish. Much-needed necessities could be purchased such as boat engine fuel, clothes and even simple tablets like paracetamol and other pharmaceuticals. There were very clean toilets here too for a 200 Kyat fee.

This is also the site where the Dragon Festival with its colourful boats takes place. We saw the very splendidly carved and painted Royal Barge and this would be pulled by many more leg-rowed boats joining in the October pageant.

At Ywama we witnessed a small home industry of weaving using the reeds from the lakes. It looked tedious and hard work, but the young girl exponent took it all in her stride. The old-fashioned bamboo treadles and shuttle flying back and forth with the weave were fascinating to watch. You could see blacksmiths at work using bellows operated by bamboo sticks.

Pigs were kept in sties on the floating weaved reed beds, fish nets were hung under the teak huts that stood on sturdy teak wood stilts. These fish enclosures were like larders for the week's meals. Tomatoes were grown alongside cucumbers, flowers and beans and indeed all sorts of produce ready for market and their own consumption.

Cheroots were being made at one location next to a silversmith and we saw umbrella makers using tree bark to create a paper covering for the frame. Various tribes were in evidence at the market as well as at the different villages on the lake, supported on reed beds and wooden frameworks of stilts. People travelled from as far as Yamwe in the Chin Hills.

We were very taken with four ladies we came across from the Paduang Kayan tribe and immediately remembered them from documentaries we had watched back home. They stood out because they wore metal coils around their necks, making their necks look really long and allegedly first used by their ancestors to protect their necks from marauding tigers, would you believe. It was not uncommon to have these coils elsewhere such as their knees. You couldn't help but stare and of course what a wonderful photo opportunity.

These ladies were obviously very used to attention from the few tourists they did see and were happy to explain that these spring coils can be taken off at will although many ladies wore them for a long time.

Many of the above sights were so intriguing we could have spent all day with them but we were also fascinated and intrigued by our next port of call, Nga Hpe Kyaung, the Jumping Cats Monastery. Full of character and history with its ancient Buddha figures, this monastery predates the Mandalay Palace by four years.

As we climbed from our boats on to a sort of solid territory, we walked past venerable old monks sitting and lying on a linoleum floor. They offered us flasks of green tea and tasty bowls of soya beans as refreshment as we waited around an open-air floor in anticipation.

Before long a lady appeared with a small hoop and some cat food followed by a few cats a black, black-and-white and a gingery-brown assortment of felines. The latter was the fattest and we saw why as the entertainment progressed. He hardly jumped but ate the lion's share of the cat food!

A monk rang a bell and the black cat jumped through the hoop, he was very sprightly and tried to steal the other cats' rewards too. It was all rather sedate and hardly Billy Smart's circus but novel. At the end of the session the monks handed sweets to the small children in the audience. A monk made a small collection from the small groups of visitors who were there, then all the monks started chanting. It really was a lovely day culminating in this activity.

We were about forty minutes' boat ride back to the Inle Princess which was bathed in the weakening light of the late afternoon as we arrived. We took advantage of the outside shower serenaded by passing ducks who were probably just laughing, before heading for a sumptuous meal in the hotel restaurant, made extra special when the staff serenaded Dianne with a guitar-accompanied rendition of Happy Birthday! As they formed a musical procession with the lights turned off and led by a chocolate sponge cake with five candles showing them the way they also bestowed on Dianne the gift of a traditional longyi, a fitting ending to a special day!

Next day we set off for the furthest reaches of the lake, a place called Sankar with our guide Mion who handed us over for the day to a delightful, attractive and bubbly young lady we picked up en route. This guide was called Phi Phi and wore black checked headgear like a turban and a black trouser suit which was typical of her Pao tribe. Starting early again at just before 8am we headed over the water to Sankar. It was a long way as Sankar is actually a separate and artificial lake at the bottom of Inle created by a dam.

We broke the journey at another market called Nampan where many Pao villagers come down from the surrounding hills to sell their produce and handmade wares. We noticed as we travelled some evidence in villages of small satellite dishes and we learned an average house back then would cost around $750, yet a longtail boat would cost $2,000.

Hills in the distance were sparsely covered in trees or vegetation, a herd of cattle on the lake shore sought refreshment amongst the water hyacinth. A man rode a beast of burden, stupas dotted the lakeshore. We took a canal through the reeds to a village passing a stand of bamboo before reaching the houses and the small, informal market.

We re-entered the main lake and headed south seeing a multi-tiered pagoda in the distance, a lone stupa on a distant ridge piercing the hazy blue sky. We were passed by a longtail boat being steadily steered and rowed by a lady, on the lakeside a lady crouches unabashed washing herself in the lake water, and above her a young girl washes clothes on the deck of the house. A blue-tarpaulined pagoda in construction leads our way to another canal where we join another wider waterway, hills converge on both sides of us as the lake narrows and locals wave to us almost in surprise on seeing westerners down this far on the lake.

The boat stopped at a few places to see a Shan tribe market at Hmawbe in the south on Inle lake. It's a really rough-and-ready local market, no souvenirs, just an authentic working market for the huge mix of local tribal humanity. We had only seen one other white face until then when we mingled with about thirty other westerners. We tasted the local home distilled liquor, which was quite potent!

We bought a stack of exercise books and pencils to give to a monastery-run orphanage on the lake. There were lots of teahouses, full of men. It was a local gathering a little like the pubs back in the UK.

We stayed for about an hour before heading on and reaching Hmawbe Bridge, a sort of checkpoint and entry to the Sankar area. Shortly afterwards we seemed to go off-piste on a short cut, the boat pilot told us, but we soon got stuck in reeds and it appeared it was too shallow, so first the pilot,

then I stepped out into the paddy-like field of reeds and manoeuvred the craft until it was freed. It was easier said than done as we sank in to the mud up to our ankles and the water hyacinth lassoed and restrained your movement, but getting going again we joined a wider waterway.

Everywhere we went we met happy, smiling, laughing and friendly people, so peaceful, incredibly helpful and resourceful, all totally at odds with the world's view of Myanmar being full of trouble and strife.

We passed a good-sized community and a large pagoda surrounded by smaller ones. A little further on was a similar sight, but this time on a hill, a place called Tarkaung. Then the hills started to get smaller, plantations appeared staked to the lake floor with tall bamboo poles to allow for the rise in water in the wet season.

Finally, after a long journey approaching six hours, we arrived in Sankar, our destination and the ancient capital of that region, a small village with interesting monasteries, pagodas and temples from the sixteenth century. Again, about thirty to forty tourists mingled throughout the crumbling temples, all seeking shade from the formidable heat of the day.

An old monastery in the centre was home to a school where a lesson was in progress. We witnessed a monk teaching them, tapping a boy on the shoulder for not paying attention, then the whole class would recite something in unison. The monastery had four monks and five novices and dusty tracks led from it around the village and through the decaying stupas. Tamarind, teak and frangipani trees dotted the urban landscape and a hospital operated for the locals.

Ladies sat low down to the floor weaving reeds and bamboo into coverings for roofs and sides of homes. We retreated for lunch at a place on the lake called Golden Island Cottage then started our journey back following a quick visit to a small saki-making cottage industry.

We stopped at the Takhaung Mwetaw Pagoda on the way back. It's simply awe-inspiring. Some impromptu stops also taken, like the time I was busting to go to the toilet made far worse as the narrow boat whizzes though the lake and all the water disperses at force around you.

We stopped at a small village settlement and I had to climb a ladder into a house which was a sort of convenience store that also sold gasoline for the longtail boats. We walked through rooms with the guide and witnessed people making cheroots, a chap weaving something which involved using his hands and twine wrapped around his big toe, before we interrupted a game of cards with big piles of kyat being wagered, the one chap tried to get me to have a game. I declined before following the guide to another location, walking through people's houses as they were going about their daily chores, and we were very surprised to find a rather nice toilet. I thought it was going to be a freefall lake toilet!

Returning to our mountain-facing sundeck we soaked up our final sunset on Inle Lake, full of the joys we had experienced. The calming tranquility away from the buzz of urban humanity was so relaxing we really didn't want to leave!

Before we could draw breath, our adventure continued to the wonderfully exotic Ngapali beach. A short transfer bright and early by leg-rowed water taxi through the morning mist took us to our taxi cab for our journey to Heho airport. From here it was a 60-minute flight to Thandwe, which serves the beach resorts of which one is Ngapali, a low-key location on the palm-fringed white sands of the Bay of Bengal.

Everything was relaxed here. We were met by a young lady from the Pleasant View Resort where we would be staying, and were shown to a very old seven-seater bus, which turned out to be a Japanese vehicle, a classic Hino-made bus that had been refurbished and was apparently a mere forty-five years old. It served its purpose well as it chugged along for the 4-mile journey to the hotel along a narrow road, lined with vegetation and occasional fields, many covered with fish drying in the heat.

We walked through the casually laid-out grounds to the beach and the Pleasant View restaurant set on a spit of rocky land which had to be waded to or rowed to at high tide. We walked the short distance to our garden view basic but comfortable villa.

To the left of the restaurant as you looked from landside was a small bay packed with local fishing boats. It was fun watching them bring in their catch of thousands of small whitebait-type fish and some exotic-looking species too like the occasional deadly puffer fish complete with poisonous spines. A cart pulled by buffalo made its way along the beach in a nod to bygone days.

During our stay we enjoyed walks and runs on the beach, seeing pop-up restaurants on the sand and witnessing the kayaking and snorkelling boats going out. We also took walks further along the coast, seeing the homes and stupas of the local inhabitants.

Leg rower on Inle Lake

Locals walked the beaches trying to sell you their wares, and we took to a lady and her daughter selling fruit. Local children asked us for 'bonbons' (sweets, which we gave in exchange for locally collected shells).One day we gave them a skipping rope which we had left over from our box of goodies we carry to give to local children as we travel.

We bought fruit from the lovely mother and daughter each day, even though we didn't need it all but Dianne would chat with them.

On the last day they brought Dianne a gift of a pearl-covered wristlet. The mother used to call Dianne 'mamma'! Zsa Zsa, who we now knew was the fruit seller's name, came along with a sprig of local flowers and a rose. We gave the little girl a furry pencil case with a few things in, and she was pleased as punch. We chatted with the mum who told us she was 37 years old, and was a lovely, tall, elegant lady.

On our last night we watched the sunset and the string-of-pearls lights appeared as darkness fell along the coastline. We waded out to the restaurant and sat at our well-positioned and favoured table number 6 for a final sundowner. We ordered a refreshing and aptly named iced cocktail called 'Tom Coolings' (nice play on words) and

Mandalay Palace

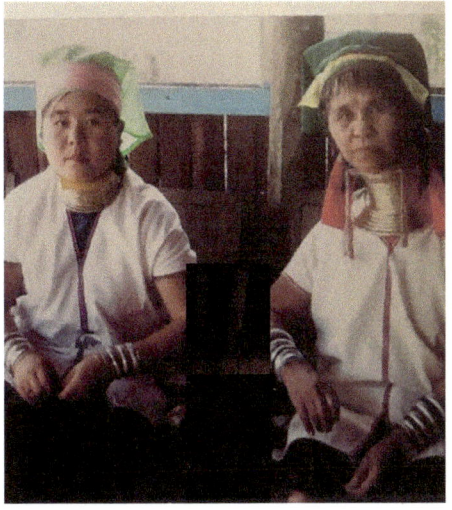

Necklaced Paduang tribe long neck women; these huge necklaces apparently defended against tigers as well as being ceremonial

This lovely waitress sang to us at breakfast time

The lovely Zsa Zsa and her daughter selling us fruit on Ngapali Beach

Local Children on beach Ngapali

Ron sampling fresh fish at a Ngapali beach pop-up restaurant

Myanmar Beer. After a delightful meal the young man showed us the way with his flashlight along the causeway and we walked barefoot up the beach to our villa. What a heavenly holiday, which sadly was coming to an end!

Next morning, we said our goodbyes, and headed back to Thandwe airport and on to Yangon for our last night in Myanmar. We had treated ourselves to the Kandawgyi Palace Hotel on Inya Lake, a grand old hotel which has since sadly been the victim of a dreadful fire. The Kandawgyi Palace Hotel was first built in 1934, not as a hotel but as a two-storey, red-brick building for the Rangoon Rowing Club. It was a regular destination for British officers who occupied Burma at the time. During the war it was used by the Japanese as a welfare department. After the war when Burma gained independence, the property became the National Biological Museum in 1948. It began operating as a hotel in 1979, when the Ministry of Hotels and Tourism took it over. The hotel featured ten bungalows made of teak. They were eventually replaced in 1993 by a large complex built in traditional Southeast Asian style

Ron, Dianne and our guide Khin Naw

and was refurbished to a high standard in 1996 complete with outdoor swimming pool.

Sadly in 2017 around 3:15am on 19 October, a devastating fire broke out. Eighty fire engines tried to control the blaze, but being a teak wood building the fire took hold irretrievably. The hotel had 141 people staying at the time. One man died and others were injured escaping, but the tragedy could have been far worse. It has never been rebuilt.

Our room had a fabulous view of the Shwedagon Paya, reminding us of its glory and that our journey had come full circle since our arrival. We recalled all the people we had met and the wonderful sights we had seen. The things we had done all cascaded through my mind like a kaleidoscope of colour.

This friendly and superstitious people are wonderful and I selfishly wanted to keep the experience to myself and not share the joy of Myanmar (Burma).

But the world needs to know and give recognition to this industrious, courteous, moralistic people. Their world is shaped by Buddhism, Animism and also Christianity, all religions and creeds in their own way encouraging people to live a good caring life.

I hope Dianne and I will return one day, because we have left a bit of our hearts there.

CHAPTER THIRTY-THREE

Four friends and two sisters

In early 2015 we had enjoyed another fantastic holiday in South Africa with two good friends, exploring Soweto and Pretoria.

When we first arrived I switched on the television and saw all the looting and riots in Soweto, which we were due to visit. Before there was time for second thoughts we met our local guide called Hastings, a larger-than-life black South African and headed into Soweto. As we entered the sprawling township, Dianne said to me, 'Is this where we've just been watching the trouble on the television?' Hastings turned around, smiled his pearly white grin and nodded. Dianne slunk down into her seat a little bemused.

Before long we came across people looting and breaking down metal shutters. As we walked towards Mandela's childhood home, the looters ignored us and just attacked the shops, which for the most part seemed to be run by incoming immigrants such as Pakistanis, Somalis and Ethiopians and in spite of the efforts of the shopkeepers the looters, many of them children, departed with armfuls of goods, mostly food and soft drinks, all watched at a distance by police officers who did nothing to intervene.

We explored the infrastructure of this sprawling township of almost 10 million people, its shebeens, day care centres and other buildings, all dominated by two huge cooling towers, highly decorated with bright lights and graffiti and a bungee jump suspended between them. There is even a campus building of the University of Johannesburg in Soweto. Shebeens proliferated under the apartheid regime as indigenous Africans were not allowed into pubs and bars. The township inhabitants created their own illicit watering holes, often run by women known as 'Shebeen Queens'! These shebeens often brewed the traditional 'umqombothi' beer made from maize and sorghum. They were meeting places, community hubs where often dancing and singing lightened the locals' spirits. Nowadays shebeens are legal.

There appeared to be a variety of levels of wealth in the township, some parts very much poor squatters' areas, very run down and ramshackle, housing mostly immigrants from the sub-continent such as Pakistan. Their sanitation was a few 'portaloos' which served a huge area of shacks. About a quarter of a mile away the area became positively middle class, nicknamed 'Beverly Hills'. Winnie Mandela and Archbishop Desmond Tutu had houses here in Venda Hills.

After a good night's sleep, we headed to the capital Pretoria via the impressive Vortrekker Monument which has numerous steps up to it and tells the story of the Dutch Boer Vortrekkers.

Vortrekkers was a Dutch term given to those who participated in the mass migration of the Boers (Dutch farmers), the Dutch speaking people

Outside Soweto Mandela's home now a museum

who had settled in the Cape Colony of South Africa before the British arrived, expanding their empire and pushing the Boers out, forcing them to find new homesteads as they migrated by wagon train to avoid conflict with the British and continue their own way of life.

Inside the Vortrekker monument it is cavernous and impressive, fortunately it does have a lift to the higher floors and viewing terrace, well worth a visit to learn about this migration and the people involved.

Pretoria is the seat of government and also the must-visit home of Paul Kruger, the nineteenth-century president of South Africa for seventeen years, a very prominent character in South African history and also the man after whom the 'Krugerrand' is named. The Union Building was impressive, fronted by statues of Paul Kruger and Nelson Mandela.

On 15 October 1962, a 'Free Mandela' campaign started as preliminary proceedings began against Nelson Mandela who was being accused of crimes against the state. He was later sentenced to five years in jail and so began his incarceration at Robben Island off Cape Town, Robben being the Afrikaans/Dutch name for 'seal island'. Little did I foresee that I would be there twenty-plus years later and eventually on a later visit meet and make good friends with one of his fellow prisoners Patrick Chamusso, sometimes known as Patrick Thidebi, immortalised in the 2006 movie *Catch a Fire!*

We met Patrick the day we left Pretoria, when we headed east to the White River and the Two Sisters Orphanage. He was a rotund chap just over five feet tall, but he definitely had a presence. Patrick Chamusso was born in 1949 in Mozambique and was a member of the African National Congress (ANC) party of South Africa, participating in the militant actions of the organisation during the apartheid era.

One of three sons, Patrick came to live in South Africa when as a teenager he followed his migrant worker father to work in the mines of South Africa. After his father's death, Chamusso's mother remarried and had another daughter. Patrick worked in various jobs in the mines and later found work as a house painter and street photographer.

At the age of twenty-eight he became an engineer at SASOL, the South African oil company, four or five hours east of Johannesburg. Patrick had no formal education, but he did well for himself and moved up the ranks becoming a driver which came with good pay. He would bring coal from a neighbouring mine to the refinery where they liquefied it to make a synthetic fuel. He played football in the local leagues and showed some talent.

However, he became disaffected with life as his family and friends were constantly persecuted, being stopped and searched, interrogated and generally terrorised by the apartheid government forces. His brother, mother and father all died in police custody. That was the last straw: he decided to make a stand against apartheid by sabotaging the Sasol refinery. I have never asked him if he did do this as he was apprehended but later released without charge. He reported that he had been tortured whilst a prisoner. He fled to Mozambique where he joined Umkhonto UM Sizwe, the militant arm of the ANC.

After military training, he returned to South Africa where he single-handedly carried out a second, partly successful, bombing at Secunda refinery, causing damage but no one was injured.

In 1980 after a manhunt, he was captured by South African special branch and brought to court on a charge of conspiracy with other ANC members to bomb the refinery. On the way, he recalled, all the lights were on green as if speeding him to his fate. In court he was told with all the other defendants to sit, but Patrick remained standing. When he was asked why he was standing he replied because you are going to give me the death sentence.

The judge adjourned and discussed Patrick's statement with his lawyer. The judge said he had contemplated the death sentence but changed his mind and gave him twenty-four years on Robben Island. He got out after twelve years and said he learned a lot from Mandela, who taught him not to hate and be bitter but to appreciate many of the simple things in life and respect and learn from people.

In 1994, after serving his time, Patrick was released under the new government's amnesty policy as apartheid was finally starting to change to a fairer society. After his release, Patrick married Conney, and the couple had three children, settling in Mganduzweni near White River in Mpumalanga province.

In 1999 Patrick and his wife fostered two sisters whose parents had died of Aids. As this epidemic took hold more children came to them, eventually they formed the Two Sisters Orphanage. Today they educate, clothe, feed and keep safe over 250 children, just while we were there an infant was left in a bundle at their gate, such is the way of life here. Ironically, Sasol, who owned refineries Patrick allegedly destroyed, now help with sponsorship of the orphanage.

Pauline and Dianne sharing fruit with the children at Two Sisters Orphanage

We arrived in White River and arranged to meet Patrick at a shopping centre, and had a drink with him before travelling with him to the orphanage at Mganduzweni, about a 20-minute drive. David and I stayed in our hire car and followed Patrick, who drove our good ladies Dianne and Pauline. When we arrived at the orphanage Patrick could not get over our trust in him by allowing our two ladies to head off into the wilds with a convicted terrorist we had never met!

Patrick told us he had been in hospital a few months before after being carjacked and injured when his car rolled over after being pursued on this very road. It sounded a bit like the wild west!

It was very humbling to meet the children and some of the house mothers and see how Patrick had set up their little school. He had created a workshop to make bedding to sell by using sewing machines, a skill he had learned in Robben Island and was teaching the house mothers to obtain a degree of self-sufficiency by making bedding like eiderdowns. We saw the little school with a map of the world painted on the outside wall and he showed us the bore hole source of their precious water. The children greeted us and were well dressed and good mannered, if a bit enthusiastic and excited at us being there.

Patrick arranged to have houses built for the formerly homeless house mothers in exchange for them looking after the children.

We have since helped Patrick a lot as we see the children at Two Sisters Orphanage as children we can help through Dreamcatcher Children, our charitable fund. Some funds were initially forthcoming and provided in the aftermath of the *Catch a Fire* film. Producer Robyn Slovo, daughter of the famous ANC cause supporter Joe Slovo, has helped us by providing film posters and copies of the DVD to auction and raffle too so funds could be raised.

Mary and Martha and *The Giver* were both filmed at Two Sisters, which enabled the orphanage to benefit from a good-sized donation.

After giving the children a selection of fresh fruit we'd picked up in White River we also shared pencils, pens and notebooks between them for their schooling needs. We sat and chatted and ate food with Patrick and he freely talked about his time in the ANC and his activities against the government. In conversation he pointed to a rocky outcrop above us and said that is where he hid with his brigade and concealed a lot of his explosives. He reminisced about a lot of terrorist activity including blowing up cars. A really fascinating day and not your usual holiday outing!

He also said he had dreamt about a vision of a white person holding the scales of justice. The vision said, 'Don't be afraid', but he replied that he was afraid as he feared death. Recently he underwent surgery in Cuba for a tumour on his brain but it turned out to be benign and they removed a tracking device from the tumour site that he claims the apartheid regime implanted to keep track of him should he escape. Madiba (Nelson Mandela) was best man at Patrick's third wedding. I just wish I could have had the honour of meeting him.

We said our goodbyes, promising to try and help in the future and travelled on to Blyde Canyon, where we stayed in garden suites at the quirky Graskop Hotel in Graskop. It was a relaxed place to stay with a great swimming pool and a good base for touring Blyde, a pretty good-sized canyon and quite verdant. We enjoyed driving the R352, starting in Sabie and passing Sabie Falls, Bridal Veil falls and Mac Mac Falls, and made it to God's Window which gave us terrific views from the escarpment. From here you can see canyons, waterfalls and all sorts of weird rock shapes and formations, a brilliant spot for photographers and painters alike. Other highlights were Bourkes Luck Potholes, naturally created by erosion from

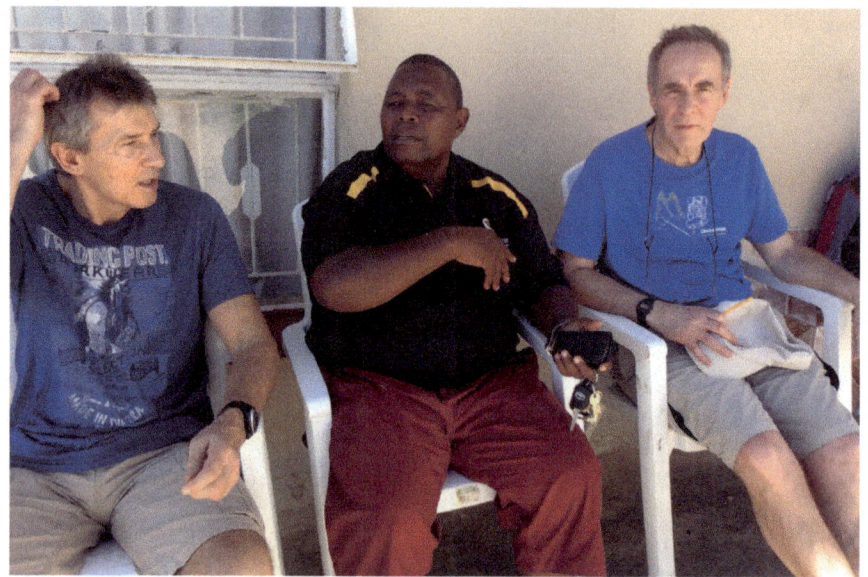

*David and Ron either side of Patrick Chamusso discussing how Dreamcatcher
Children can help the orphanage*

the river and elements with all sorts of round boulders and gaps in the rock
and pools of water, well worth the diversion from the road, and the Three
Rondavels, three rounded mounds reaching a slight point at the top.

About 3 miles north of the potholes there is a great view over the canyon,
strangely named Lowveld View at 1,219 metres above sea level. You see below
you the Blyde River wending its way, carving a path through the valley with a
dam in the distance. This was probably our favourite view of the canyon.

Our next port of call was the amazing Kruger National Park, formed
in 1926, named after Paul Kruger, the third president of the South African
Republic, and inaugurated during his presidency. Just south of the Tropic
of Capricorn, its two million hectares (7.577 square miles) is home to over
500 species of bird and 147 species of mammals. Of course, most people
want to see the 'Big Five': buffalos, elephants, lions, leopards and rhinos,
black and white. We have been fortunate to see them all on more than one
occasion in the Kruger and elsewhere. We enjoyed the hot weather and
the relative lack of people, it being February. We were surprised it was so
quiet, but then were told October to March is the rainy season. We were
fortunate to have the best of both worlds, as the park was rich with foliage
yet we saw most animals, usually best seen in the dry winter months of

this hemisphere.

We drove our own vehicle into the Kruger and saw an amazing assortment of animals on our own including a leopard, which was panting away in a dry stream bed just off the road, trying to find some cover under the foliage on the banks of the stream. We also took a safari and saw an incredible stalking of gazelles by some leopards one morning, who walked right in front of our open Toyota Land Cruiser, totally oblivious of us. One jumped up on the bonnet for a better view and then headed off towards the kill. We were told to keep very still!

We then headed south through Swaziland with its own currency called Emalengeni and on to the amazing St Lucia wetlands, where we witnessed hippos wandering around the residential neighbourhood grazing and occasionally invading people's gardens if they failed to fence them off securely. It was one of those double-take moments as you drove slowly around the municipality.

We visited a nearby beach at Cape Vidal to be greeted with a sign stating beware of hippos, sharks, estuarine crocodiles and rip tides, yet they failed to mention one or two other more rampant and common blights to enjoying the beach. David and I had walked to the water's edge and paddled in the water whilst we talked to a local fisherman who was fishing from the beach. All of a sudden my foot felt really sore and looking down small jellyfish were all around our feet and marks relating to the strands of their stingers were tracking up our feet and ankles. Soon David was similarly affected, so

we beat a hasty retreat, heading back to the girls, who had camped under an overhanging tree for shade on the edge of the beach, and we searched for the equivalent of dock leaves or some sort of vegetation that might ease the pain and irritation of the jellyfish stings.

As we settled down the next little incursion from the local inhabitants was a group of monkeys, who had spied that a bag had been hung on a tree branch and made a rapid foray to steal it, baring their teeth as we tried to fight them off, which fortunately we were able to do. After that we kept a close eye on their whereabouts, which was at a suitable distance. Before long, however, we observed them moving along the beach where other monkeys were prowling the dunes and trees, waiting to pounce on people's belongings when they were at the water's edge or swimming. Sure enough, we saw a family head into the water for a swim leaving their belongings on towels and the monkeys took advantage, stealing everything they could carry and heading to the safety of the tangled vegetation all pursued by the victims. A very interesting and tense day.

We spent some time in Durban with the busiest harbour in Africa, at the fabulous Anchors Rest, a guesthouse with honour bar and comfortable accommodation in delightful grounds including a swimming pool at Umhalanga Rocks, a nearby location to Durban. On a trip to the Natal Shark Board building, we realised the local rugby team is not called the Durban Sharks for nothing. We learned all about the sharks local to the area in a very interesting and informative presentation when they showed us the size of some of the jaw skeletons they had.

Durban was very cosmopolitan with a long and attractive beachfront and walking path, from which we spotted the Juma Masjid Mosque, the largest in the southern hemisphere. We were fascinated to learn that the province Durban sits in Natal, which was named after Vasco De Gama in 1497 when he sighted land on Christmas Day and thus the Portuguese name for Christmas became the province name.

Leaving Durban, we headed up to the stupendous Drakensberg Mountains, en-route seeing where Gandhi's passage of inequality started in Pietermaritzburg, where he was thrown off a train for being coloured in a white compartment even though he had a valid ticket. Gandhi has a fascinating life to ponder, South Africa being a bigger part of it than some think. He had studied law at University College London, being called to the Bar in June 1891 before returning to India. A couple of years he went

Ron looking down on Blyde Canyon

to South Africa to help a distant cousin who needed a lawyer.

Gandhi formed a group of stretcher bearers to help the British army in the Boer War, to show how valuable Hindus could be. Gandhi and his 1,100-strong Natal Indian Ambulance Corps moved to the front line at the Battle of Spion Kop (a term used for the famous Liverpool football ground supporters' terrace named after this battle). The British fought to capture this hilltop, but 300 men died in the attempt, most of whom were from Lancashire and many of these Liverpool itself. Gandhi's Ambulance Corps had to carry wounded soldiers for miles to a field hospital because the terrain was too difficult for the ambulances to negotiate. Gandhi and thirty-seven other Indians received the Queen's South Africa Medal.

He entered politics by forming the Natal Indian Congress to try and galvanise support and legislation to stop the discrimination and bullying in the country. Of course, as we now know, it was 1994 before blacks were allowed the vote in South Africa. All this political activism and understanding of racial discrimination stood him in good stead for his future efforts in his homeland of India.

Lady and child in Lesotho Kingdom

In Lesotho after ascending the Sani pass Highest pub in Africa

Before long we reached the famous battlefields of Kwazulu Natal, we knew it was our guide when he strode purposefully across the restaurant towards us; he was short and somewhat rotund, a swagger stick under his arm, he wore a khaki slouch hat and most definitely portrayed your archetype of a Boer. We went on to learn his ancestors had fought the British in the Boer Wars. He was very knowledgeable and brought the history of the area to life. Drinking in the history of the Zulu and Boer Wars, we visited Isandlwana and the infamous Rorke's Drift, scene of the heroic defence against the local tribes, as portrayed in the film Zulu. This was quite a humbling visit and you have to be in awe of the Zulus and their efforts to protect their land and culture. The word Zulu, incidentally, means heaven!

On our way to Spion Kop we stopped at Ladysmith where Lt General Sir George White and his men resisted the Boers at the famous siege of Ladysmith for close to 4 months from November 1899 to the end of February 1900 they were finally relieved by advancing British forces having lost 800 men dead and 800 prisoners were taken by the Boers who only lost 52 men.

We also took in a a small town called Underberg. We stayed nearby in some accommodation at a trout hatchery, which we used as a base to visit Lesotho (pronounced Lee-sootoo). Our local guide Vivien picked us up in his four-by-four and took us the relatively short distance to the border where we filled in some documentation and had our passports inspected.

The scenery soon became dramatic as we ascended the narrow, rough dirt track full of potholes, very bumpy indeed made worse by recent rains. We passed protea bushes with their large flowers and alive with colourful birds called malachite sunbirds, who feed on the protea and other bushes and plants like aloe. First, we saw these small birds with their long tail feathers on the wing and then feeding and resting by the side of the road.

The adult male is metallic green when breeding, with blackish-green wings with small yellow pectoral patches. The underparts plumage is yellow, flecked with green. The female has brown upperparts and dull yellow underparts with some indistinct streaking on the breast.

The drop into the valley was perturbing, made even more so when you witnessed the wreck of a vehicle that had left the road and descended the rock-strewn valley sides, so we were happy to stop at a viewpoint a couple of hairpin bends from the top, looking down the valley at the tortuous bending track we had just come up. A group of men descended past us herding sheep. Soon we dropped over the lip of the valley at the top to a small settlement at 2,893 metres. We learned about the village and villagers, how the huts were built, how they baked bread and about their beliefs and life and they fed and watered us with some basic local foods. Their clothing seemed to consist of a cloth dress over which were wrapped colourful home-made blankets. Because of the harsh climate at this altitude they also wore balaclavas. Their three strands on the national flag are called khotse, pula and nala, meaning peace, rain and prosperity and they regularly recite this greeting when they shake hands with you in their native Sesotho language.

They are proud of their history and told us about them defeating the Zulu and holding the British at bay. Yet in 1869 they asked the British for protection from the advancing Vortrekkers who were expanding their empire. The British army duly obliged and Basutoland, as it became known, became a British protectorate. To symbolise her support, Queen Victoria sent the King of Basutholand a blanket and to this day the men in-particular wear a blanket in memory of this. Women pin the blankets on young girls before marriage, one on top of another. A statement by the king at the time was 'we are protected like wearing the Queen's blanket', thus the symbolism which endures today.

A collage of five battlefields pictures from the Boer and Zulu wars including Isdahlwana, Rorkes drift and Spion Kop

We returned to the border post before heading back down the track and had a drink in the highest pub in Africa where they had good food and very clean toilets, always a bonus in Africa! On the way down we witnessed bearded and Cape vultures riding the thermals and a delightful Cape robin and then passed a herd of grey antelope on the hoof.

We had enjoyed a fabulous day, though tempered by our appreciation of the harsh life these people have in Lesotho. At the time their life expectancy was in the late forties, which happily has increased in the last few years. In 2017 life expectancy for women was 56.7 years and for men 52.2 years. The major problems, as in the rest of South Africa, were Aids and malaria.

Our friends headed home to Blighty and we continued onwards, flying north to Inhambane in Mozambique to visit friends who ran a beachside lodge on the idyllic Barra beach, often referred to colloquially as 'Barradise'.

As we flew home with lots of great memories and with our souls and spirits warmed by the experience as well as having been infused by some glorious winter sunshine, little did we know that following a routine colonoscopy two weeks later our lives would be changed immeasurably and indeed for ever.

It's true I had not felt 100 per cent for a couple of years, but put this down to the upset of losing my father and also my mother not coping with life as she approached her late eighties and the constant support and attention we needed to give her, which ultimately led to her going into a residential home, and all this, together with clearing our family home of over fifty years was a very tough time physically and emotionally. Also, I was now in my sixties and figured this drained, strange feeling came with old age, especially the uncomfortable but not painful feeling in my right lower back.

THE RAINBOW NATION

From the melting pot of Soweto
To capital Pretoria
Paul Kruger's residence
Voortrekker Monument euphoria
Two Sisters Orphanage
Ride with a freedom fighter
No time for lunch

Four photos of Mozambique 'Barradise' and arriving at Inhambane airport

So, a few pounds lighter
Graskop sundowner
Refreshing swim
Glasshouse restaurant trout
Full to the brim
On to the Kruger
After Pilgrim's Rest
Witnessing majestic leopards
Totally impressed
Swaziland experience
From Pike's Peak to Malendala
Then the weavers' colony
Around the pool at Bayala
Umhalanga Rocks
A stroll along Marine Parade
Welcoming Anchors Rest
Top place we have stayed
Pietermaritzberg
Where Gandhi left the train
To stand where his life changed
Both humbling and insane
On to Underberg
Nights under the thatch
Views so resplendent
There were trout here to catch
An epic journey ensued
Up the Sani Pass
The highest pub in Africa
Where we raised a glass
Long journey to the battlefields
Once under Zulu rule
Just in time for a sundowner
and a dip in the pool
This wonderful nation
And colourful tribes
Dutch influence
Its inspiring vibes

From Table Mountain
And the wines of the Cape
To the wildlife and bush
You can plan your escape
This beautiful country
From mountains to sea
Full of surprises
Chequered history
Abundance of resources
Coastline for miles
When harmony reigns
Mandela smiles.
What is the future?
That no one can know
If you get the chance
You really must go!

CHAPTER THIRTY-FOUR

Becoming an author

So, having been slowed up by cancer and not being able to do my overseas tours any more, I read a lot more as I tried to deal with the effect of chemotherapy and the recovery from major operations. This soon led to my desire to fulfil an ambition: to write a novel.

I had been to some creative writing classes and acknowledged I was not a natural but, undaunted, I started with a plot and gradually developed the discipline to write. I had been inspired by lots of books I read on the Second World War and in-particular the Special Operations Executive, some fiction some not.

I was lucky enough to have a grammar school education at the Wakeman School, but had no university opportunity, having worked from the age of seventeen. I did have a two-year further education course crammed into one year at St Austell, and latterly a PGCE course when I was fifty, but my education has been from life, and the thirst for knowledge especially social, economic and political history satisfied by copious reading and of course travel has taught me so much about different cultures, religions, people and indeed myself! For the most part anything I have learned has been from the oft-used cliché, the 'university of life'!

One historical fiction writer, Fiona McIntosh, was very important in galvanising me to write. Having written to her and commented on how much I had enjoyed a couple of her books, *The Lavender Keeper* and the sequel *French Promise*, I told her that I was planning to write my own novel to help with my rehab and she generously offered me a free place on her residential writing course in Adelaide to give me a target to get well and travel to Oz, to do the course and go and see my boys in Sydney.

After a couple of setbacks and further cancer procedures I didn't make it, but this offer and regular encouragement from Fiona led me to complete my first novel, *Don't Bring Me Flowers*, which proved a lot harder to write than I thought it would. Even though I finally published in November 2018 after six rewrites and lots of editing, I am determined to make my next one nearer to perfect and look on this one as a practice.

I have to tell you the joy of writing once you manage to come to terms with the discipline of actually sitting down and putting pen to paper or in my case hitting the keyboard, is so liberating and intoxicating.

As my knowledge and skills developed, I found that even though you may have a plot, or even have written a synopsis of how you see the story developing and the characters involved, it is a real learning curve. Initially you tend to include a lot of 'exposition', perhaps spending too long on setting scenes but you soon learn to edit that and cut to the chase to try and keep the story flowing and of course intriguing the reader enough to maintain their interest. Pace, hooks and developments in the story are very important.

It was amazing to me how the plot changed and it took me on twists and turns I had never envisaged, so much so I re-edited and rewrote at least six times over three years and that was before it was professionally edited.

Characters explode onto the scene in some cases, in others they slide in almost unnoticed before becoming a major player. Before you know it you go down a new avenue to a new destination

Of course, you also learn that you need to keep note of the plot changes and twists and simple things like the colour of someone's eyes or hair as you do not want them to change when you refer to them later in the book. You keep a note of dates you have quoted and so many other things to make sure everything rings true and is accurate, especially with historical fiction.

This brings me to another point: an awful lot of research takes place before you put pen to paper and start the book proper. Indeed, the plot and details get tweaked and morph as you go along, for sure, because it fits better or you learn something new. The old story of having a break from the manuscript and then revisiting it often helps with clarity and structural editing.

My day tends to start with breakfast about 7.30am, then an hour or so checking emails, social media and the news and of course football scores and news, usually Shrewsbury Town related. After a shower and shave it's time to work, usually about 9am. I tend to work until 12 midday, taking regular breaks to stretch and clear the mind, and of course copious cups of tea. Time seems to fly most days when I'm writing as I'm excited about progressing the story, even I want to know what's going to happen. I then tend to take some lunch for an hour and perhaps have a short walk. Then back to work mid-afternoon until about 5pm when I might walk again to get some fresh air.

It all depends on the season. In summer I might change my pattern. I work Monday to Friday and have the weekend off as a rule so I can enjoy time with family and of course take in sporting events. This is where summer sometimes messes up my schedule. During the cricket season I might go and watch England or even a local team to break things up and enjoy the fine weather while we have it. I guess I am a loose disciplinarian. In many ways I also govern my writing with word targets of a minimum of 500 per day, sometimes I write 2,000 or more, at other times it's difficult to get going.

I already have a plethora of other literary projects, some born out of ideas from our travels, some from visits to influential places and events like the Hay Literary Festival, others just ideas that pop into my mind. The Hay Festival is a great event which we have been to over the last five years, both the summer and winter festivals. We have met many interesting people and been able to share a few moments with luminaries such as the wonderful surgeon David Nott and newscaster, journalist and author George Alagiah with his wife Frances Robathan.

Victoria Hislop, one of our favourite authors, was very pleasant and gave me literary encouragement with my various projects. We've also met Tahmina Amam, who has written

Ron and Victoria Hislop discussing books at Hay Festival

Ron signing copies of his first novel
Don't Bring me Flowers

some fascinating books around her Bangladeshi family background and Luke Harding, who wrote *Collusion*, a very intriguing exposé of Trump's long-standing business relationship with Russia before he was president.

We listened to Gina Miller at the festival offer a rather balanced approach to the result of the Brexit referendum, yet in the months following her actions were anything but. Initially impressed with her diatribe, we ended up disappointed with her actions, which, after all, speak louder than words, or so they say!

Comedians at the festival often portray the attendees as all Guardian-reading lefties, most with a good education and from a wealthy background. It's food for thought but rather a generalised swipe.

Apart from this memoir, upcoming projects include a book investigating the spiritual side of Shrewsbury, my home town, and some children's books based on Dennis and Roy, stories I told to Tom, my youngest, when he was a small child. All these tales are loosely based on my dad and uncle's childhood growing up in the village of Penwyllt in the 1930s and 40s. A wonderful development means a publisher called Orphan's Publishing has been very helpful and supportive in making this dream come true. The real bonus is that Dianne has done all the illustrations, so the project is a husband-and-wife joint effort.

I'm also working on the life story of a woman who worked as a de-cryptographer in Egypt during World War II. I would still love to write another novel, for which I've already written a first draft. Like the biography, some of this novel is set in Egypt and we had planned a tour to research Cairo in particular and visit some of the venues from that era. Alas, Covid-19 has put paid to that for the time being.

For me writing has been a real silver lining to having cancer. There are so many projects to bring to fruition!

CHAPTER THIRTY-FIVE

House sitting

Given we have been quite confined to barracks over the last three to four years because of my need for treatment, scans and operations, I came up with the idea of house sitting and joined an operation called Trusted Housesitters, a web-based company.

We had hoped to travel worldwide to house sit years ago but we have started more locally and have now looked after houses and pets in Wales and England and one little excursion to Cyprus. It's a great experience staying in some new communities and some lovely houses meeting some splendid people and some terrific animals. We have looked after pigs, donkeys, sheep, chickens, cats and dogs, not forgetting the goldfish.

Here are some of our tales.

To start with we enjoyed a lovely time in one of our favourite playgrounds, Snowdonia, with the delightful Flicky, a golden retriever, and his sidekick, the goldfish Nemo. The lovely Walters family were off to the Canaries and we had the joy of exploring the beauty that is the Welsh coast and the mountains and lakes of Snowdonia.

We continued with plenty of house sits in the UK, lots in Wales, some in Shropshire close to home – such as with the wonderful Prideaux Family at Butler's House, and the lovable Tarka and Toggle, a black cocker spaniel and black Labrador, well trained dogs who melt your heart, true companions – others around the country such as Yorkshire. Before long, my health had

Tarka and Toggle

improved to the level where I had received some consecutive clear scans from the cancer so we decided to spread our wings and head for Cyprus.

Late one evening I saw the opportunity so visit the south of Cyprus to look after a couple of dogs and a cat whilst the owner went to explore for a week. We received a rapid reply and had some email exchanges, which resulted in us flying to Paphos just a couple of months later. Part of the submission by the homeowner was very frank in that she pointed out the accommodation was very basic and, although billed with a swimming pool, that was going to be empty.

We took the precaution of booking a reasonably cheap two-star hotel down the road. Our thinking was we would use it for a few nights prior to the house sit then just use it for the lovely pool after that. It was wonderful to be jetting off to the sun again! We had arranged a transfer with the tour company and were duly met and delivered to our lovely little hotel on the north-west coast of Cyprus. Travelling through the olive groves and surprisingly green landscape boosted by the winter precipitation was a very inviting journey as we entered the Cypriot spring.

Service, food and facilities proved to be well above the star rating of this family-run hotel, which had been built at the far end of Latchi on the beach. We really enjoyed the few days of sunshine and hired a car to travel around the area and explore before our house sit responsibilities began. One day we drove as far as we could along the coast until we came to the border which divides the island between the Turkish and Greek zones. There seemed to be few people around in the small fishing town before the border and no one immediately in evidence at the control point with its custom barriers lowered. We decided to return to our hotel via the Troodos mountains and headed off into the pine-clad mountains where we hardly saw another vehicle or person for hours.

We had thoroughly enjoyed our mini vacation and now went to meet our house sit owner and the lovely creatures Zoran, Zoritsa and Lillie. Christina had told us we could use her car, so we arranged to be collected from the car hire depot just down the road from our house sit, which was on the outskirts of the town of Polis.

Christina was a lovely lady, somewhat hippy-like and a web designer, so she was able to work from home. Her plan was to be away in northern Greece for a week where she planned to buy some land and set up an eco-camping operation.

Meanwhile we were left with the responsibility of looking after the three bedroomed, two-bathroom villa with pool and with the use of a vehicle. Well, that is as grand and aspirational as I can be. The vehicle we were collected in was a really battered old Toyota whose paintwork was well worn, the doors did not lock and the windows wound down and stayed that way. No working air conditioning or radio, or any comfort, for that matter. The seats were so worn they gave no support to your derriere and the seatbelts didn't work.

On the upside, the dogs Zoran and Lillie were delightful and we only saw the tortoiseshell cat Zoritsa around feeding time twice a day. A window was left open for her to come and go as she pleased, so even though we locked the property it was a waste of time. Anyway, there was nothing anybody would want to steal.

There was a large area of lawn and fruit trees and bushes to the side of the house, but we were told to beware of snakes. Just then the general job gardener appeared and proceeded to mow the lawn, hopefully scaring off the snakes. We stared down into the debris-filled, six-feet end of the pool with its dirty rainwater and abundant toads and just hoped we did not stumble over the other detritus which was scattered around the area and end up in there.

The house was generally not very clean and when Dianne went to perch on the end of the settee and moved a pillow only to find a dead mouse It was clear Dianne had made her mind up that she would be returning to the hotel and not staying a night here, so the pressure was on me.

We waved Cristina off and settled into life with the pets taking them for walks along country roads and to the beach and along paths beside the beach. Once we became used to it, we had a brilliant time. We found the cost of living quite reasonable and enjoyed our walk into Polis. Here we met a jeweller and craft shop owner who made their own gifts and had links to Shropshire.

Mosquitoes were a small issue but not bad at all. The strangest day was walking along a road we had used a few times when Zoran, a large, strong dog, yanked on the lead and tried to chase a kitten it had spotted. The kitten duly shot up the nearest olive tree after traversing the overgrown roadside ditch. It was then I spotted movement in the brush and saw an olive-green snake with a huge girth slither slowly towards the tree and slide upwards after the kitten.

Boy, were we glad we had the dogs on their leads, which initially was just in case traffic came barreling along the winding road. We had been warned about snakes but not seen them and it turned out this was one of the most venomous snakes we could wish to meet, the blunt – nosed viper. We took the dogs home then returned to see what could be done for the kitten, fearing the worst, but fortunately it had reunited with its siblings and mother further up the road.

One day walking home we were stopped randomly on a country lane by a man wanting to see our dog licence. We thought it was a ruse but, true enough, he came home, checked our licences and called the office before approving our legal status for the canines. We were quite pleasantly surprised Cyprus was so hot on this ... well done, them.

While we were in Cyprus Dianne checked her Facebook and couldn't believe my daughter Rhian and our grandson Owen were staying in Paphos. We sent them a message and hooked up with them, spending a great day with them. Strangely, Owen's half-sister Drew, who is a model, was also there in a manner of speaking, as she had done some work for Accessorize and her posters were all over Paphos airport.

All in all, we had a fab time, and hope we can do more overseas sits.

I had been feeling a lot brighter and more confident following the first really big long-haul trip in four years and so set about arranging house sits and future plans with gusto! As I will say again there is no time like the present because, as the song goes, 'if tomorrow never comes' ... enough said.

One evening I was checking my emails when a list of home sits came in from Trusted House-sitters. Amongst the possibilities as I skimmed through half-interested I saw the picture of an azure swimming pool and then looked at the heading 'Dubai'. I lazily shouted down the stairs to Dianne, 'house sit with a nice pool here', knowing Dianne loves to swim. In a trice she shot up the stairs, checked it out and before I knew it, I was applying for the sit in a place neither of us had ever wanted to visit! Within a day or two we were skyping the family and within a week we had our flights booked to Dubai with a cheeky little side trip for a week to Oman.

Arriving in the small hours of the morning to the very smart and organised Dubai International Airport we collected our bags and headed for the taxi rank, which was sporting an enormous queue but to our pleasant surprise the constant flow of taxis and quick departures meant we only waited about 15 minutes and we were on our way to the housing development close to Montgomerie Golf Course in the Emirate Hills area. A balmy 32°C hit us as we stepped out of the car at 2am.

Following an overview of our duties and meeting the delightful Saluki hounds, we settled into life in Dubai rather well. We found the grocery shopping a lot more expensive than at home, but then they import just about everything in Dubai. We soon got into the rhythm of very early and late dog walking to avoid the worst of the heat and made full use of the pool.

The taxi app Careem was well used, as was Yalla, meaning 'let's go', both a very efficient and cheap way of mastering the exploration of the few delights Dubai has to offer. Sure, we were impressed by the wonderful engineering feat of the Burj Khalifa, even if we despaired of the Disneyesque queueing and procession to the higher floors 124/125 and above only to be greeted by

a hazy skyline … but it was air-conditioned and a respite from the temperatures in June which reached 43°C.

We tried the souk in Dubai Creek, only to be what can only be described as mugged when trying to buy a piece of Arab headgear. The physical pushing and intimidation lost the growing crowd of souk sellers a sale. I was relieved to get out

Dubai dog walking with this elegant Saluki Hound

of there, by far the worse experience in overseas bazaars and souks I have encountered and I have been in a few … they seemed desperate. It was so hot in the late morning sun we eventually sought respite in a McDonalds. I know, much against the grain but an oasis that morning, as most eating and other establishments were closed. Did I fail to say? It was Ramadan and we were here in Dubai until the penultimate week.

We loved the dogs and their owners were lovely too. The poor dogs slept about eighteen hours a day and were never keen to go for a walk, yet on the day of our arrival we had witnessed them off the lead in the evening and they could certainly move fast. Most of the time a tortoise could have overtaken them.

We did take a cruise around the Palm but overall, we just enjoyed reading a book by the pool and looked forward to getting out of Dubai …

Oman was a contrasting tale, the short flight with Emirates lasting just forty minutes and out of respect for our fellow flyers observing Ramadan, we were given the most delightfully wrapped and presented gift boxes of snacks and drinks so we could discreetly eat or, better still, wait until we had disembarked and we were in our own accommodation.

On arrival, a rarity for us, our suitcase was rather badly damaged (yet usable). We had to spend time reporting the damage and as such we reached arrivals a little late and initially could not find our host guide. Finally a rather tired, wan-looking chap in a Kuma (their headgear like a cap) and dishdasha approached us with a greeting sign that was facing into his chest, only when he reached us and turned the sign the right way did we realise Majid was our new best friend … he was obviously looking forward to 'iftar', the breaking of the day's fast at sunset.

Over the next week we were awed at the dedication to prayer five times a day and the fast from food and drink and other abstentions, and became far more aware than we were in Dubai of the prayer, reflection and devotion to community.

Our first hotel was situated on the corniche, the wonderful arc of deep-water harbour wall backed by many old buildings and of course the souk at Moutrah. The blue onion-domed mosque with its minaret was a highlight on the city skyline, which refreshingly barely reached three stories anywhere, thus allowing the minarets of the mosque to stand out against the Muscat hills which provided a fabulous backdrop and divided the ocean from the hinterland of Oman.

Ron's new Omani headgear!

In the evening, the pontoon and promenade were busy with those wishing to go for an evening stroll, especially after the long fast to iftar and the subsequent feast. Of course, the whole area took on a new persona as it became illuminated against the dark brooding mountain peaks, with Al Jalali Fort built by the Portuguese on a rocky outcrop standing sentinel at one end counterbalanced by the modern fish market in front of the Marina Hotel and excellent restaurant Bait Al Luban, a former guesthouse, at the other. The new fish market built in 2017 is shaped, some say, like a wave, others say like a fish. This is a very busy area from sunrise as the fishermen drop their catch and buyers pour in.

A real must when visiting Muscat is the contemporary Grand Mosque, built and funded by Sultan Qaboos to celebrate his thirtieth year as sultan. It is absolutely fabulous and you cannot help but be impressed, particularly by the main prayer hall, which sports a Persian carpet 70 x 60 metres and hand made by over 600 Iranian ladies. It took them over four years, such is the craftsmanship. This prayer hall with a capacity of 6,500 is just for the men although there is a closed-circuit television link to a smaller prayer hall for up to 800 women. It is said that, including the outside areas for prayer, the complex could cope with 20,000 coming to pray.

We enjoyed learning more about the worship and beliefs of the Muslim faith in the visitor centre and met a very lovely lady from Marrakech who was volunteering and explaining about this fabulous structure and enlightening us about the Quran and its teachings and the belief of Jesus as a prophet and his mother Mary referred to as Maria or Mariam in the Quran. On our departure they kindly gave us some more descriptive literature in English and a gift to take away, some frankincense and an ornate cup in which to burn it.

We learned more about the abaya, the long, usually black, garment the ladies wear, and also the hijab, the veil which covers the head as well as the niqab, which covers the face, often referred to as a burqa (or burka). Men tended to wear the traditional long white (sometimes coloured) dishdasha, a loose-fitting robe.

Men wear a checked rectangular or square headscarf, usually with a rope band on top to fasten it in place. The ghutra (headscarf) can be white but is often red and white check or black and white check and called a Shemagh or Keffiyeh. It is worn slightly differently in Oman like a turban and called a Muzzar'. I also found a hat in Muscat called a Kumma in Oman.

Next morning we took a quick excursion around old Muscat, hidden in the next fold of the mountains on the coast and home to a fabulous museum, the Bait Al Zubair, which is essentially a restored old house full of Omani handicrafts and furniture and other related ephemera and therefore a reflection of Omani culture. The Opera House looked wonderful but again was closed for Ramadan. The embassy quarter and Parliament building, the latter built in 2013, were shining examples of the new meeting the old. We were finding the heat somewhat oppressive but it's surprising how you manage to acclimatise.

Before long we were heading south along the well surfaced road which follows the Arabian Gulf coast the hinterland is protected by the Eastern Al Hajar mountains which gradually gives way to a low lying barren parched and rocky land yet this scrub dotted terrain of the Arabian Peninsula is attractive in its own way. After about 45 minutes we pass Quriyat, a small attractive town known for its dhow building and a great old fort, but just what we needed on this stiflingly hot day was the Bimmah Sinkhole, a natural hole in the limestone just off the highway, which has toilet facilities and picnic shelters. There's a long, steep concrete staircase to the bluey-green seawater pool, but it's well worth the effort.

We were lucky as the benefits of it being Ramadan and midday were that it was almost deserted, just twenty or so tourists and non-Muslim residents enjoying some respite from the heat. After a bit of a picnic away from our driver so we didn't feel guilty about eating and drinking in front of him when he couldn't partake, we pushed on past the quiet sandy beach at Fins (white beach) before reaching Wadi Shab. Here you can take a boat to a footpath up the wadi leading through the canyon which is normally full of little pools you can cool yourself in, but we arrived close to 2pm when it was far too

Sunset at Sama Al Wasil in the Wahiba sands of the desert as we entered the Empty Quarter Very hot at 45°C

hot (43°C) to contemplate any walking, so we took some photos and moved on.

Returning to the main road, Highway 17, we headed on south past Qalhat, now a small village but originally Oman's first capital city as its port was very important in its day. Reaching Sur, we strayed from the highway and checked out the historical old town and its beaches, viewpoints and working dhow shipyard, the last of its kind in Oman, where the labour-intensive process is a sight to behold.

We opted not to head to Ras Al Jinz for the turtle reserve as we had witnessed similar in Tobago in the Caribbean as well as other places and instead made our overnight stop Sama Al Wasil in the Wahiba Sands area of the desert. We stopped at the small desert border perimeter town of Al Wasil and watched as our guide Majid chatted with locals and a young man who had a garage barely the size of a hut let down the tyres on our Toyota Land Cruiser so that Majid could drive more easily in the sand we were about to encounter. So with a wider area of tyre to move in the dunes we headed off past the old Al Wasil Castle, some of which had been renovated and contrasted nicely with the weather- and time-worn ruins of the wind patterned walls, all set against the backdrop of verdant date palms reaching for the clear blue sky and the Omani flag lifeless atop the castle's tower. A few bearded old men sat and talked their characterful faces worn like the walls of the castle, as if carved by time.

The men were dressed in their dishdashas an ankle length, long sleeved collarless gown and 'Muzzars' a type of turban although some men prefer a traditional 'keffiyeh' which is fixed on top of the head for warmth or to keep cool, indeed this often small check fabric is versatile and can be lifted and wrapped across the face to protect against sandstorms.

As we entered the Empty Quarter, Majid revelled in his chance for some off-roading, admitting his love of rallying. About 18 kilometres in, flanked by two huge dunes, we arrived at the oasis of Sami Al Wasil, its chalets built in a circle like a wagon train and additional accommodation in tented glamping units outside this. No Wi-fi, but a pleasant eating area under canvas and also a communal area where music and other entertainment can be had in the main season but sadly lacking as the two of us and our guide were the only guests that night. We settled in, then went up to see the sunset, Majid driving us up the undulating dunes of red and white sand. As the sun set, we returned and made friends with the camels. Sadly, the guide fed them a chicken burger we had failed to eat but we added some grass forage, so at least they had some salad. On the way out to see the stars later we were walking along the stone path they had created and Majid startled us telling us to stop as an angry scorpion blocked our path, its stinger at the ready. Majid took off his shoe and threw it at the scorpion, which disappeared rapidly, but our guide insisted on following, trying to get a photograph. Nuts!

The stars were okay but we always say nothing will ever live up to our 'Tom Price' moment at the settlement of Tom Price, a time we spent in the Aussie outback, where the stars were so thick you could barely pick out the sky between.

Following a decent supper buffet and a good night's sleep, we had breakfast, took a few last-minute pics and headed for the former ancient capital of Oman, Nizwa. It's about a three-hour drive through quite flat scrub-like terrain. We had planned to visit Wadi Bani Khaled but once again the hot summer season defeated us as they had closed the area for maintenance. We stopped at a little rest area which was well off the road and we had to drive through a stream to get to it, it was like a mini oasis and the facilities were modern but predominantly hole-in-the-ground toilets. A bigger problem was the horde of flies who were like a winged amoeba of black, buzzing menacingly. Dianne couldn't face it; I braved it and wished I hadn't but, heh, it's a hot, dusty country.

On we drove in our comfortable air-conditioned Toyota Land Cruiser and soon neared Nizwa. We first visited the delightfully verdant and hidden valley of Birkat Al Mouz with its stands of date and banana palms below

Nizwa Fort respite from the 40°C sun!

the now forgotten original town with its ruins you can still explore, ancient falaj irrigation channels still bringing the elixir of life to the area … water. We took a stroll and soaked up the tranquility. We then visited a recently opened and much restored fort called Bait Al Rudaida.

Soon we headed for Nizwa and our delightful accommodation, the Heritage Inn, perfectly placed close to the souk. It is actually a few old houses of merchants and politicians now restored for use by tourists, very quirky, we even had a well in our room, sealed with a glass top I hasten to add, and the weirdest shaped bathroom as our shower and toilet were in a corner sealed off almost from the sink and entrance door and as you looked up you could see light coming in from a natural skylight left in the ceiling. Generally, the room was dimly lit, great for a night perhaps a little too claustrophobic for longer.

We had a great meal in the restaurant opposite, Al Aqr, belonging to the hotel. The food and service were excellent and a buffet very reasonably priced.

Time then for a little retail therapy in the souk, where I purchased my 'Massar'. People were really friendly whether they were selling gold, vegetables and produce or garments, no one was pushy. It was very relaxed, the souk has been restored and sprawls a little but is relatively compact. It is not really targeted at tourists as you can buy goats, cattle and birds as well as the usual but it makes it more interesting on market day, for sure. It seemed to be much busier first thing in the morning.

We loved the tour around Nizwa fort which has been well restored and made very interesting by the splendid guide Esma. We had visited a fort the day before which was newly opened and was more like a military museum

Ron contemplates skydiving into The Grand canyon of Arabia!

with all its weaponry of rifles and muskets on display. It was interesting for a while but being hot, we just wanted to chill. We were made very welcome as the only guests and given ripening dates off their trees.

Next day we headed via the impressive adobe fort at Bahla and Jabrin Fort, all very well restored but we were 'forted' out! We were more than happy to head up into the mountains and to Jebel Sham, the highest mountain in the Al Hajar, the Green Mountains range, about an hour or so from Bahla Fort. We were told it would be a lot cooler and the temperature certainly reduced but it was still close to 30°C even near the summit.

After a good tarmac road of switchbacks, then a few kilometres on a dirt track we reached our accommodation, Sama Heights, where we enjoyed a fabulous mountain view room which we loved, with a big four-poster bed, a great terrace, all mod cons and very reasonably priced. Before sunset we took a drive to the Grand Canyon of Arabia. You might think that an over-the-top description, but having seen a few canyons in my lifetime including the Colorado Grand Canyon, I can tell you this is not overrated. It is amazing and well worth the visit to gaze into the abyss with the Wadi Gul way below and the steep sides of the Hajar Mountains rising towards the clear blue sky. It truly deserved the accolade of 'awesome'. We loved the place, and watched a cool sunset, to boot!

Next day, for a change of pace we visited the old mountainside villages of Misfah and Al Hamra. One of the real benefits of being there in the searing heat of summer and at Ramadan was that we had the places to ourselves. Misfah is really quirky as the narrow paths between mud-built

Living like the locals for an afternoon

tenement houses wind their way up and down the hillside, revealing ornately painted and designed doors. Some of the houses seemed to be built into the hillside, some appeared to be hanging off it. They were like our old wattle and daub houses in the UK but these were made of mud and other materials like straw and small bits of gravel. On the lower areas we spied some date palms and some terraced verdant areas served by the irrigation system which cascades down the mountainside with clear, cool water. One area is for bathing for ladies only. Dianne checked it out and said it was cool, neat and had the occasional frog!.

We then descended to Al Hamra, a 400-year-old small town, more a village really, with many quite well-preserved old houses and, like Misfah, built by Yemeni refugees. One house, called Bait Al Safah serves as a living museum, where we had the undivided attention of the guide, who was from the Nile delta in Egypt. The ceilings were made of palm beams, with palm fronds to top off the mud and straw. We were escorted up storey after storey and were quite surprised how tall the building was. We passed many living rooms and bedrooms and even one full of beehives, would you believe. The rooftop terrace gave us a fabulous view over the town and valley.

Our host was very excitable and wanted to show us everything. He eventually settled down and made us tea and shared some dates with us before encouraging me to wear a dishdasha and Kuma whilst Dianne was clothed in an abaya and hijab and we posed with Dianne holding an earthenware pot on her head. It was great fun.

Finally, we headed back to Muscat. We had thoroughly enjoyed this snapshot of Oman and really felt like we had experienced or at least got a feel for how Arabia once was pre-mobile phone and internet: a very simple and relaxed pace of life.

Back in Muscat we took one more swing past the Muttrah souk for some frankincense and had a parting evening meal after 'Iftar' was in evidence at the lovely Bait Al Luban restaurant with its authentic Omani food and ambience close to the fish market. If you visit, get the balcony table if you can, looking out over the corniche to the fort, past the royal yachts and towards the huge incense burner in Al Riyam Park where frankincense is burned and dispersed by the sea breezes. Definitely try the ginger and mango drink as there is no alcohol. Thanks, Oman. Shukran!

CHAPTER THIRTY-SIX

Back in the saddle

After some decent scan results and with my energy and confidence growing, 2018 had been a good year. I had started to join Dianne in some of her dance classes, thanks to Sarah Bright and Bryony at I Can Dance Shrewsbury. I loved the Charleston, doing theme evening sessions like Peaky Blinders. We actually went to the Black Country Museum for a Peaky Blinders themed night, all dressed up, a fab evening! The dancing was great fun and certainly was a good workout, too. I highly recommend it for the camaraderie, the exercise and the feelgood factor it brings, and this comes from someone who struggles to conform to learning a routine and has two left feet!

Having learnt the basics of the Charleston, Di and I joined in with a world record attempt for the most people to dance a Charleston routine. We broke the Guinness Book of World Records' previous record on 22 September 2018 as a grand total of 1,096 people danced their way into the record books in Shrewsbury Quarry Park, raising £5,025 for Severn Hospice into the bargain!

With another clear scan in October 2018, we felt confident enough to make plans to visit the boys in Sydney during January 2019 and finally meet our new grandson Benji who was nearly two years old. Flights booked we were all excited and made plans to catch up with friends by flying back together via Kolkata, Bhutan and Darjeeling before going our own way home via Mumbai.

Charleston world record holders, Ron on far-right

In August we heard that the boys' mum had been suffering with backache and after various tests it appeared it was more complicated than just pulled muscles or lumbago. In spite of our best efforts and that of the boys to get to the bottom of the problem, news was conflicting and continued to deteriorate until Wend invited the boys over for a week each in November 2018. It was quite clear by then that some biopsies had shown up cancer without a primary tumour and things had gone too far. Very sadly

By order of the Peaky Blinders!

Wend passed away on 21 December 2018. I couldn't believe it and even now find it very hard to come to terms with. Earlier that year she had visited Australia and had been to a university reunion in the USA in June and spent time at their apartment in the south of France. Life was great for her and her husband Ronnie. How quickly life can change.

Despite a pall of depression hanging over us, we still travelled via Singapore and arrived in Sydney on 1 February 2019. It was great to see the lads but I could feel their grief and I felt like a guilty intruder.

I reflected on the first time I had travelled to Oz. I was with some Australians and travel agents from the UK. We started talking about where words came from, actually the conversation started with a young lady travel agent saying what a weird name for an airline QANTAS was, and was it an aboriginal word?

She was soon advised by some of us old moustache Pete's, that it was actually an acronym, which stood for Queensland and Northern Territories Air Services, the original name of the company, an airline which then of course branched out and became an international favourite.

This chat then led to why Aussies call us 'Poms'. A couple of ideas were put forward. One was that the original, shall we say tourists, from the UK, came on prison ships as convicts. On their prison clothes was stamped POHM, which stood for 'Prisoner of Her/His Majesty', so Poms we were!

One of the Aussies with us also suggested it was because we had light skins like a pomegranate, so the shortened version pom. You have to give him points for imagination. Someone else's contribution was that we were sometimes all called 'Jimmy Grant' as rhyming slang for immigrant. Who knows? It was fun chatting about it.

Ron and Dianne at Sydney Harbour

I love Australia and New Zealand and more so now my two lads Peter and Tom live in Sydney and have made a life there and now I have an Australian grandson. Benjamin Thomas Lafourcade Morgan born 26 March 2017. A real live Aussie!

You may hear it said that Brits and Americans are divided by a common language, but Americans and Australians have their moments.

Dianne and I were once travelling to Ayers Rock by car one day. It was a pretty hot and dusty journey, even though we were lucky in our Holden hire car to have a novelty for us, air conditioning that had separate controls for each of us. Di was rarely cool and wanted the a/c on full blast but as that was too fierce for me I could regulate my side of the car, which made it perfect. Nevertheless, we needed a break from the long journey and stopped at a road station as others did.

I sidled up to the busy counter where a large and rather rough-around-the-edges Aussie bloke was preparing meals and snacks as fast as he could, perspiration on his brow. A lady, maybe his wife, was taking orders. The American chap in front of me perused the menu board behind them. 'Excuse me,' he said to the lady, 'what's in the bacon and egg sandwich?'

Without looking up or missing a beat the bloke preparing the food said, 'What the **** do you think is in a bacon and egg sandwich?' before continuing his work.

Rather embarrassed and taken aback, following a short pause the American said, 'I'll take two coffees and two cheeseburgers.'

Seeing Benji and holding him for the first time was magical as other grandparents will, I'm sure agree.

We checked out Peter and Tom's homes and places of work, met their partners and in the case of Peter his Chilean in-laws who were lovely and in spite of us not speaking Spanish and them having no English we managed to get on really well. Tom also moved apartment while we were there and we managed to explore a little of Sydney we had not been to, like Neutral Bay and the Kirribilli area and Balmoral Beach as well as Bilgorla, before we all joined together for a bit of family time at the beautiful and relaxing Jervis Bay. We stayed in a fabulous house with hot tub in Vincentia, a short walk from a necklace of beaches.

On the first evening there was a tumultuous thunderstorm and Di and I could not resist heading out into it, watching some young ones still swimming in the sea before seeking cover as the thunder and lightning increased

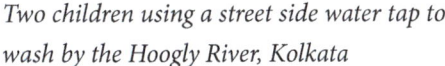

Two children using a street side water tap to wash by the Hoogly River, Kolkata

in ferocity. We had waterproofs and shorts on and just loved the warm rain cascading over us and the sea becoming wilder. There is nothing better than being at one with nature. Meanwhile the rest of our party were starting to become concerned, funny how the tables turn from parent to child and back again. When we turned up a little damp but full of buzz from our experience, we remembered what it was like waiting for our children to come home, wondering what may have befallen them ... the relief was great but a little chastisement was in order!

Finally, it was time to say goodbye. I cracked up at the airport as usual, but then we met our friends David and Pauline and headed off on our sub-continent adventure to India via Singapore.

Our flight to Singapore passed quickly and we were soon heading through Singapore passport control on the way to reclaiming our luggage. My passport control officer barely acknowledged me as I gave him my passport and ticket for onward travel. After a short while he put the passport and ticket on the counter and put his two thumbs up. I copied him, put my thumbs up and grabbed my documents to walk off before he said 'Wait!' What he had meant was for me to put my thumbs on the print scanner to the side of the kiosk. Duly done, he let me go without any further show of emotion. Smiling, I recounted my story to my fellow passengers. We all had a laugh and went to check in for yet another airline, the biggest Indian carrier, Indigo.

At 2.30am we arrived in the bustling Kolkata Airport and were supposed to be met by a driver arranged by our hotel, but no one showed up so we queued at a rather shabby kiosk outside the terminal and booked a very reasonable taxi. Somehow we managed to get four suitcases and four people into the car plus the driver for a hair-raising drive of braking and weaving in and out of traffic, including an emergency stop to avoid a stalled tuk-tuk. We were so relieved to reach our somewhat dated Peerless Inn.

We had a much-needed 5-hour sleep before we were met by the very co-ordinated and accommodating Mr Mukherjee along with the very knowledgeable guide, Krish. India is not everyone's bag as it certainly can get under your skin for different reasons to different people. We just love the country and its people. Mr Mukherjee and Krish were brilliant in showing us as much of Kolkata as possible. Kolkata was the capital of India until 1911 and situated in the huge state of Bengal when the British decided they would change the capital to Delhi as they were concerned about a groundswell of support for the anti-British movement.

The decision would ultimately have the effect of the eventual partition of Bengal which was decided by Lord Curzon then the Viceroy of India. The partition took place on 16 October 1905 and separated the largely Muslim eastern areas from the largely Hindu western areas. The Hindus of West Bengal who dominated Bengal's business and rural life quickly pointed out that the division would make them a minority in a province including Bihar and Orissa. Hindus were incandescent with rage at what they saw as a 'divide and rule' policy, but Curzon insisted it was a purely administrative decision.

Although the damage was done Bengal was reunited by Lord Hardinge in 1911, following many riots in protest against the policy and the growing belief among Hindus that East Bengal would have its own courts and policies and potentially independence.

In 1947, Bengal was partitioned once again as India became independent and Pakistan and India became different countries based on religious grounds, one predominantly Hindu and Sikh India and Pakistan for the most part Muslim. In 1955 East Bengal became East Pakistan and in 1971 became independent as the country now called Bangladesh.

Even today Bengal is a hotbed of political movements with left-of-centre politics currently dominating.

Over the two days we had in Kolkata we visited many former colonial areas. There were plenty of buildings remaining from that era, but some, like the fabulous Writers' Building, stood empty. This former East India Company secretariat building was the workplace for junior clerks the administrative backbone of the company.

But Kolkata is far more than a former bastion of the British empire. True, we were impressed by the hugely impressive Victoria Memorial which is currently a museum. This marble edifice was proposed by Curzon and is fabulous. Ironically, the gardens are full of Indians strolling and taking photographs of the statue of a former Empress and Queen of England who never even visited India.

It was the grittier places that made a lasting impression. Mother Teresa's house, of course, was a mecca for us being brought up on stories of her work. It was only being there that we learned more about this amazing lady. Her obvious continuing legacy was what she has done for the poor children of Kolkata. She was born in Albania in 1910 and joined a convent in Ireland before becoming a missionary in India. After her death in 1997 she was laid to rest in a mausoleum in the Mother House, whilst a museum extols the life she led and her simple room is open to the public. We visited St Mary's School and talked to two of the Italian volunteers there who gave up time each year to help these children all at their own expense, truly humbling.

We headed down to the Hooghly River, its banks rather dirty and litter strewn alongside the polluted river. It didn't seem to be the place to be bathing but it was! The Flower Market was full of scents and a rainbow of colours as blooms were weaved into garlands for celebration and praise. It was a very busy market, so we headed up the bridge steps onto the prominent Howrah Bridge to scan the horizon.

An Armenian church, a synagogue, the Parashnath Jain Temple, St. John's Church and St Paul's Cathedral were just some of the diverse religious places we visited. Strangely, it was the simplicity of the Anglican St John's, over 150 years old, that stayed with me, not just because of the Black Hole of Calcutta memorial tucked away in a corner of the churchyard. It had an understated beauty and was formerly the cathedral until St Paul's took over that mantle.

College Street was full of booksellers, lots of the tomes in English. We walked to see Tagore's house. Many have not heard of Rabindranath Tagore, but he was the national poet very influential at the beginning of the twentieth century and won the Nobel Prize for Literature. The whole day was incredibly informative as we also visited the home of Surandra, brother to Netaji Subash Chandra Bose (Netaji Bhawan), the revolutionary freedom fighter.

Bose was a real anti-British Indian and courted Hitler and the Japanese to get support to develop a revolutionary Indian Army against the British and allies. Not only did Bose succeed in training an army to fight against the British he started the first female fighting regiment, the Rani of Jhansi, named after an Indian princess.

Our first visit of the day had been to the Marble Palace, built by Raja Rajendra Mullick, one of the wealthiest merchants of Kolkata in the nineteenth century and still owned by the family. We enjoyed some coconut milk bought from a street seller outside the gates of the palace before drifting on.

Getting in to see the international cricket ground, the Eden Gardens Stadium was great. Fort William didn't really exist as such now and the so-called 'Black Hole of Calcutta' used to be marked by a plaque by the General Post Office, which was where the former cell had been.

We opted not to go in the Victoria Museum but instead spent time wandering about the eighteenth-century South Park cemetery with its huge mausoleums to many of the early colonial settlers who struggled with the climate and disease. Most surprising was Walter Landor Dickens, son of English novelist Charles Dickens. He was originally buried in Bhowanipore, a war cemetery, and the tombstone was later moved to here in 1987.

Whilst in Kolkata we had a lovely Bengali meal at 6 Ballygunge Place with very tasty food. We enjoyed simple things like the 'loochi' bread to mop up the delicious sauces. We also visited the atmospheric and renowned Indian coffee house, which was staffed by uniformed waiters in somewhat stained whites with green belts. They

Indian Coffee House College Street Kolkata Existed since 1876 and from 1947 was the meeting place for the poets, artistes, literati and people from the world of art and culture. Iconic restaurant but a very dubious Chicken sandwich.

wore splendid green jewelled headgear, more reminiscent of the guards on palaces India-wide. It was a place buzzing with chat and people eating and drinking, under the whizzing fans suspended on a metal framework above. We were given chicken sandwiches, which totally went against the grain for me as I usually stay vegetarian in India, but I survived!

Kolkata was brilliant, but off to Bhutan we flew, heading for the Himalayas. A short flight on Druk Air, the national carrier of Bhutan, brought us to the Land of the Thunder Dragon. We received a Bhutanese welcome with silk scarves placed around our necks by our guide Dolay, a native of the Haa valley, and Taewan, our driver.

So started a wonderful adventure in this kingdom of smiles, giving us a deeper insight into the lives of the diverse people of Bhutan. We experienced many activities including farmhouse visits, special Buddhist ceremonies. We saw many of the country's cultural highlights including many forts (Dzongs), such as Punakha Dzong and Rinpong Dzong with its wooden cantilevered bridge, and the infamous Taktsang Monastery (Tiger's Nest) in the Paro Valley, the National Memorial Chorten, the twelfth-century Changangkha Temple, a huge Buddha above Thimphu and Chimi Lhakang in the sub-tropical Punakha Valley. We also did some fantastic ridge walks.

It was the people who really shone through for us, though, always happy and smiling and wanting to help to improve our experiences. Tipping didn't even cross their minds as motivation to help us.

Our first night was after a 3- to 4-hour-drive over the 10,000-ft high Dochula Pass, where it became rather cool and damp with even a little sleet blowing in the wind as we took in the 108 choten stupas that had been built here to symbolise the 108 books of Buddha. We enjoyed a cup of hot tea all gathered around a wood-fired stove and, a regular companion to every drink of tea, a plate of dry Jacobs crackers.

On the road to the capital Thimphu we had checked out our first ancient bridge and in Thimphu itself had our first visit to a lovely little restaurant overlooking the river called the Chha. We wandered around the local produce market full of lots of lovely fresh organic vegetables, fruit and dried fish. We were expecting to see fresh fruit and meat, but there was none at this time of year. Any meat is imported from India but because of the people's Buddhist faith, they believe in kindness towards animals, value citizens' happiness and respect the environment and so there are no slaughterhouses in Bhutan.

Roads in Bhutan were not motorways other than some better sections near the capital, but all were pretty well maintained considering the wilderness and mountainous terrain they twisted through.

We soon descended towards our first overnight in Bhutan, the delightfully rural Drubchhu resort outside the small town of Punakha. Set amongst green paddy fields on the site of an old granary, it was a lovely sanctuary of peace where we enjoyed good food and drink and where the staff were very friendly and always keen to help with any enquiries. Our spacious, comfortable room for two nights looked out over the terraced rice fields, the hotel gardens and vegetable patch and across the valley to the mountains. It is such an attractive and relaxing location and we very much preferred it to downtown Punakha. Breakfast was good with made-to-order omelettes and dinner enjoyable too. Our favourite red wine was 'Vintria', a Bhutanese wine.

The next day our guide Dolay took us a few kilometres past the confluence of the two rivers Pho Chu and Mo Chu, the father and mother rivers, which meet at the Punakha Dzong. We headed to a small car park by the pristine waters of the river before crossing the old bridge on foot to ascend to the distant temple on high. Our hike was to Khamsum Yulley Namgyal Chorten, about a 40- to 50-minute hike through a beautiful terrace of paddy fields. This temple was completed in 1999, built by the Queen Mother in honour of the King. Located on top of a hill it commands a very beautiful view of the valley below. Like most forts and temples, you cannot take pictures inside but we climbed the various staircases to the top and took photos of the views.

After an enjoyable walk we headed for the Punakha Dzong. It was a very special day as the annual Punakha Tshechu festival was on. There was a throng of people witnessing traditional dances and music reliving legends and folk tales, but the colour was provided just as much by the very decorative and vivid hues of the splendid costumes of visitors, who all seemed to be in their best national and regional dress. Tshechu's are large social gatherings at which you can clearly see the social bonding among people from remote and spread-out villages. Equally, people are happy for you to take their photographs. The much-vaunted black hat dance was akin to a very slow, sombre Morris Dance!

We loved the day here but after a good night's sleep, it was a warm bright morning, so Dolay took us for a walk down through the nearby village of Lobesa with its basic shops and bars, many of which seem to be festooned

(Top Left) Two young boys at Punakha Tshechu Festival; (Centre Left) Two Ladies and (Top Right) Characterful man at Tshechu Festival

with what became a recurring theme, lots of wooden phallus. It must be something religious. We started our walk across the fields here, which seemed quite dry.

We spotted a game of 'khuru', which is like darts back home but played with very large projectiles with flights which are hurled around 25 metres to a very small archery-style target, of which there was one at each end of the field of play. These guys were good, we established they were teams from Punakha and Thimphu, all dressed in 'goh, their traditional dress which is knee-length robe made of cloth, wrapped around their bodies often patterned with stripes held in place by a belt. Its design allows a pouch like part of the garment to store a wallet or mobile phone and similar small objects. There were about eight members on each team facing off on this delightful sunny Sunday morning. Between rounds they would perform a ritual dance and this became especially animated if someone achieved a bullseye and the dance was accompanied by chanting and singing. It appeared that if they

(Top) Gangtey Goemba in the snow before our trek down the Phobjikha Valley to our hotel for the night and (Bottom) The 'Goh' – Bhutan National Dress

managed to get a dart within a hand's width of the target it is one point, two points if they hit the wooden target and three for hitting the bands of colours. There is a village in the Punakha district called Khuru, so perhaps this is where the game originated.

We pushed on a little way to Chimi Lhakhang, which is dedicated to Lama Drukpa Kunley (also known as the Divine Madman) and is popular among childless Bhutanese couples as a temple to seek blessings for fertility. There were returning couples giving donations to the monks here, others asking for blessings. It was all very interesting, being in our sixties we felt we were pretty safe visiting here.

We headed further east and higher into the mountains where it started to get a lot colder. One thing that probably keeps tourism in Bhutan optimum and Bhutan unspoilt is they insist on a minimum spend each day and a tour including full board is arranged for everyone, thus putting off the backpacker squads who spend little. It also explains the lack of tourists we came across. Most food in Bhutan was vegetarian and tasty desserts were invariably fruit or a piece of cake. One day we were treated to an apple pie and pomegranate seeds.

All men seemed to wear the traditional Goh, which seemed to have a pouch round the mid-riff where they kept belongings like their phone and wallet. Striped or checked sleeves were rolled back and always looked impeccable. When cold they rolled the sleeves down to protect the lower arms and hands like a muff. Underneath they wear a T-shirt, shorts and knee-length socks.

We crossed over the pass and dropped towards Gangtey, passing a herd of yaks and also some langur monkeys. We visited Gangtey Goemba, a sixteenth-century monastery, still active and inhabited by monks, where there was sleet in the air and the remains of some overnight snow on the ground. Here we headed off on a pleasurable two-hour walk which kept to the ridge and gave us a good feel for this wide Phobjikha valley which opened before us. It is supposedly the most beautiful and shortest of the existing nature trails in Bhutan. From the small hilltop overlooking Gangtey Goemba, we headed downhill through flower meadows to Semchubara village and from here through beautiful pine forests and into the open valley. After passing a chorten and Khewa Lhakhang, the trail ended just below our overnight hotel in Gangtey Dewachen. Before we called it a day, we drove down towards valley end and saw some black-necked cranes.

Our Bhutan Guides: Dolay and Driver Taewan

We were ready for a good feed and rest. Our rooms had wood-burning stoves and, though it made them cosy, we felt it messed up our breathing at night with the fumes and when the fire went out it was darn cool. We loved the resort, though, and next day were greeted by a light sugar-coating of snow and saw five cranes in flight, maybe heading to Tibet.

We returned over the Dochula Pass to Thimphu and the modern Hotel

Bhutan Suites. We were most impressed in this city by the Great Buddha Dordenma. We took a morning drive up into the hills to this large statue of Lord Buddha, then took a walk for about two hours on an easy nature walk along a narrow footpath, enjoying the view of the Thimphu valley. This was the most impressive Buddha I have ever seen, and I have seen a lot on my travels over the years.

In the afternoon we visited the folk heritage museum, a replica of a traditional farmhouse; the weaving centre to see the textiles of Bhutan; and the most surprisingly interesting of the lot, a very basic paper factory which was very labour intensive and produced paper very different to that we are used to at home. The visit takes about twenty minutes. You start by seeing the Jungshi tree bark used to make the paper soaked and cooked in an iron cauldron over a wood fire. You then enter the factory, where everything is done by hand. You see the process up close, as women shred the soaked bark, which is then further shredded into a pulp, then made into paper in vats using bamboo screens. It's then pressed to get the water out, and dried. There's a lovely gift shop where you can buy notepaper, notebooks and so on.

We had a walk around Thimphu which was not as appealing as other towns we'd visited and full of stray dogs which was a common theme throughout Bhutan. They didn't really bother you as you walked around but at night, they drove you nuts with their barking.

We were lucky to see a group of locals practising the national sport of archery and were most impressed with their skills in getting close and hitting targets 140 metres away. We saw the traditional wooden bow and the very technical bow used in the Olympics nowadays. I will certainly look out for Bhutan in the next Olympics archery competition.

Next day, it was on to the main reason for our trip to Bhutan: Taktsang Monastery. My health hadn't been too good whilst we'd been away. Unfortunately, I was battling an attack of acute prostatitis.

After breakfast we headed to Paro and our hike to the monastery. The walk of about two hours uphill takes you almost a kilometre above the Paro valley floor. (For those who cannot hike you can arrange a horse for a transfer up to a cafeteria midway, but our guide, like us, felt it was cruel on the horses.)

The view of Taktsang Monastery, known as the 'Tiger's nest', built on a sheer cliff face 900 metres above the valley floor is a spectacular sight.

The monastery is an important pilgrim site for Buddhists. The great Guru Rimpoche is said to have flown here on the back of a tigress when he brought the teachings of the Buddhist Dharma to Bhutan in the eighth century.

It was a very warm day, so we divested ourselves of most of our heavy walking gear, took some water and headed upwards. It was a well-worn track and frankly we could have guided ourselves up. We regularly stopped to look out over the valley as we ascended the 3,000 feet from the valley floor, which was already 7,000 feet above sea level. After about an hour, passing prayer wheels, some water-powered, some of the most beautiful views towards the mountains opened up through trees festooned with prayer flags. We reached the cafeteria, a welcome oasis where we had a cup of tea with the usual dry crackers.

We chatted with a couple from Melbourne who had walked this far as part of a group but were not going as far as their other friends who had gone

ahead. It was clear this walk needs respect. Whilst it is no Kilimanjaro it is tiring and add a little altitude into the equation it certainly needs a modicum of fitness.

Ron & Dianne on the way to the Tiger's Nest in Bhutan!

We pushed on but the combination of heat, altitude and the climb was taking its toll and the pace was slowing. We reached a spot where we watched a llama being born before the really great viewpoint of the Tiger's Nest. We were within touching distance and the elation started to flow.

Not far now, or so it seemed, but it was wishful thinking, it was still on the far end of the deep valley as the trail turned into steep descending steps before it climbed up again, over 700 steps in all. It was draining and I really felt it as we headed down and looked up that the monastery was getting further away, making us all slightly despondent.

The stairway dropped towards a bridge across a lovely waterfall, partly iced over, that, according to local beliefs, fell into a sacred pool. From here, the steps started to climb up to the monastery. After over 100 steep steps we were finally at the monastery entrance.

It was so exhilarating to finally reach Tiger's Nest Monastery. Even though photography is not allowed inside I have my own mental images of this wonderful monastery and its location. We left our shoes and phones at the entrance.

Seemingly small, this 'tardis' of a monastery disclosed several temples within its caves. We visited three where statues of Guru Rinpoche were evident alongside statues of other Buddhist images and beautiful religious paintings. The meditation cave of Guru Rinpoche is sealed off behind a gilded shutter in a small temple. It is only opened once a year so we were unable to view it.

Our guide Dolay left us to our thoughts for a short while and Dianne and myself held one another and felt blessed that we were in this sanctuary of peace and calm high above the valley, almost as if we had been carried up away from the turmoil of pain and fear that the cancer had brought to our lives over the previous four years and we felt truly blessed for this deliverance from evil.

Needless to say, the journey down from the monastery was full of elation and a wonderful feeling of achievement and good fortune having been bestowed upon us. A buffet at the cafeteria, a last close look at the Taktsang Monastery and the views from the lofty toilets on the side of the valley made us smile. Our souls were renewed with joy and we felt a true sense of calm.

On our return to the hotel, Dolay had arranged a body massage for us all, which was more than welcome and eased our aching bodies, for just £15 each.

After a good sleep and great breakfast we learned it was the king's birthday and through our local guide we were invited to a celebration at the school in Paro, where the local governor and all the local dignitaries and dancing groups set about celebrating. Everyone seemed so invested in extolling the virtues of the king and queen. A fabulous start to the day.

One little girl spoke to us in perfect English and talked about the king's 39th birthday, we were shown to a special tent where we had guest seats and then given butter tea and snacks. We had a photograph taken with the young girl's class, then headed for Paro Dzong and learned about the fascinating 'Circle of Life' before walking down to the river crossing and lunch with a tasty cheese momo (like a dumpling) at Charo and Doro. We drove along the road out of Paro to take a distant view of Jumolhari, a 7,500-metre peak, before doing a bit of shopping in Paro.

We didn't think the day could get any better but after discussion with Dolay we had arranged for a birthday blessing for Dianne and Pauline at the beautiful Gangtey Palace altar where we were staying, This fine hotel, built in the late nineteenth century, was once a very important and strategic place for Bhutan because of its geographical location when defending against Tibetan invasion. Home of many 'penlops' (governors), and once home to the Royal Grandmother (Queen Mother) Nehru. The first prime minister of India visited here in 1958 too. Bear in mind Bhutan has kept itself quite isolated over the years.

The palace sits on a ridge, overlooking the valley and facing the Paro Dzong across the Pachhu River. 'Gangtey' is the Bhutanese equivalent of 'raised land'. A pair of mythical snow lion statues guard each side of the palace's entrance.

We were ushered into the altar room at the palace and the ladies Pauline and Dianne were given the honour of lighting a yak butter lamp as they dedicated a wish. We then realised the hotel staff including the manager were behind us. We were ushered into an ornate room adjoining the altar room and served tea before being presented with two birthday cakes, saying Happy Birthday Dianne and Pauline respectively. It was a very emotional time as we all sang Happy Birthday and shared the cake and tea with these wonderfully generous people. An unforgettable birthday, for sure.

We had enjoyed the food with the fab view from the restaurant across to the floodlit dzong each night and had enjoyed the walk around the dzong, probably my favourite, I would say. Then we enjoyed a visit to a homestay

(Above and Left)
King's birthday
Parade and festival
where we were
welcomed by the
locals

Wood burning stove in hotel room

Dianne, David, Pauline and
Ron above Gangtey

and to the hot stone baths they prepared for tourists, almost on a commercial scale now.

Our accommodation was quirky. They had managed to blend the history of the building with modern conveniences, but the emphasis was on the history and original feel of this super property.

Next day it was time to leave this wonderful country and its people but not before travelling to the seventh-century Kyichu Lhakhang, one of the 108 temples constructed by the Tibetan King Songtsen Gampo. Here we enjoyed a blessing of longevity and safe travels before travelling the short distance to Paro airport.

We said our goodbyes to Taewan and our guide Dolay. We promised if we made it back again, we would visit Dolay's home valley of Haa. With that we checked in for our flight.

As we walked out to the Druk Air flight in the morning sunshine for our flight to Bagdogra, I looked at Dianne and surveyed the surrounding mountains. We had loved our time here and we both became quite emotional. Thank you, Bhutan! Kaadinchhey La!

Gangtey Palace Altar Birthday celebration

Our plane took off towards the west. As if to wave goodbye to Bhutan, it dipped its wing to the right to move around a range of hills ahead of us before dipping its left wing and heading up a wide valley as it ascended to cruise altitude and headed for India.

Sad to leave Bhutan at Paro airport

As much as we loved Bhutan, we were now travelling to a destination we had always wanted to visit. We flew just twenty-five minutes across the border into India and the bustling city of Bagdogra. The orderly small airport of Paro seemed a world away from the old-fashioned hectic greeting we received at Bagdogra, but we were happy to be on our way to the tea capital of the world, Darjeeling, home of the Toy Train. Passing a number of military bases we took on Rohini Road, the most common route these days. We soon left the plain and started to weave uphill as we headed for Kurseong where Rohini Road meets the National Highway NH-55 (i.e. the Hill Cart Road) which then leads to Darjeeling. Very green vegetation had taken over from the drier plain of the Bagdogra area. We had passed through Silguri where we could have taken the train but it would have taken hours longer to Darjeeling than the three we were expecting by car.

It might have been less stressful, however, as we constantly overtook vehicles, sometimes on bends, but somehow, we survived the horrendous drops which descended to the valleys below. As we reached Kurseong the bright sunshine we had been greeted with at Bagdogra had disappeared to be replaced by a misty, cool atmosphere.

Just before dark we were dropped at the small two-star hotel Villa Everest, a small former private home, now a low budget but comfortable boutique hotel with twelve rooms and, the reason we chose it, views of the Himalayas from our bedrooms … well, there would have been if this thick mountain mist wasn't present.

After freshening up we went out for a walk having decided to return for dinner in the hotel. We walked down Gandhi Road past an old church and clock tower, you could feel the former, now faded, grandeur of many

Darjeeling Tea at Glenary's

buildings here. We dropped in to Riddi Siddhi, a money exchange, and received a great exchange rate for sterling before doing a little perambulation around the little town of Darjeeling which clings to the steep hillside. The mist and light rain made the town feel like a run-down mill town in industrial northern Britain as darkness descended. The busy honking traffic was almost annoying and depressing by now, as we dropped past the post office then up a narrow road with lots of shops to avoid the choked main roads before re-joining Gandhi Road and seeking solace for the first of many times in Glenary's. Of course, a Darjeeling tea was de rigueur, but a fantastic apple pie really brightened our day.

A good rest saw us ready to explore Darjeeling. We descended the hill for five minutes down to Darjeeling station and booked on the 10am Toy Train to Ghum. We had to complete a booking form then pay just £13 each for a two-hour return trip. We stopped at the small station buffet and a few chocolate bars and drinks were very cheap, very different to a British station where you pay a premium. Needless to say, the train was full of tourists.

The small engine steamed away, pulling us on the 'Himalayan Princess' up the track, slowly passing the shops and other buildings, many within

Himalayan Princess Darjeeling known as the Toy Train!

touching distance as the train at times zig-zagged up the incline and traffic worked its way around the train, waiting until we occasionally met a traffic jam ourselves. Views are mainly from the right of the train going up. We have a quick stop at Batasia Loop where there is a pretty garden and a war memorial.

In Ghum we decided to look at the very good but small museum. Fortunately, the sun had come out for our return to Darjeeling. We headed for a well-deserved cheese sandwich on the rooftop terrace of a rather rundown Keventers restaurant. There was not much of a view as the mist still hung over the valley sides.

We decided to walk to the zoo and Mountain Institute which was a bit further than we thought but we were soon at the gates, where we were charged the princely sum of 100 rupees each (about £1) to enter. Whilst I hate the idea of animals being penned in, we saw the amazing snow leopard, Himalayan bear and the majestic Royal Bengal tiger, which took great delight in reversing towards the fence and spraying urine over the watching visitors, which was probably the most fun it had each day and gave its opinion on captivity.

The main reason we were here was to see the Himalayan Mountain Institute. This was the legacy of Sherpa Tenzing, a hero of mine, for sure. His mausoleum stood outside the institute. The museum itself had references to all the greats, especially Tenzing and Hillary, had pictures of Steve Venables and in the book of comments and signatures of those affiliated to mountain climbing were those of Chris and Jo Briggs from Pen-Y-Gwyrd.

A stroll back to Glenary's for another apple pie and tea was compulsory. We returned for dinner at Glenary's that evening

Tenzing Institute in Darjeeling

as it was Dianne's birthday. We had a fabulous night listening to music by a local chap who played lots of favourites like 'Blue Ridge Mountains' and 'Fields of Barley'. We met an Aussie group who had been trekking for days in nearby hills.

Next morning the sun was up and we could see the Himalayas from our room. We walked in the warm sunshine to the Happy Valley Tea Plantation, about 25 minutes' walk past a very busy bus station. We enjoyed a tour of the dormant tea factory as we were in close season. We had some tea tasting before walking up the steep hill right to Chowrasta, down to the Tibetan refuge then back up to the Heritage Resort just below an observation point where we had a lovely cup of Darjeeling and danced the Charleston on the roof to be sent over Facebook.

We then found the large house of Sherpa Tenzing on A J Bose Road, just below the Villa Everest, called Gangla after his father's family name. Villa Everest was formerly home to the *Darjeeling Times* where it was first printed. Vivien Leigh was born in the convent just below. Mother Teresa came on retreat to Darjeeling and said she heard God in the sound of the train, encouraging her to minister in Kolkata.

Our stay in Darjeeling over, Ahmed, our young driver, picked us up

Famous Taj Palace Hotel in Mumbai

by jeep and took us back to Bagdogra Airport via roads off the beaten track, a rather scenic route. We took off on time with Jet Airways for our next destination, Mumbai, warmed by a lovely in-flight biryani.

The flight was just over three hours and arriving in Mumbai Airport was a bit of a culture shock after the old-fashioned facilities at Bagdogra. It was ultra-modern and as we emerged it was like arriving in Orlando, Florida with lots of exotic bushes and palm trees. We were supposed to be met by a female driver and unmarried mother, one of a group of people our destination hotel tries to support.

Sadly, the woman didn't have her sign ready initially and so it was down to us to try and spot her. This rather slovenly and awkward lady proved rather strange and not at all friendly or helpful. We tried to cut her some slack, but after leading us to the car park she then expected us to carry our heavy cases down flights of steps whereas normally porters would be engaged to help and

Ron and Dianne at the Gateway of India, Mumbai

Crawford Market Mumbai

be tipped well for their services, no massive hardship but she just wasn't switched on to visitor's needs.

The vehicle was new and we settled in and enjoyed the vista from a new causeway bridge crossing a waterway through Mumbai, formerly Bombay and before that named by the Portuguese 'Bom Baia'. Our lady driver made us very nervous, though, as she almost crashed three times and let out a huge belch which made us laugh but then added a regular repeat of burping. Then out of the blue she pulled into a fuel station to refuel. Everything was very unorthodox, we tried to make conversation but her English was poor and our Hindi non-existent. Finally, we were relieved to arrive at the perfectly situated Hotel Abode in Colaba.

After a pleasant greeting from a young lady and help up the stairs with our bags we checked into our simple, clean and characterful room, showered and went next door to the Café Colaba where a great choice of snacks and teas and coffees set us up for a pleasant evening stroll in the immediate environs.

We did a little shopping in the nearby streets and stopped at the famous Leopold Cafe, once a victim of the Mumbai terrorist attacks of 2008. It was buzzing but quite expensive, so we decided we wouldn't repeat this visit.

After a good night's sleep, we were ready for a full day's sightseeing with our local tour guide, a young fellow called Sachin, and driver Turvinder. Sachin was a typical guide with an agenda and although armed with a list of our preferred sights to take in he took control from the beginning, failing to engage with us, just facilitating the various sites starting with the Gateway of India and view back to the Taj Palace Hotel. It was wonderful to be there on a slightly hazy morning. We tried to engage him and occasionally got through, but for the most part he was not listening, just instructing the driver which way to go, occasionally being rude or harsh to him. He was constantly on the phone and answering calls and we felt we were being fitted into his day. In fairness by the end of day, we had spoken to him about his small family and treated him and the driver to lunch. We had requested a visit to a dress shop for Dianne and we know how this works with kickbacks and, whilst feeling a little pressured and undervalued, we accepted that. Later we asked to go to an alternative shop, only to be taken to a high-end Kashmiri silk shop, so we called the tour to an end. We tipped well, perhaps we should have reduced that tip but we stuck with our pre-planned amount.

We managed to see a lot during the day, including Manai Bhavan where Ghandi stayed with a good friend at his home and was actually arrested, and the Dhobi Ghats where the Dhobi wallahs were fascinating to watch, handwashing linens from the city's hotels and hospitals. We were intrigued to see they even had a school in the complex for their own children.

We had lunch at the Captain Cook restaurant, which was quite clean and well-presented and the regional selection of small dishes typical of the area suggested by Sachin was perfect and as always too much. It included plenty of spicy dishes, breads and chapati and some desserts.

We enjoyed the visit to the historic and vibrant Crawford Market full of fruits and vegetables of all colours and sizes. We were talked into buying £15-worth of spices. How did that happen? Another kickback for the guide, no doubt.

St Thomas Church is well worth a visit, now being refurbished and cared for by UNESCO. It has a large impressive nave, and has seat plaques at the front where King George V and Queen Mary sat and also where Mother Teresa visited. We also went to see the bungalow where Rudyard Kipling was born. His father was Dean at a school now known as the Sir JJ Art School.

We looked down on the harbour from a close-by viewpoint called Malabar Hill, the place to live with lots of expensive properties in that area, and visited the Hanging Gardens and Kamala Nehru Park along the way. The Hanging Gardens were interesting as underneath were huge reservoirs, but not far away was an amazing place where people who die are laid out and birds circle and swoop down and eat the carcases, part of the belief system of the Zoroastrian religion. The Parsis have built stone towers like open-air auditoriums containing three concentric rings of marble slabs — an outer ring for men's bodies, a middle ring for women's and an inner ring for the children. For centuries, bodies left on the slabs were consumed within hours by neighbourhood vultures, with the bones left in a central catchment to leach into the soil.

We were in the park close to evening, so local people were out in their fine clothes to enjoy time with their families. The Shoe Garden has a large shoe construction. As we stood and watched children playing in and around it like a climbing frame, memories of my own childhood and the nursery rhyme, 'There was an old lady who lived in a shoe', came flooding back. The Walkeshwar Temple and Banganga Tank, a sacred pool over 800 years old were both inspiring, and taking in Mumbai's Marine Parade known as the Queen's Necklace was a wonderful way to appreciate the city!

We passed Watsons Hotel, now abandoned and run down, which is ironic and almost kismet as Tata, the owner of the steel company in addition to many other businesses, was once thrown out of it. He with others opened the Taj Hotel, which now flourishes whilst the once main hotel of Mumbai has gone.

Chhatrapati Shivaji Station, formerly the Victoria terminus, is a fine station, one of the platforms of which was used for filming *Slumdog Millionaire* scenes. The ticket hall is a fabulous architectural sight to behold.

I loved watching a cricket game on the Oval Maiden just under the clock tower and Bombay Court House, the former looked very much like 'Big Ben' in London this clock tower was designed by Sir George Gilbert Scott.

We loved Mumbai it was relaxed and not as intense as some of the beggar-sieged cities of India. Dharavi is a large slum area but very well organised in the businesses and activities that operate there. Fisherman Village on the beach was interesting, with fish being dried on bamboo frames to

Shoe Garden Mumbai

475

become the well-known Bombay duck but also where the terrorists arrived by boat to commit the atrocities at so many iconic Mumbai institutions in 2008.

We rounded off the day with a sundowner in the harbour bar of the Taj Palace Hotel. We were encouraged by the servant of forty-seven years, Mr Mistry, to have a large Kingfisher beer, which was on special offer and served with an attractively presented cascade of nibbles. We watched the hubbub around the Gateway of India as the sun went down and ferries returned from Elephanta Island.

Mr Mistry had witnessed many evenings with clients as the sun dropped low on the horizon but now the sun was setting on his own career as he takes away many memories of his life that has been the Taj! We wondered if he had been on duty the day the Taj was assaulted by terrorists in 2008 when so many people lost their lives and considerable damage was done to the hotel.

For the only time in Mumbai, we were harassed on the way back to our hotel by two separate beggars. We will always try to help but these guys tried to drag us to stores to buy them food. They would not accept money saying storekeepers would not take money from them. The first lad disappeared, then returned with a babe in arms. We eventually gave in and went to a store but when he wanted a huge pack of baby milk for £20 and more we said no way and gave him 50 rupees. He refused to accept until we walked away. We felt a little under siege but they eventually gave up. We felt so bad. We had been told by guides and others not to encourage them and I hear the beggar population has been reduced but what does that mean, how are they cared for and supported?

We are so lucky to have been born in the United Kingdom, with all the blessings of plenty, a postcode lottery for sure.

Our same taxi lady took us back to the airport from our delightful lodging the Abode Hotel, perfectly run by Nilesh and his team. They cannot do enough for you and give great local advice and can arrange just about anything for you. Our lady stopped again on the way to the airport to get herself some snacks but was kind enough to give us a small local delicacy as a gesture.

As the warmth of the Mumbai breeze wrapped around us, we said our goodbyes to the Indian Sub-continent and entered the harsh air-conditioned environment of the airport terminal. And so, ended this fabulous adventure, all of us feeling happy and blessed.

CHAPTER THIRTY-SEVEN

———

Just when you thought it was safe to go back in the water!

On our return from the Middle East and the joys of Oman in June 2019 we went about our normal summer activities, back to the Park Run and some house sitting to cover a few days at Harnage with the delightful Tarka and Toggle. Then we were surprised to receive a letter asking us to go and see the oncologist. The scan taken in April that Birmingham showed as all clear, had finally after two months been reported on in Shrewsbury and an area of concern identified in my para-aortic lymph glands, which were unusually enlarged. A further CT scan was arranged.

We tried to stay busy and, as good fortune would have it, we were house sitting at Criccieth with the donkeys Rosie and Gracie and sheepdog Murphy again. This part of the Welsh coast had become synonymous with our respite when trying to get our heads together and relax.

Early one morning we rose and headed for the Watkin Path figuring we would take a walk up to perhaps Yr Aran, but as it was a lovely sunny and dry morning before we know it we were heading towards the summit of Snowdon. It was the first time we'd climbed this route, long and of course rather steep higher up, but it is in our opinion the most attractive approach. We saw the massif of Snowdon rising to the left of us and it looked rather foreboding with steep scree slopes and the summit shrouded in mist. Before

Tyddyn Iolyn - One of our favourite house sits with Rosie and Gracie the donkeys and Murphy the sheepdog

long we were heading in that direction as the path twisted this way and that and became rockier and more obscure. We stopped, took a boost from our favourite Velaforte energy bars and drank plenty of water for the assault on the summit.

Snowdon did not disappoint. As we entered the mist occasionally a window opened to the valley below, then closed around us. We could not see a distinct path and being early in the day there were not a lot of people on the path. After a fair time of scrambling and trying to ascertain the route upwards we could hear the occasional voice above us and decided to move more to the left as an overhang had appeared above us.

The mist was making the hand and foot holds a little damp and therefore more precarious and rocks were crumbling below our feet. After about twenty minutes a few souls were heard, then seen, coming down and we started to feel more at ease. A steep and challenging scramble soon brought relief as above us we could see the Snowdon café and railway. We were there. It took us 2 hours and fifty minutes from valley to summit, not too bad for a couple of senior citizens.

We chose to take the Coca-Cola route down the Llanberis Path, a long path but really easy, before reaching the Sherpa bus stop in 1 hour 50 minutes and catching the fabulous Sherpa bus service back around to our car. We felt we had achieved a lot and all before lunch!

Another meeting with the oncologist ensued on the Monday evening, our weekend exploits soon a distant memory. Dr Nasim said their concern was confirmed, which was not great news as the lymph nodes are conduits to the whole body. Two of them were enlarged and one growing quite quickly, three in just three months, which indicates a strong likelihood that the cancer is returning.

Dr Nasim was very good in explaining the way forward: firstly a PET scan in Stoke, probably the following week, to establish the metabolic qualities of what was happening and by injecting a radioactive glucose-based dye they would be able to see if cancer was evident and also see if it had spread; then either stereodactic radiotherapy with the dramatically named 'cyber knife' in Birmingham mid- to end of August, or possibly the knife first and possibly followed by chemotherapy if that was unsuccessful.

I signed forms for my chemotherapy to start in three weeks hence Irenotican and 5FU for probably twelve cycles over at least three months. The 'Ready Brek kid' would be riding again that Autumn!

It had been wonderful to feel as normal as anyone else for almost two years and to be able to make plans.

However, it was now time to batten down the hatches to get through the viciously debilitating chemo, think of the positives and hope for an extension to my life in reasonable shape!

People will never understand totally if they have not been through it … films and documentaries can educate but unless you have walked in the footsteps of a 'cancer dancer' you will never really know the discomfort, pain, reduction in operation of your faculties, the fear, the depression at times and the hopelessness. You lose the sense of taste, sometimes lose your hair, feel incredibly bilious and it seems worse for me as I cannot seem to be sick to have some relief so it's a constant state of nausea. I had forgotten 'chemo brain', but then I would, wouldn't I?

In the twelve cycles the build-up effect of the chemotherapy means the two-week cycle of recovery time becomes less and less, finger ends would crack to the extent they were so sore and then would bleed, your mouth becomes so sore you want to eat but struggle to do so. Nausea and fatigue are constant bed friends and the horrible metallic taste of all your food adds to the degradation of one of life's remaining joys, made worse by the fact that I love to cook, yet couldn't even gauge the taste of my own cooked food.

Poor sleeping patterns, punctuated and hindered by acid reflux eased by Gaviscon and sitting up all night, leg pain from my thread-veined, road map legs made trying to even nod off very difficult. Two-hour stints of dozing were only possible by having regular pain relief, which became stronger and stronger and caused other problems. Omeperazole dosage was upped to twice a day to avoid stomach damage, causing more likelihood of an avenue for cancer to proliferate. All the time you are becoming more weary, wondering if it is all worth it.

Quality of life is poor. You are just essentially surviving outside the fire of retribution which is slowly burning towards you. What do you say when people say how are you? You just respond okay and use clichés like 'no pain no gain!' Trying to put on a brave face whilst holding on to your hope and faith, but thinking how this must be affecting your loved ones, who feel helpless. It is not just about you so you try and share their optimism of a better future as the Diana Jones track 'Better times will come' plays on constant reload inside your mind … it tears everyone apart.

People will never understand how low and depressed you can get and never witness all those times you just burst into tears at the total futility of it all when you fall to the bottom of the well and sense in the darkness the unscalable sides of your situation, greased with slippery slime that stops you reaching the fading light of hope above you whilst waiting for someone or something to rescue you, but hope, faith, courage and positivity you must have to survive.

Your ileostomy bag, which disappointingly to friends' quips, does not match my shoes, starts to split at night as your stoma changes shape and becomes inverted under attack from the chemotherapy-infused faeces. Like a sea anemone under attack it expands and retracts, the toxic contents of your bowel making the stoma bleed and become very sore as it also attacks your abdomen skin, the acidic faeces burns and erodes the sticky seal of the stoma bag and faeces escape onto your clothing and bed sheets, making another mess as you try and escape to the bathroom, wash and cleanse yourself before replacing with a new bag and securing strips before cleaning up all the mess as you stand naked in the freezing cold of night. Is chemo tough? You can bet your darn life it is, but it might just give you some more quality time with your family and friends, so it's a gamble most people take.

I had feared having an ileostomy but I actually got used to it very quickly and I found it was far less arduous than I feared and actually improved my quality of life compared to the ulcerative colitis I had suffered with for twenty years. It is embarrassing at times, especially as it takes on a life of its own as the bowel opens when it feels like it, and can sound like an out-of-tune kazoo band member. I have to just smile sometimes in company as on tricky days it's like having an ileostomy with Tourette's.

You can be embarrassed and not go out in public, but it can also be amusing and a talking point if you are strong enough to cope with that viewpoint and most people are understanding.

My return to an almost vegetarian diet of vegetables and fruit ends up giving me a lot of wind, so much so it's as if my ileostomy is a free spirit and occasionally has a very loud will of its own, whether in company or not and to try and pass it off as playing the bagpipes does not fool many!

Once in a course of chemotherapy simple things can become difficult as your thinking can become befuddled and makes you think you are experiencing some sort of dementia. One morning I stood in the kitchen with a table napkin and a tray in my hand and couldn't remember where to put them for a split second. I struggle to remember things so write them down but then forget to look at the list. Some of you may be smiling and thinking that's old age and part of it might be. Who knows?

My first chemo included a platinum-based treatment called Oxilaplatin. It had a strong effect on my central nervous system, with my nerves tingling like crazy and extremities suffering from neuropathy. I wasn't even able to use my fingers to button up my shirt. I just fumbled, becoming more and more annoyed and frustrated with myself.

You need to have hope and faith and the love of those around you, who suffer with you! When you come through this it makes you a far more thankful and appreciative individual who thanks whoever they believe in for the smallest of mercies.

It does beg the question, though: 'Why is cancer like an epidemic?' A search is on for cures but why not the search to prevent and find out what causes the cancers, such as maybe pesticides in our food?

You never know how many chapters life will hold for you, but I feel you must keep turning the page while you can. When I look back at the people I have met and those that have had a big influence on my life I did not foresee the road ahead and from an early age was always happy to let life play out and understand fate has its ways. As enjoyable as life has been at times it is countered by the hard and sad times, but through all of them I have grown as a person and gained experience and knowledge and the ability to be more prepared for future events.

Never did I believe I would have travelled so much in my life, nor did

I ever think I would be in my forties before I met my true soul mate and companion, but I guess the road of life was preparing me for each eventuality as it will continue to do so for when I eventually leave the bonds of earth and safari to new horizons beyond our current understanding.

And so full circle from four years before when I felt life was close to an end, we have returned from a wonderful voyage of life and experience in Australia, Singapore, India and the wonderful Bhutan, feeling totally blessed and so, so lucky!

At 11am on 6 December 2019, I witnessed the blowing up of the iconic Buildwas Power station cooling towers, a blot on the landscape to many, a familiar landmark to others. All through my teenage years since 1969 when man landed on the moon. I passed them regularly visiting my grandparents at Ironbridge and later Madeley. The towers' destruction was a reminder that nothing is forever, it felt like they were destroying the images that were part of my life and a new landscape will be the reward for our children and grandchildren ... I had long since been aware that time moves on and is never on your side so make the most of every day and every opportunity. Time will not come again!

This has been highlighted by the outbreak of the killer pandemic Covid-19. Who would have thought something like this would change the world and the way we lived? Ironically, lockdown was like my life already as I spend a lot of time working at home, writing and raising funds for charity.

On the subject of raising money for good causes we took advantage of the pandemic to set ourselves a challenge. It all started because in 2019 we signed up to climb Ben Nevis and raise funds in aid of Marie Curie, but unfortunately my chemotherapy started and went on for five months.

We duly signed up with Marie Curie Fund to climb Ben Nevis in September 2020 and started doing a few hills at the end of July. We thought we had better get some training into our legs and had the idea we would get up early one day and climb Snowdon. One Monday morning we headed for Pen-y-Pass on Snowdon. The weather forecast was awful with winds gusting to 40 mph accompanied by torrential rain. I suggested to Dianne that it was perhaps a bad idea to go, but she lamented, 'I've made the sandwiches now!' So that was it, we were going!

After a 5 am start to get there early, thus making sure we secured a parking place in the small car park, we arrived to find a deserted parking area. We were the only ones there, which should have rung some alarm

Ron has recently qualified as a Snowdonia Gold Ambassador an area he loves dearly!

bells.

Nevertheless, we decided to push on up the Miners' Track, reassured that we had good all-weather gear. True to the weather outlook, the rain was relentless and as we rounded each bend the wind blasted us as if we were being hosed down. Dianne had made the faux pas of wearing trekking shoes instead of boots, which most days would have been okay but at times the wind blew us backwards and sideways, forcing us to keep our heads down. Before long we thought we had gone off-piste as we were walking up a stream but then realised the path had become a raging torrent of water, not good news for Dianne in her trekking shoes.

We pushed on, not a soul in sight, it seemed a lot longer and harder than the many times we have climbed Snowdon before. We knew the cafe was closed at the top, so we carried all our food, water and any supplies we might need, including emergency gear.

Finally, after about two hours, we reached Llyn Glaslyn and started up the steep ascent to the ridge and then the summit, but the water was cascading down the rocks and gulleys, and the mist meant we could barely see a pathway and even our top-notch weather gear was starting to be penetrated by the driving rain and wind.

We took a view. We figured the gusts of wind were going to be worse on the top, so much against the grain, we decided to retreat, having gained only about 360 metres in height in about two hours. Crazy!

Returning to Pen-y-Pass it was a whole hour before we saw another soul

Ron on Snowdon Climb in severe weather with no other soul in sight

on the route. As the weather was clearing and easing, one or two people were starting to climb up. We reached the car park, saturated and aching, but a cup of tea made a difference, and we started to be philosophical and tell ourselves we had enjoyed the experience, especially having the mountain to ourselves.

I spoke to a Snowdonia Park warden who was incredulous that we had tried to summit in this appalling weather. His raised eyebrows said it all. 'Don't worry, we were socially distanced!' I reassured him.

When we arrived home that evening, we had an email from Marie Curie to say this year's Ben Nevis walk had been cancelled. We just had to laugh. But then the seed of another fundraising plan germinated. We decided to climb hills and mountains in the UK to the aggregate height of Mount Everest, 8,848 metres/29,029 ft to raise much-needed funds for the Lingen Davies Centre and Severn Hospice in Shrewsbury to thank them for the kindness and care shown to us, family and friends. These establishments are essential components of our caring community and do a fabulous job!

As our plan developed, we began our climbs in earnest with the Shropshire hills. Lawley was trekked with great friends Graham and Jan Croft. Caer Caradoc, Stiperstones, Corndon, the Wrekin and many more were summited before we progressed to the higher Welsh Peaks. For a change we headed to the highest southerly peak in the United Kingdom, which happens to be Pen y Fan in the Brecon Beacons, close to where my father was born and raised. Penwyllt is dear to my heart, so we visited Dad's home village, which is now a ghost community, the industry having gone, and just a handful of dwellings and people remain here.

The weather forecast was again abysmal for the four days we were to spend in the Brecon Beacons, but it had to be climbed. After an awful night's sleep, we drove thirty minutes to the foot of Pen y Fan for a 7am start. We were blessed with a couple of hours of dry, sunny weather which was the only window in 96 hours.

We headed off on the steady but not technical walk, although naturally it became steeper as we trekked upwards. We soon reached the ridge when the paths divide and the mist was so thick you could barely see a couple of metres in front of you.

We instinctively took the only path we could see to the left which quickly ascended. After a while we found ourselves scrambling over rocks, so I used an app on my phone to check our position. It showed us we were heading for the second-highest peak, Corn Du. We reached it in a strong 50-mph wind and traversed it to Pen y Fan at 886 metres. We were surprised to find for ten minutes we were the only ones there. It was cold, misty and windy so we soon came off the summit and headed back down into the valley. Mission accomplished.

Next on our list were more Shropshire hills, including the highest, Brown Clee at 540 metres. It was a gloriously sunny day and a walk in the park after Pen y Fan. Apparently, there is no hill higher going due east until you reach the Ural Mountains in Russia. The hill is dotted with all sorts of interesting remains, such as iron and bronze age cairns, the remnants of iron ore smelters, industrial ruins where once coal was mined. Most poignant is a memorial to the crews of aircraft from World War II. Twenty-three allied and German aircraft met their demise on the Clee Hills.

On our walk we met a lovely couple who were taking a break from renovating a house locally and were accompanied by their lively dog Pip. They had recently climbed Ben Nevis and pointed out to us it was a rather long, hard climb but, subject to the weather, quite straightforward, even though the top was a little precarious if the mist was down.

Armed with that information, in early September we headed north for our final push to reach the height of Everest. Our first stop was the Lake District where we were joined by friends David and Pauline. The weather forecast once more was dreadful with lots of rain. I guess that should be no surprise in the Lake District. We decided to have a warm-up by trekking some Western Fells and one morning drove to Bowness Knott where we parked and strolled along the side of Ennerdale water.

As we started to ascend towards Starling Dodd and Red Pike, the weather took a turn for the worse and the rain just fell in sheets. We passed a couple of walkers coming down who were saturated. This was a very hard uphill slog on a poorly defined track through marshy ground. We eventually followed the gulley through which the beck ran off the mountain and finally we emerged through tussock grass to the ridge in the thick mist then, lo and behold, the mist cleared and the sun came out and we had fabulous views across the Solway Firth into Scotland. There were very few people walking these fells, although it's fair to say most hills and mountains we walked were a lot quieter than most summers would have been due to Covid.

We reached the summit of Starling Dodd at 636 metres and continued east to Great Bourne at 613 metres before taking a sharp descent back down towards Ennerdale Water. This walk exhausted us all, we'd not just climbed a total of 785 metres but had walked 14 miles into the bargain and for the first few hours the weather had been abysmal. Still, another summit or two and we were still going.

After a few days' recovery we were ready for the highest mountain in England, Scafell Pike. We were blessed as the weather gods were on our side for once and whilst not a hot sunny day it was dry for our early morning start from Wasdale Head. It was starting to get busy and about thirty or so of us were trekking up the well-made path on a long winding ascent to the summit.

The ascent was pleasant and when the morning mist blew away we had some fabulous views back down to Wastwater. We really enjoyed the experience and little quirks like having to remove our boots to gingerly ford Lingmell Gill, as the stream was quite full given the recent rains. The water was cold but refreshing especially on the descent, rejuvenating tired feet.

By 11am we were at the top at 978 metres, the summit covered in mist and the wind bitterly cold. We returned to the National Trust base camp in a total of five hours, which we were very pleased with. We duly celebrated with a Ritson Blonde ale in the Wasdale Head Inn, once run by Will Ritson. Will, who has now passed away, was a real tall tale teller and in fact his tall tales have inspired an annual competition to find the 'World's Biggest Liar' which is held here!

After a great week in the Lake District, we said goodbye to our brilliant friends, Pauline and David, who headed to the sunshine of southern England and we progressed north to the decidedly more erratic weather of Scotland.

Ben Nevis conquered at the culmination of Climbing Everest in the U.K. fundraiser

After a long drive we arrived at our super guesthouse in the lee of Ben Nevis at Fort William. We made a reconnaissance that evening to the Ben Nevis Inn at Achintee where we planned to start our final ascent of the challenge. This is a super little inn, very contemporary compared to its historic roots, and a great place for a meal and to listen to stories of other trekkers and outdoor pursuits people who were here to enjoy the Scottish wilderness.

Ben Nevis, 'Beinn Nibheis' in Gaelic, has more than one meaning. 'Ben' means mountain but 'Nevis' can be interpreted as malicious or venomous, and it certainly lived up to its name for us. But there is another interpretation: 'mountain in the clouds or heavens', so we will go with that more poetic description.

Next morning, we rose early for a 7am start. The weather forecast the day before was for rain turning to snow, but fortunately that morning the forecast had changed to a drier spell and rain later. I don't know why I bothered checking, as we concluded we should just take gear for all seasons and get on with it because the weather is so changeable. Armed with our electrolyte and water drinks, plenty of high-energy Veloforte and Clif bars but skipping breakfast, we set off in just above freezing temperatures on our last mountain climb in our quest to reach the height of Everest.

The track winds its way steadily at first, before the real uphill battle takes place. It is truly a long and rocky path. After twenty minutes it started to rain heavily so we stopped and donned our waterproof gear. I'd stopped at a bench and was struggling to get my waterproof trousers on. Dianne had spotted some gear on the bench and when I turned, she had it stuffed up under her waterproof jacket to keep it dry, making her look pregnant. As I turned towards her, she let out a huge sigh of disbelief and asked, 'Why do you bring so much other gear you are not going to wear?' I looked at her nonplussed as she produced a heavy pair of leggings and a fleecy top from under her waterproof. Quite bemused, I started to laugh, then explained these were not mine, someone must have discarded them. We both had a chuckle and pushed on.

It became harder as the temperature seemed to plummet as we trekked higher. By the time we were approaching the summit, visibility was terrible, there were some people ahead of us but we couldn't make them out as they disappeared into the ether of the swirling mist.

It was very difficult to pick a route on to the top of Ben Nevis as dangerous drops into oblivion seemed to draw you away from any central path to the summit. The mist and wintry storm blasts of sleet and hail hindered our progress. Driven on by the thought of raising money for our local charities, we persevered. On a bitterly cold summit we were battered by strong winds, snow, hail and freezing temperatures with a -9°C wind-chill factor, too cold to take many photographs. But we did it, 1,345 metres or 4,413 feet. It had taken us 6 hours 45 minutes up and down and thanks to all the wonderful people who supported and donated, this fundraising challenge, we hope, will succeed in raising over £11,000, to be divided equally between Severn Hospice and the Lingen Davies Centre.

Never, ever give in, you just do not realise with a little luck and lots of positivity and grit what you can achieve. With the advances in medicine and love and support from your friends and family you too can overcome the trials and tribulations of life and embrace the good days and joys and wonderful people you have been blessed to know in your life.

Edith Eger's fabulously inspirational book *The Choice* has helped me greatly. I took away a lot of thoughts and mantras, some of which I have adapted to my own needs and thoughts.

One such thought is you cannot change what has happened in your life, you cannot change what you did or what was done to you, BUT you can choose how you live now today and in the future. Choose to be free.

Equally, healing is not about recovery, it is more about discovery. By

Cheque Presentations to Severn Hospice and Lingen Davies Centre

that I mean you can discover hope in hopelessness, an answer where there doesn't seem to be one, discovering this answer is not what matters, it's what you do with it.

Don't stay remote from people and keep your concerns and troubles to yourself, share your feelings because expression is better than depression, which is what can happen when you keep things bottled up. I learned that by internalising my worries and problems it led to stress and illness, first in the shape of ulcerative colitis and ultimately cancer. By sharing your worries and concerns you liberate yourself from an ever-growing prison cell full of negativity and despondency. Essentially sharing your woes means expression leads to lightening the load and making you free.

Every day we are faced with choices. You either take the positive trail or descend down the negative avenue.

CHOICES!

I will/I won't
I can/I can't
Lessons learned or past regrets
I did/I didn't
I could/I couldn't
I would/I wouldn't
I should/I shouldn't

You can ask why me? What have I done to deserve this? Or you can endure, enjoy today and embrace the future. Achieve and fulfil what dreams you can in a time frame that you are unsure of. Think of your news not as doom and gloom but that you have been blessed with an awareness of your mortality and a chance to live every moment.

And make the best of every one of those moments rather than be disgruntled and negative with every little setback or occurrence that annoys or hinders you.

Looking back is natural and often needed as we learn from the past but looking forward is necessary as the future is the only road to hope!

As a chapter ends, a new one begins, this is the book of life. No one knows how long the book of life is, so fill each chapter with new experiences, try not to pass up opportunity because of fear or uncertainty but embrace the possibility of success.

We do have a choice, gloating perpetuates division, bitterness, anger and blame breeds despair, whereas understanding, encouragement, tolerance, positivity and optimism garners hope. Which will you choose?

Instead of focusing on your troubles, be thoughtful, show kindness and patience to others. It is far more liberating.

There is no right or wrong in life as we learn, if we embrace every failure and success as the bridge of learning we may improve the future!

One thing I learned quickly after a terminal diagnosis was to be more aware of my senses. Listen to the birds, smell the flowers, be aware, witness the beauty there is in life. Simple things like taste your food, savour every mouthful instead of eating too quickly Make sure you live, laugh and love at every opportunity you have.

Every day you have a choice, you can be bitter, or you can be happy. By doing what you feel is right helps you shed your fears, helps you accept life and death for what they are and you can leave a legacy to your loved ones and those around you in the knowledge that you are at peace, unencumbered by guilt. Your loved ones can then celebrate your life rather than thinking what else could they have done to make life better for you.

Embrace the future and be kind. I heard a saying once, I'm not sure who originally said it, but it goes like this: 'If you have to choose between being right and being kind, choose being kind, you will always be right.'

We are all human beings no matter the colour of our skin, or what our political and religious beliefs are. Be enriched by the many cultures in our world, learn from people by listening instead of imposing your opinion, be tolerant of everyone because you never know what cross they are having to bear each day, be understanding and patient never arrogant, haughty or rude and you can make the world a better place! Smile and you will be surprised how that makes others smile!

Think of the word HOPE as an acronym: Harnessing Optimistic Positivity Every-day!

For now, until the music stops, I shall continue with HOPE and carry on being a 'cancer dancer', learning new steps and routines all the way!

CHEMO MAGIC!

A recluse by choice stares at the world outside
A prisoner exists where once there was pride
Nausea, fatigue and abdominal pains
Torture inflicted to achieve a gain
Silently the elixir spreads
Cell by cell
A living nightmare endured

To make me well
Three days of toxic infusion
My body lies wracked by poisonous confusion
A week elapses
I begin to feel brighter
Body's ills subside
My mood becomes lighter
I begin to work up an appetite
I no longer fear
The depth of the night
I have lost my sense of taste
Some lose their hair
As our bodies waste
Peripheral neuropathy
Tingles in my fingers and feet
In night's silence I hear my heart beat
Twelve cycles on
Judgement draws nigh
The rewards for the suffering
must surely be high
This is it, the results of the scan
Will I be doomed
Or am I Peter Pan?
My mind's in a blur
As the consultant talks
Will we leave the room
Destined for long walks
A nice pot of tea?
It will make you feel better
Someone on the periphery
Discusses the weather
I look at the sky
All I see is shrouds
Where is the sun?
Hidden by clouds
Then one two three
I am back in the room

Those magic words
have lifted the gloom
'I'm pleased to say
Your scan is clear
I won't need to see you until the New Year'
The brutal treatment
Has been the answer
For the foreseeable future
I'm a true cancer dancer

Acknowledgements

A huge thanks goes to Rob Sutcliffe, a fabulous liver consultant at the Queen Elizabeth Hospital in Birmingham, aided and abetted by a fabulous surgical team and wonderful secretaries in Jo Stevens and Andrea McCaffrey, who gave us hope and without whom this book would not have been written. Ongoing medical care in Shrewsbury with Rob Clark, Dr Nasim and the Lingen Davies Chemo Day Centre has also been of vital importance.

Sincere thanks to my editor Deb Renshaw, who has been as diligent and patient as ever, also to Rich Evans Design who once again has created a fabulous cover design. Gratitude and appreciation to the team at Matador my fabulous self-publishing company.

A debt of gratitude to the late Brian Bass who gave me my first job in travel at Luxitours in Shrewsbury. Of course, a heartfelt thank you goes to all my wonderful clients over the years, many of whom became friends and some may even recognise themselves in some of the tales in this tome!

Heartfelt thanks also to our many friends and family, especially David and Pauline Turner and Gra and Jan Croft, who have been there for me most of my life. My wonderful friend and therapist Janet Nicholls has helped me through the hard yards of treatment with her care, kindness and therapeutic holistic abilities.

Special mention to Edith Eger, who wrote an inspirational book *The Choice* based on her amazing life of survival against the odds and whose words helped support my positivity.

Lastly, thank you to my dear wife and best friend Dianne, my trusty companion and fellow traveller on many of these adventures, who has shepherded me through very testing times for better or worse.

In 1995 following the death of his father, Billy meets Gwen, an old friend of his parents, and so begins an odyssey into their mysterious wartime past, bringing surprising consequences. Billy's quest for the truth leads him to Wales, Scotland, Myanmar, Singapore, Australia and Japan, where he unearths heart-wrenching stories of bravery, sacrifice, love and survival during World War II in Southeast Asia.

Will the roller coaster of discovery, new relationships and unexpected changes in his life bring Billy the contentment and fulfilment he craves?

REVIEWS

The story twists and turns with quite a few 'I didn't see that coming' twists in the tale. The book is moving, and in some parts sad, however it too is an uplifting book and certainly one I found hard to put down. Highly recommended. EMMA H

An intriguing story which unravels to reveal the horrors of war through personal hardships. Twists and turns right to the end to keep the reader fully engaged. A thoroughly good read!! Pauline T

A wonderfully poignant first novel, written with the panache of a seasoned author. Having all the ingredients of a best-seller, its cleverly interwoven story and memorable characters linger long after the final page is turned. Brilliantly researched and beautifully descriptive, this book will take you on an unforgettable journey. Isobel S

A CHRISTMAS MIRACLE' is a perfect gift to lift the spirits. Get this heart-warming festive tale for your children and Grandchildren! It is a short children's book set in the 1930s in a South Walian hill village and tells the tale of a heroic act and the wonderful community spirited efforts of two small boys Dennis and Roy and their faithful dog Lucky!

REVIEWS

It's a heart-warming tale...we loved it! Rhian T

Absolutely beautiful story! My Grandchildren loved it. Gill P

Eve and Mia are really enjoying this book and I love reading it to them every night. Sally B

 Matador

For exclusive discounts on Matador titles,
sign up to our occasional newsletter at
troubador.co.uk/bookshop